Springer Series in Statistics

For other titles published in this series, go to
http://www.springer.com/series/692

Christiane Lemieux

Monte Carlo and Quasi-Monte Carlo Sampling

Christiane Lemieux
University of Waterloo
Department of Statistics & Actuarial Science
200 University Avenue W.
Waterloo ON N2L 3G1
Canada
clemieux@math.uwaterloo.ca

ISSN: 0172-7397
ISBN: 978-1-4419-2676-0 e-ISBN: 978-0-387-78165-5
DOI: 10.1007/978-0-387-78165-5

Printed on acid-free paper

springer.com

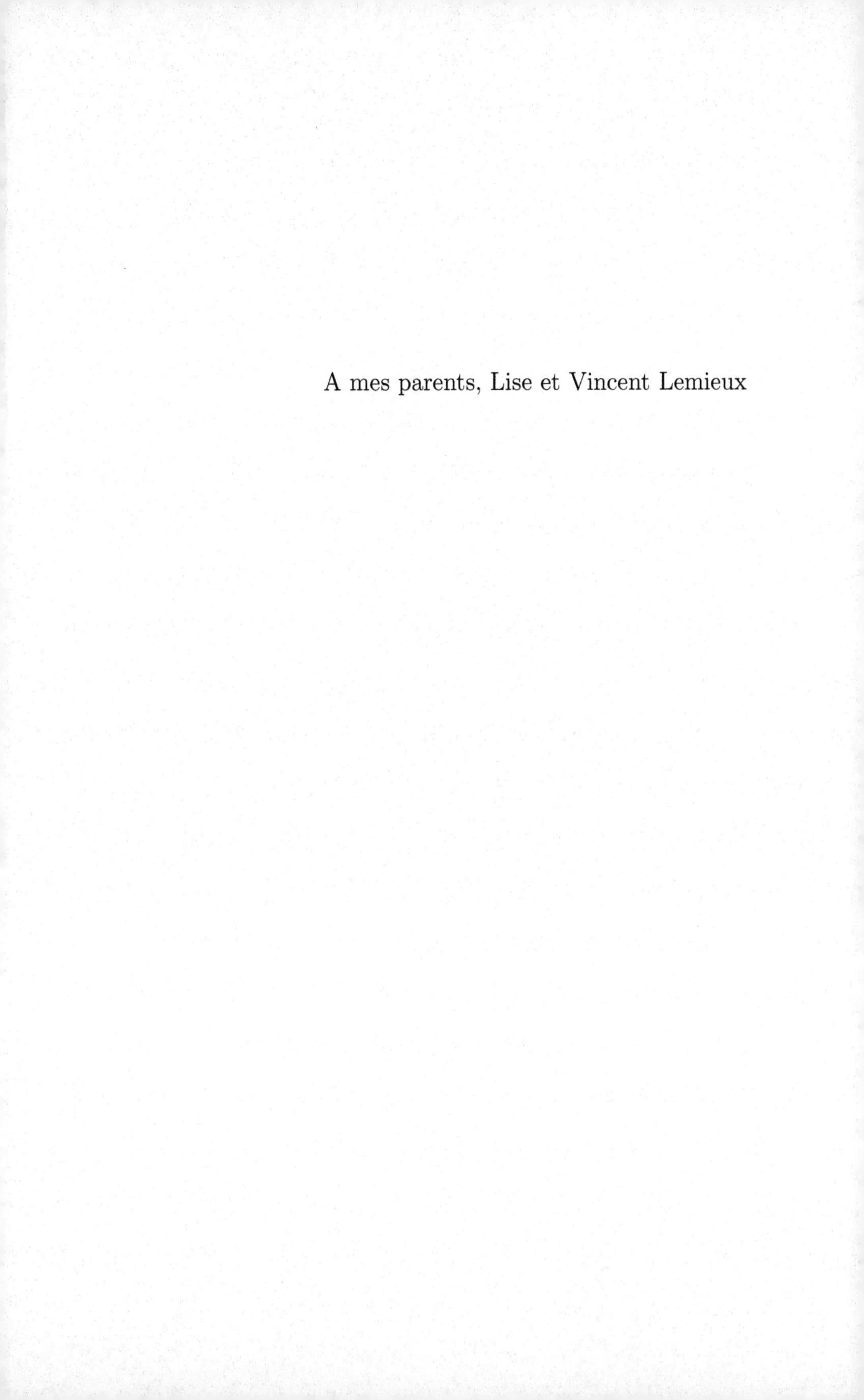

A mes parents, Lise et Vincent Lemieux

Preface

The goal of this text is to provide a self-contained guide to Monte Carlo and quasi–Monte Carlo sampling methods. These two classes of methods are based on the idea of using sampling to study mathematical problems for which analytical solutions are unavailable. More precisely, the idea is to create samples that can be used to derive approximations about a quantity of interest and its probability distribution. In the former case, random sampling is used, while in the latter, *low-discrepancy sampling* is used.

Quasi–Monte Carlo sampling methods are typically used to provide approximations for multivariate integration problems defined over the unit hypercube. They do so by creating sets or sequences of vectors $(u_1, \ldots, u_s)$, with each u_j taking values between 0 and 1, that sample the s-dimensional unit hypercube more regularly than random samples do, hence mimicking in a better way — with less *discrepancy* — the uniform distribution over that space. For this reason, most of the theory that underlies these constructions has been developed for problems that can be described as integration problems over the s-dimensional unit hypercube.

On the other hand, random sampling — via the use of Monte Carlo methods — has been developed and used in a variety of situations that do not necessarily fit the formulation above, which makes use of a function defined over the unit hypercube. In particular, stochastic simulation models are usually constructed using random variables defined over the real numbers, the nonnegative integers, or other domains that are not necessarily the unit interval between 0 and 1. However, the computer implementation of such models always relies, at its lowest level, on a source of (pseudo)random numbers that are uniformly distributed between 0 and 1. Therefore, at least in principle, it is always possible to reformulate a simulation model using a vector of input variables defined over the s-dimensional unit hypercube.

Being able to perform this "translation" — between the more intuitive simulation formulation and the one viewing the simulation program as a function f transforming input numbers $u_1, \ldots, u_s$ into an observation of the output quantity of interest — is extremely important when we want to successfully replace random sampling by quasi-random sampling in such problems. For this reason, we will be discussing this translation throughout the book, referring to it as the "integration versus simulation" formulation, with the understanding that by "integration" we mean the formulation of the problem using a function defined over the unit hypercube.

Because integration is the main area for which quasi-random sampling has been used so far, a large part of this text is devoted to this topic. In addition, simulation studies are often designed to estimate the mathematical expectation of some quantity of interest. In such cases, the translation of this goal into the formulation that uses a function f, as described in the preceding paragraph, means we wish to estimate the integral of that function. Hence these problems also fit within the integration framework.

A number of books have been written on the Monte Carlo method and its applications (especially in finance) [120, 121, 137, 145, 165, 211, 236, 293, 314, 386, 391, 418, 424], stochastic simulation [45, 175, 217, 218, 243, 389], and quasi–Monte Carlo methods [128, 308, 339, 441]. The purpose of this text is to present all these topics together in one place in a unified way, using the "integration versus simulation" formulation to help tie everything together. After reading this book, the reader should be able to apply random sampling to a wide range of problems and understand how to correctly replace it by quasi-random sampling. The selection of topics has been done in that perspective, and I certainly do not claim to be covering all aspects of Monte Carlo and quasi–Monte Carlo methods or surveying all possible applications for which these methods have been used. A very good source of information that contains the most recent advances in this field is the biannual *Monte Carlo and Quasi–Monte Carlo Methods* conference proceedings by Springer.

This book is organized as follows. The first chapter introduces the Monte Carlo method as a tool for multivariate integration and describes the integration versus simulation formulation using several examples. The more general use of Monte Carlo as a way to approximate a distribution is also studied. The second chapter gives an overview of different methods that can be used to generate random variates from a given probability distribution, a task that needs to be done extensively in any simulation study. This material comes early in the text because of its relevance in understanding the integration versus simulation formulation. Chapter 3 contains information on random number generators, which are essential for using random sampling on a computer. Methods for improving the efficiency of the Monte Carlo method that fall under the umbrella of *variance reduction techniques* are discussed in Chapter 4. A description of quasi–Monte Carlo constructions and the quality measures that can be used to assess them is done in Chapter 5. Several connections with random number generators are done in that chapter, which is the reason why their presentation precedes our discussion of quasi–Monte Carlo methods. Chapter 6 discusses the use of quasi–Monte Carlo methods in practice, including randomized quasi–Monte Carlo and ANOVA decompositions. The last two chapters are devoted to applications, with Chapter 7 focused on financial problems and Chapter 8 discussing more complex problems than those typically tackled by quasi–Monte Carlo methods.

This text can be used for a graduate course on Monte Carlo and quasi–Monte Carlo methods aimed either at statistics, applied mathematics, computer science, engineering, or operations research students. It may also be useful to researchers and practitioners familiar with Monte Carlo methods who want to learn about quasi–Monte Carlo methods.

The level of this text should be accessible to graduate students with varied backgrounds, as long as they have a basic knowledge of probability and statistics. There is an appendix at the end explaining a few key concepts in algebra required to understand some of the quasi–Monte Carlo constructions. Problem sets are provided at the end of each chapter to help the reader put in practice the different concepts discussed in the text.

There are several people whom I would like to thank for their help with this work. Radu Craiu, Henri Faure, Crystal Linkletter, Harald Niederreiter, and Xiaoheng Wang were kind enough to read over some of the material and make useful comments and suggestions. The anonymous reviewers from Springer also made suggestions that greatly improved this text. The students in my "Monte Carlo methods with applications in finance" course at the University of Calgary in the winter of 2006 used the preliminary version of some of these chapters and also tested some exercises. Lu Zhao worked on the solutions to the exercises for a subset of the chapters. Although their help allowed me to fix several mistakes and typos, I am sure I have not caught all of them, and I am entirely responsible for them. If possible, please report them to clemieux@uwaterloo.ca.

I would also like to thank various persons who helped me get a better understanding of the topics discussed in this book. These include Carole Bernard, Mikolaj Cieslak, Radu Craiu, Clifton Cunningham, Arnaud Doucet, Henri Faure, David Fleet, Alexander Keller, Adam Kolkiewicz, Frances Kuo, Fred Hickernell, Regina Hee Sun Hong, Pierre L'Ecuyer, Don McLeish, Harald Niederreiter, Dirk Ormoneit, Art Owen, Przemyslaw Prusinkiewicz, Wolfgang Schmid, Ian Sloan, Ilya Sobol', Ken Seng Tan, Felisa Vázquez-Abad, Stefan Wegenkittl, and Henryk Woźniakowski. In addition, I would like to thank John Kimmel at Springer for his patience and support throughout this process. The financial support of the Natural Sciences and Engineering Research Council of Canada is also acknowledged.

Finally, I would like to thank my family for their support and encouragement, especially my husband, John, and my two wonderful children, Anne and Liam. Also, I am very grateful for all the wisdom that my father has shared with me over the years in my academic journey. He has been my greatest source of inspiration for this work.

Waterloo, Canada, October 2008 *Christiane Lemieux*

Contents

1 **The Monte Carlo Method** ... 1
1.1 Monte Carlo method for integration ... 3
1.2 Connection with stochastic simulation ... 12
1.3 Alternative formulation of the integration problem via f: an example ... 20
1.4 A primer on uniform random number generation ... 22
1.5 Using Monte Carlo to approximate a distribution ... 25
1.6 Two more examples ... 27
Problems ... 34

2 **Sampling from Known Distributions** ... 41
2.1 Common distributions arising in stochastic models ... 42
2.2 Inversion ... 44
2.3 Acceptance-rejection ... 46
2.4 Composition ... 48
2.5 Convolution and other useful identities ... 50
2.6 Multivariate case ... 51
Problems ... 55

3 **Pseudorandom Number Generators** ... 57
3.1 Basic concepts and definitions ... 58
3.2 Generators based on linear recurrences ... 60
3.2.1 Recurrences over $\mathbb{Z}_m$ for $m \geq 2$ ... 61
3.2.2 Recurrences modulo 2 ... 64
3.3 Add-with-carry and subtract-with-borrow generators ... 66
3.4 Nonlinear generators ... 67
3.5 Theoretical and statistical testing ... 68
3.5.1 Theoretical tests for MRGs ... 70
3.5.2 Theoretical tests for PRNGs based on recurrences modulo 2 ... 75
3.5.3 Statistical tests ... 80

Problems ... 85

4 **Variance Reduction Techniques** ... 87
4.1 Introduction ... 87
4.2 Efficiency ... 89
4.3 Antithetic variates ... 89
4.4 Control variates ... 101
4.5 Importance sampling ... 111
4.6 Conditional Monte Carlo ... 119
4.7 Stratification ... 125
4.8 Common random numbers ... 132
4.9 Combinations of techniques ... 135
Problems ... 136

5 **Quasi–Monte Carlo Constructions** ... 139
5.1 Introduction ... 139
5.2 Main constructions: basic principles ... 143
5.3 Lattices ... 146
5.4 Digital nets and sequences ... 153
5.4.1 Sobol' sequence ... 157
5.4.2 Faure sequence ... 161
5.4.3 Niederreiter sequences ... 163
5.4.4 Improvements to the original constructions of Halton, Sobol', Niederreiter, and Faure ... 164
5.4.5 Digital net constructions and extensions ... 170
5.5 Recurrence-based point sets ... 174
5.6 Quality measures ... 179
5.6.1 Discrepancy and related measures ... 180
5.6.2 Criteria based on Fourier and Walsh decompositions ... 187
5.6.3 Motivation for going beyond error bounds ... 197
Problems ... 197

6 **Using Quasi–Monte Carlo in Practice** ... 201
6.1 Introduction ... 201
6.2 Randomized quasi–Monte Carlo ... 202
6.2.1 Random shift (or rotation sampling) ... 204
6.2.2 Digital shift ... 206
6.2.3 Scrambling and permutations ... 206
6.2.4 Partitions and Latin supercube sampling ... 209
6.2.5 Array-RQMC ... 210
6.2.6 Studying the variance ... 211
6.3 ANOVA decomposition and effective dimension ... 214
6.3.1 Effective dimension ... 216
6.3.2 Brownian bridge and related techniques ... 222

6.3.3 Methods for estimating σ_I^2 and approximating $f_I(u)$ 225
6.3.4 Using the ANOVA insight to find good constructions 228
6.4 Using quasi–Monte Carlo sampling for simulation 229
6.5 Suggestions for practitioners 237
Problems 239
Appendix: Tractability, weighted spaces and component-by-component constructions 241

7 Financial Applications 247
7.1 European option pricing under the lognormal model 247
7.2 More complex models 256
7.2.1 Heston's process 257
7.2.2 Regime switching model 258
7.2.3 Variance gamma model 260
7.3 Randomized quasi–Monte Carlo methods in finance 260
7.4 Commonly used variance reduction techniques 273
7.4.1 Antithetic variates 273
7.4.2 Control variates 273
7.4.3 Importance sampling 275
7.4.4 Conditional Monte Carlo 279
7.4.5 Common random numbers 281
7.4.6 Moment-matching methods 282
7.5 American option pricing 283
7.6 Estimating sensitivities and percentiles 288
Problems 298

8 Beyond Numerical Integration 301
8.1 Markov Chain Monte Carlo (MCMC) 303
8.1.1 Metropolis-Hastings algorithm 305
8.1.2 Exact sampling 310
8.2 Sequential Monte Carlo 312
8.3 Computer experiments 320
Problems 332

A Review of Algebra 335

B Error and Variance Analysis for Halton Sequences 341

References 347

Index 369

Acronyms and Symbols

$\Rightarrow$	convergence in distribution
$\lceil x \rceil$	the smallest integer larger than or equal to x
$[x]$	integer nearest to x
$\lfloor g(z) \rfloor$	polynomial part of a formal Laurent series $g(z)$
ant	antithetic
AWC	add-with-carry
CDF	cumulative distribution function
CI	confidence interval
cmc	conditional Monte Carlo
crn	common random numbers
CUD	completely uniformly distributed
cv	control variate
Eff	efficiency
$\mathbb{F}_m$	Galois field with m elements
$\mathbb{F}_m((z^{-1}))$	field of formal Laurent series over $\mathbb{F}_m$
gcd	greatest common divisor
HW	half-width
I_d	the $d \times d$ identity matrix
i. i. d.	independent and identically distributed
ind	independent
IPA	infinitesimal perturbation analysis
IS	importance sampling
LCG	linear congruential generator
LFSR	linear feedback shift register
LR	likelihood ratio
MC	Monte Carlo
MCMC	Markov chain Monte Carlo
MRG	multiple recursive generator
MSE	mean-square error
$\mathbb{N}_0$	the set of nonnegative integers
$N(0,1)$	standard normal variable

OA	orthogonal array
$\Phi(x)$	CDF of an $N(0,1)$ evaluated at x
P_n	$\{\mathbf{u}_1, \ldots, \mathbf{u}_n\} \subseteq [0,1)^s$
$P_n(I)$	projection of P_n over $I = \{j_1, \ldots, j_d\} \subseteq \{1, \ldots, s\}$, given by $\{(u_{i,j_1}, \ldots, u_{i,j_d}), i = 1, \ldots, n\}$
pdf	probability density function
PRNG	pseudorandom number generator
pst	poststratification
$\rho(X,Y)$	correlation coefficient between X and Y
roa	randomized orthogonal array
RQMC	randomized quasi–Monte Carlo
SAN	stochastic activity network
scr	scrambled
SIS	sequential importance sampling
str	stratification
SWB	subtract-with-borrow
$\mathbf{A}^{\mathrm{T}}$	transpose of the matrix $\mathbf{A}$
$\mathbf{1}_A$	indicator function for event A; that is, $\mathbf{1}_A = 1$ if event A occurs and is 0 otherwise.
$U(a,b)$	the uniform distribution over $[a,b]$
$U([0,1)^s)$	the uniform distribution over $[0,1)^s$
$\mathbf{u}_{-I}$	the vector $\mathbf{u}$ without the coordinates u_j with $j \in I$; that is, $\mathbf{u}_{-I} = (u_j : j \notin I)$.
$\mathbb{Z}_n$	the ring of integers modulo n
$\mathbb{Z}_n^*$	the integers modulo n without 0
z_α	$100(1-\alpha)$th percentile of the $N(0,1)$ distribution

Chapter 1
The Monte Carlo Method

The Monte Carlo method is a widely used tool in many disciplines, including physics, chemistry, engineering, finance, biology, computer graphics, operations research, and management science. Examples of problems that it can address are:

- A call center manager wants to know if adding a certain number of service representatives during peak hours would help decrease the waiting time of calling customers.
- A portfolio manager needs to determine the magnitude of the loss in value that could occur with a 1% probability over a one-week period.
- The designer of a telecommunications network needs to make sure that the probability of losing information cells in the network is below a certain threshold.

Realistic models of the systems above typically assume that at least some of their components behave in a random way. For instance, the call arrival times and processing times for the call center cannot realistically be assumed to be fixed and known ahead of time, and thus it makes sense instead to assume that they occur according to some stochastic model.

The Monte Carlo simulation method uses random sampling to study properties of systems with components that behave in a random fashion. More precisely, the idea is to *simulate* on the computer the behavior of these systems by randomly generating the variables describing the behavior of their components. Samples of the quantities of interest can then be obtained and used for statistical inference.

For instance, Monte Carlo simulation of the call center above would be done by performing the following steps: (i) Choose a model describing the system, including a description of the probability distributions for the random variables in the system (arrival times of the calls, types of calls, processing time per type of call, etc.); (ii) write a computer program that implements this model and can thus simulate the behavior of this call center over a certain period of time; (iii) use the program to create a sample of observations for

C. Lemieux, *Monte Carlo and Quasi–Monte Carlo Sampling*,
Springer Series in Statistics 692, DOI: 10.1007/978-0-387-78165-5_1,

the average waiting time experienced by the customers with and without the additional service representatives; and (iv) perform statistical inference on these samples to determine if the service representatives added significantly help to reduce the waiting time.

In addition to this *stochastic simulation* formulation, the Monte Carlo method can be used for problems that have no inherent probabilistic structure, for instance for the computation of multivariate integrals [165, 339, 391, 418] — discussed heavily in this text — and for solving systems of linear equations [125].

The development of the Monte Carlo method as a statistical computing tool goes back to the mid-1940s, when the first electronic computers were built. More precisely, it was John von Neumann and Stanislaw Ulam who first worked on the idea of using random numbers generated by a computer in order to solve problems encountered in the development of the atomic bomb. The name *Monte Carlo* — used in the title of the 1949 paper [320] by Metropolis and Ulam — refers to the famous casino in Monaco, where randomness is also used in a repetitive way. Early papers on the topic are [319, 320], and historical accounts can be found in [95, 165, 318].

In this chapter, we first review the Monte Carlo method in the context of integration. We then explain how estimation problems typically tackled by stochastic simulation can be formulated in that context, thus revealing the larger and more general scope of Monte Carlo methods. We also present a simple example illustrating the nonuniqueness of the integration formulation that corresponds to a given estimation problem. Then we discuss the use of Monte Carlo methods to estimate a distribution, going beyond the more traditional goal of estimating the mean. We conclude with two additional examples to illustrate further the integration versus simulation formulation.

Before going further, we provide below a description of the different key concepts discussed in this book.

Monte Carlo method: The use of random sampling as a tool to produce observations on which statistical inference can be performed to extract information about a system.

Monte Carlo integration: Special use of the Monte Carlo method, where we randomly sample uniformly over some domain $V \subseteq \mathbb{R}^s$ and use the produced sample $\{\mathbf{x}_1, \ldots, \mathbf{x}_n\}$ to construct an estimator for an integral of the form

$$\int_V f(\mathbf{x})d\mathbf{x},$$

where f is a real-valued function defined over V. Note that we can usually recast the problem so that $V = [0,1)^s$, an assumption that we make throughout this book. Hence, for our purposes, we think of integration as being defined over $[0,1)^s$ and as being tackled by producing a sample $\mathbf{u}_1, \ldots, \mathbf{u}_n$ of points with each $\mathbf{u}_i$ in $[0,1)^s$.

Stochastic simulation (or Monte Carlo simulation): The application of the Monte Carlo method to problems where the goal is to study properties of systems having stochastic components. Typically, it results in a sample from a random variable of the form $Y = h(\mathbf{X})$, which represents some output measure of interest. The vector $\mathbf{X}$ contains random variables modeling the system's stochastic components and are the ones that are simulated in order to obtain a sample from Y. An example illustrating this definition will be described in Sect. 1.2.

Quasi–Monte Carlo sampling (or *quasi-random* or *low-discrepancy* sampling): Method used to produce sets $\{\mathbf{u}_1, \ldots, \mathbf{u}_n\}$, with each point $\mathbf{u}_i$ in $[0,1)^s$, that sample the unit hypercube $[0,1)^s$ more uniformly than a random sample of n independent points does.

1.1 Monte Carlo method for integration

To explain how to apply the Monte Carlo integration method, we start with a discussion of univariate functions. We chose this one-dimensional setting simply to ease the presentation. As we will see later in this section, the advantage of the Monte Carlo method over other numerical integration schemes typically holds for larger dimensions, say at least 4 or 5.

Suppose we are given a function $f(x)$ defined over an interval $A \subset \mathbb{R}$. The goal is to compute the integral

$$I(f) = \int_A f(x)dx.$$

If f is simple, chances are that we can easily integrate it and thus give a closed-form solution for $I(f)$. For example, if $f(x) = x^2$ and $A = [0,1]$, then from calculus we know that $I(f) = 1/3$. However, in some cases it is not possible to find a closed-form solution for $I(f)$. A simple example is when $f(x)$ is the probability density function (pdf) of a standard normal random variable and $A = [0,c]$ for some real constant $c > 0$. More precisely, the problem here is to compute

$$I(f) = \int_0^c f(x)dx, \tag{1.1}$$

where

$$f(x) = \frac{1}{\sqrt{2\pi}} e^{-x^2/2}.$$

In this case, $I(f) = \Phi(c) - \Phi(0) = \Phi(c) - 1/2$, where $\Phi(\cdot)$ is the cumulative distribution function (CDF) of a standard normal random variable. Since no closed-form expression exists for $\Phi(c)$, this implies that $I(f)$ has to be approximated.

One possible approach to construct an approximation for $I(f)$ is to use the Monte Carlo method. To do so on the problem above, we first need to generate an i.i.d. random sample of n numbers $x_1, \ldots, x_n$ uniformly distributed between 0 and c and then form the approximation

$$Q_n = \frac{c}{n} \sum_{i=1}^{n} f(x_i). \tag{1.2}$$

To construct the sample $x_1, \ldots, x_n$, we assume for now that we have an algorithm `Rand01()` that outputs independent random numbers uniformly distributed between 0 and 1. By calling `Rand01()` n times, we can construct a sample $u_1, \ldots, u_n$ of i.i.d. random numbers and then transform them using $x_i = cu_i$ for $i = 1, \ldots, n$. Since each $u_i \sim U(0,1)$, clearly each $x_i \sim U(0,c)$ because

$$P(x_i \leq x) = P(cu_i \leq x) = P(u_i \leq x/c) = \begin{cases} x/c & 0 \leq x \leq c \\ 0 & x < 0 \\ 1 & x > c. \end{cases}$$

One way of understanding what the approximation (1.2) does is to look at Fig. 1.1, where Q_n is interpreted as the area of a rectangle of base c and height

$$\frac{1}{n} \sum_{i=1}^{n} f(x_i)$$

that approximates the mean value of f over its integration domain.

Alternatively, since the variables x_i are random, we can think of Q_n as a random variable, compute its expectation, and verify that it is equal to $I(f)$.

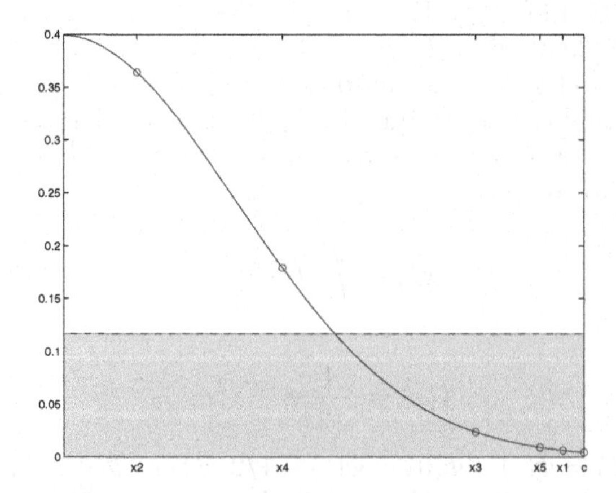

Fig. 1.1 Monte Carlo method in one dimension with $n = 5$ points. Q_n corresponds to the surface area of the shaded rectangle, whose height is given by the average over the five evaluation points.

That is, we have

$$\mathrm{E}(Q_n) = \frac{c}{n}\sum_{i=1}^{n}\mathrm{E}(f(x_i)) = c\int_0^c \frac{f(x_1)}{c}dx_1 = I(f), \tag{1.3}$$

where the second equality comes from the fact that each x_i is uniformly distributed over $[0, c]$, and thus their pdf is $1/c$. Hence, from (1.3), we have that Q_n is an unbiased estimator of $I(f)$. In addition, the strong law of large numbers implies that Q_n converges to $I(f)$ almost surely with n. In other words, we are guaranteed that if we are willing to take n large enough, our approximation Q_n can become arbitrarily close to the desired quantity $I(f)$ with probability 1.

This simple example illustrates the basic idea of Monte Carlo. As we mentioned at the beginning of this section, for functions of one variable such as the one above, there exist (deterministic) numerical methods that can provide much more accurate approximations than Monte Carlo [73]. For instance, based on the *trapezoidal rule*, the integral $I(f)$ given in (1.1) is approximated by

$$Q_n = \frac{c}{N}\sum_{i=0}^{N}\frac{1}{2}(f(x_i)+f(x_{i+1})) = \frac{c}{2N}(f(x_0)+f(x_n)) + \frac{c}{N}\sum_{i=1}^{N}f(x_i),$$

where $N = n-1$ and $x_i = ci/N, i = 0, \ldots, n-1$. Thus, here we approximate $I(f)$ by the sum of the area of n trapezoids of width c/N, with the height of their sides determined by f. Figure 1.2 illustrates the process.

The trapezoidal rule is part of a family of numerical integration methods called *Newton-Cotes formulas* that use equally spaced points to evaluate the integrand. Another member of this family is *Simpson's rule*, where a piecewise-polynomial function (rather than a linear function, as in the trapezoidal rule) is fitted through the function evaluations. For a given odd integer n and by setting $N = n - 1$, this is achieved by using the weights

$$c/3N, 4c/3N, 2c/3N, 4c/3N, \ldots, 2c/3N, 4c/3N, c/3N$$

for the values $f(x_0), f(x_1), \ldots, f(x_n)$ rather than $c/2N, c/N, \ldots, c/N, c/2N$ as in the trapezoidal rule. These methods are particularly useful for well-behaved, smooth functions, and their error usually depends on the value of second- or higher-order derivatives of f. Another family of deterministic numerical integration methods are the *Gaussian quadrature methods*, which use evaluation points given by the roots of a certain polynomial rather than using equally spaced points.

While it is true that, for functions of one variable, methods like Newton-Cotes or Gaussian quadrature can easily outperform the Monte Carlo method, the situation is different in the multivariate case. More precisely, consider the general multivariate integration problem where the goal is now to estimate

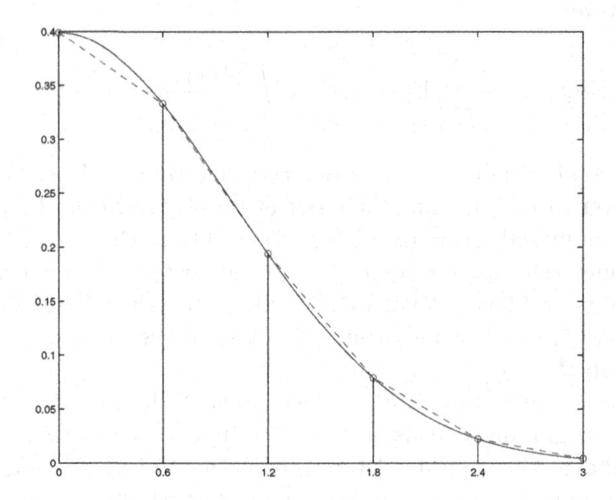

Fig. 1.2 Trapezoidal rule with $n = 6$ points. $I(f)$ is approximated by the sum of the area of $N = n - 1 = 5$ trapezoids.

$$I(f) = \int_{[0,1)^s} f(\mathbf{u})d\mathbf{u}, \tag{1.4}$$

where $\mathbf{u} = (u_1, \ldots, u_s)$ is an s-dimensional vector in $[0,1)^s$ and $f : [0,1)^s \to \mathbb{R}$ is a real-valued function. (At the end of this section, we come back to our choice of fixing the integration domain to be the unit hypercube $[0,1)^s$.) For instance, in the forthcoming Example 1.1, $s = 2$ and we consider the function

$$f(\mathbf{u}) = f(u_1, u_2) = u_1^3 + \frac{2\sin u_2}{1 + u_1}.$$

When the integral $I(f)$ given in (1.4) cannot be evaluated analytically, a general approach to approximate it is to use a quantity of the form

$$Q_n = \sum_{i=1}^{n} w_i f(\mathbf{u}_i),$$

where $P_n := \{\mathbf{u}_i, i = 1, \ldots, n\} \subset [0,1)^s$ is a point set in $[0,1)^s$, and the weights w_i satisfy $0 \le w_i \le 1$ and $\sum_{i=1}^n w_i = 1$. In other words, the approximation Q_n is obtained by taking a weighted average of n function evaluations of f made at the n points in P_n. The extension of methods like the trapezoidal rule or Simpson's rule to this case consists in defining P_n to be a *product rule.* That is, P_n is defined as the Cartesian product of some fixed one-dimensional point set. For example, the multivariate version of the trapezoidal rule would be to choose some N and use the point set

$$P_n = \{(i_1/N, \ldots, i_s/N), i_j = 0, \ldots, N, j = 1, \ldots, s\}$$

with $n = (N+1)^s$ and associated weights

$$w_{i_1,\ldots,i_s} = v_{i_1} \ldots v_{i_s},$$

where

$$v_l = \begin{cases} 1/N & \text{if } 1 \leq l < N, \\ 1/2N & \text{if } l = 0 \text{ or } l = N. \end{cases}$$

On the left-hand side of Fig. 1.4, we see an example of a point set used by the trapezoidal rule when $s = 2$, $N = 31$, and thus $n = (N+1)^2 = 1024$. In addition, Example 1.1 illustrates how to construct an approximation for $I(f)$ using the trapezoidal rule with $s = 2$ and $N = 4$ for a total of $n = (N+1)^2 = 25$ evaluation points.

Example 1.1. When $s = 2$ and $N = 4$, the trapezoidal rule consists in using the approximation

$$\begin{aligned}
&\frac{1}{64}(f(0,0) + f(0,1) + f(1,0) + f(1,1)) \\
&+\frac{1}{32}(f(0,1/4) + f(0,1/2) + f(0,3/4) + f(1/4,0) + f(1/2,0) + f(3/4,0) \\
&+f(1,1/4) + f(1,1/2) + f(1,3/4) + f(1/4,1) + f(1/2,1) + f(3/4,1)) \\
&+\frac{1}{16}(f(1/4,1/4) + f(1/4,1/2) + f(1/4,3/4) + f(1/2,1/4) + f(1/2,1/2) \\
&\quad +f(1/2,3/4) + f(3/4,1/4) + f(3/4,1/2) + f(3/4,3/4)).
\end{aligned}$$

In Fig. 1.3, the hollow circles are points with a weight of $(1/2N) \times (1/2N) = 1/64$, the hollow squares have a weight of $(1/2N) \times (1/N) = 1/32$, and the black circles have a weight of $(1/N) \times (1/N) = 1/16$.

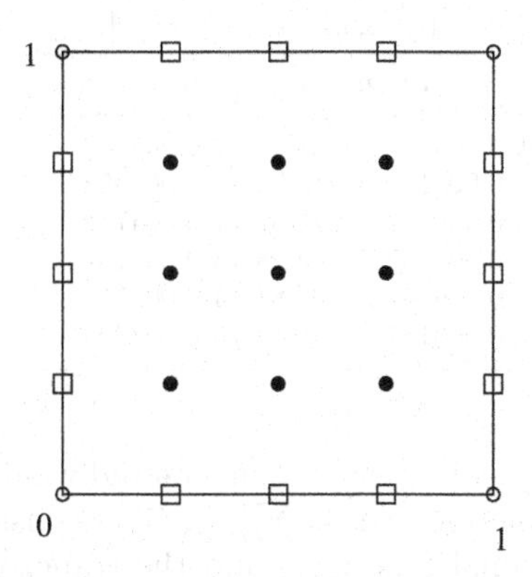

Fig. 1.3 Weights for trapezoidal rule with $s = 2$ and $N = 4$.

For the function $f(\mathbf{u}) = u_1^3 + 2\sin(u_2)/(1+u_1)$, the trapezoidal rule with $N = 4$ yields the approximation 0.8447, whereas the true value $I(f)$ is given by $I(f) = 0.25 + 2\ln(2)(1 - \cos(1)) = 0.8873$, so the error is 0.0426.

The problem with these product rules is that the number of sampling points n must grow exponentially fast with the dimension s in order to keep the error bounded. This is due to the fact that, for these rules, the order of magnitude of the error bound is the sth root of the order of magnitude of the one-dimensional rule's error [339]. For instance, the error of the trapezoidal rule when $s = 1$ can be shown to be in $O(n^{-2})$ under certain conditions; on the other hand, the s-dimensional version of this rule has an error in $O(n^{-2/s})$. In Table 1.1, we show (in the second column) the error obtained by the trapezoidal rule when approximating the integral of

$$f(\mathbf{u}) = \frac{1}{s}(\sqrt{u_1} + \sqrt{u_2} + \ldots + \sqrt{u_s})$$

over $[0,1)^s$ when $N = 10$ as s goes from 1 to 6. As expected, the error remains constant although the total number n of evaluation points increases from 11 to 11^6 as s increases from 1 to 6. Equivalently, if we keep n approximately equal to 11^3 by using $N = [11^{3/s} - 1]$ (where $[x]$ denotes the integer closest to x), we see that the error increases substantially as s goes from 1 to 6 (third column). We also show in the last column of this table the behavior of the error when $s = 4$ and N increases from 10 to 15. For the corresponding sample of values of n, we can use regression to estimate the exponent α such that $cn^{-\alpha}$ fits the behavior of the error $|I(f) - Q_n|$ best. Doing so, we find $\alpha = -0.4$, which is not too far from the rate $-2/s = -1/2$ predicted by the theory.

Table 1.1 Behavior of the trapezoidal rule for $\sum_{j=1}^{s} \sqrt{u_j}/s$.

s	$\|I - Q_n\|$		$s = 4$	
	$N = 10$	$n \approx 11^3$	N	$\|I(f) - Q_n\|$
1	0.006157	0.000004	10	0.006157
2	0.006157	0.000970	11	0.005354
3	0.006157	0.006157	12	0.004712
4	0.006157	0.016928	13	0.004189
5	0.006157	0.035384	14	0.003756
6	0.006157	0.063113	15	0.003393

For this simple example, it is easy to understand what goes wrong with the trapezoidal rule: The function $f(\mathbf{u}) = \sum_{j=1}^{s} \sqrt{u_j}/s$ considered there is simply a sum of s one-dimensional functions, and the trapezoidal rule is designed so that only $N + 1 = n^{-1/s}$ *distinct* evaluation points are used for each of these one-dimensional functions, although in total we are using $n = (N+1)^s$

function evaluations. Hence, if N is fixed, the error remains constant even if $n = (N+1)^s$ increases. Alternatively, if n is fixed, then $N \approx n^{1/s}$ decreases with s, and thus the error increases with the dimension.

More generally, we can attribute the inadequacy of product rules as the dimension s increases to a phenomenon called the *curse of dimensionality*, which was coined by Richard Bellman [28] to describe the fact that each time s increases by one, the "size" of the space $[0,1)^s$ to be sampled "increases" (in some sense). This "curse" is especially harmful to these rules because as s increases they continue to use a set P_n that is built in a purely one-dimensional fashion and thus fails to recognize the increase in the size of $[0,1)^s$. As s increases, this results in larger and larger gaps in $[0,1)^s$, where there are no points from P_n, but, on the other hand, more and more points map to the same place on any given one-dimensional axis.

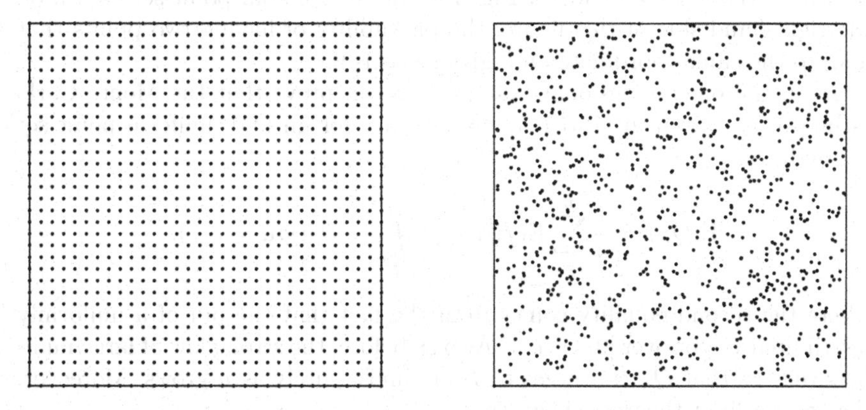

Fig. 1.4 Left-hand side: 1024 points for trapezoidal rule; right-hand side: 1024 random points for Monte Carlo.

One possible approach for constructing a point set that avoids the rigidity of the rectangular grids utilized by product rules is to use a purely random sample of n points uniformly distributed in $[0,1)^s$, which is precisely what the Monte Carlo integration method does. Indeed, with Monte Carlo, the set P_n is formed by n i.i.d. uniform points over $[0,1)^s$ and the weights w_i are all set to $1/n$. More precisely, the integral $I(f)$ in this case is estimated by the Monte Carlo estimator

$$Q_n = \frac{1}{n}\sum_{i=1}^{n} f(\mathbf{u}_i), \tag{1.5}$$

where the points $\mathbf{u}_i$ are i.i.d. uniform over $[0,1)^s$. The pseudocode given in Fig. 1.5 shows how to use Monte Carlo for the two-dimensional function $f(\mathbf{u}) = u_1^3 + 2\sin(u_2)/(1+u_1)$ given in Example 1.1.

An example of a random two-dimensional point set is given on the right-hand side of Fig. 1.4. Each point on this figure may correspond to one of the n evaluation points (u_1, u_2) used in the pseudocode given in Fig. 1.5 in

```
Evalf()
    Q ← 0
    for i = 1 to n
        u1 ← Rand01()
        u2 ← Rand01()
        Q ← ((i − 1) × Q + u1^3 + 2 × sin(u2)/(1 + u1))/i
    return (Q)
```

Fig. 1.5 Pseudocode to approximate the integral of $f(\mathbf{u}) = u_1^3 + 2\sin(u_2)/(1+u_1)$ using Monte Carlo. Q is updated at each new evaluation point.

the case where $n = 1024$. Note that by contrast with the rectangular grid shown on the left-hand side of Fig. 1.4, for the random point set shown on the right-hand side of this figure, the probability of having two points that map to the same coordinate on a given axis is 0.

As in the one-dimensional example, we can prove that the Monte Carlo estimator Q_n given in (1.5) is unbiased since for an i.i.d. uniform point set P_n we have

$$\mathrm{E}(Q_n) = \frac{1}{n}\sum_{i=1}^{n} \mathrm{E}(f(\mathbf{u}_i)) = \int_{[0,1)^s} f(\mathbf{u})d\mathbf{u} = I(f),$$

where the second equality comes from the fact that the pdf of a uniformly distributed vector over $[0,1)^s$ is 1. Also as before, the strong law of large numbers tells us that Q_n converges to $I(f)$ almost surely as n grows. Moreover, the central limit theorem shows that

$$\frac{Q_n - I(f)}{\sigma/\sqrt{n}} \Rightarrow N(0,1),$$

where $\Rightarrow$ means convergence in distribution and σ^2 is the variance of $f(\boldsymbol{U})$. That is,

$$\sigma^2 = \int_{[0,1)^s} (f(\mathbf{u}) - I(f))^2 d\mathbf{u}.$$

Thus, approximate confidence intervals of the form

$$\left(Q_n \pm \frac{\hat{\sigma}}{\sqrt{n}} z_{\alpha/2}\right)$$

can be constructed for $I(f)$, where $\hat{\sigma}$ is the sample standard deviation given by

$$\hat{\sigma} = \left(\sum_{i=1}^{n} \frac{(f(\mathbf{u}_i) - Q_n)^2}{n-1}\right)^{1/2}$$

and $z_{\alpha/2}$ is the $100(1-\alpha/2)$th percentile of the standard normal distribution. Hence the probabilistic error of the Monte Carlo estimator is in $O(1/\sqrt{n})$, which is independent of the dimension s. The variance of Q_n can be estimated by $\hat{\sigma}^2/n$ and is often used as a benchmark when the Monte Carlo method is compared against other (stochastic) integration methods.

Table 1.2 gives results for the Monte Carlo estimator similar to the ones presented in Table 1.1 for the trapezoidal rule. That is, the second to fourth columns of the table show the Monte Carlo error as s goes from 1 to 6 and n is 11^s, $[11^{3/s}]^s$, and 1331 for the second, third, and fourth columns, respectively; the sixth column shows the Monte Carlo error when $s = 4$ and n goes from 11^4 to 16^4. In comparison with Table 1.1, we added the column $n = 1331$ to see what happens when n is fixed, while the value $n = [11^{3/s}]^s \approx 11^3 = 1331$ used in the third column varies with s due to the rounding operation that was necessary in order to apply the trapezoidal rule.

Table 1.2 Behavior of the MC error for $\sum_{j=1}^{s} \sqrt{u_j}/s$.

s	$\lvert I - Q_n \rvert$				$s = 4$
	$N = 10$	$n \approx 11^3$	$n = 1331$	N	$\lvert I(f) - Q_n \rvert$
1	0.070262	0.000415	0.000415	10	0.000528
2	0.015078	0.000050	0.000741	11	0.000739
3	0.000254	0.000254	0.000254	12	0.000461
4	0.000528	0.000510	0.000779	13	0.000416
5	0.000373	0.000839	0.000340	14	0.000253
6	0.000081	0.001424	0.000137	15	0.000097

The results for the Monte Carlo error are quite different from those obtained with the trapezoidal rule, which were given in Table 1.1. First, when s increases from 1 to 6 and $n = 11^s$ (second column), we see that the error decreases with s and eventually becomes much smaller than the one obtained with the trapezoidal rule. When n remains constant as s increases (third and fourth columns), the error stays more or less the same and does not have the same upward trend as the trapezoidal rule. Finally, when $s = 4$ and n goes from 11^4 to 16^4, the error is at least 10 times smaller than with the trapezoidal rule and decreases in a slightly more erratic way. These results support the suggestion — based on the comparison of the convergence rates of $n^{-2/s}$ versus $n^{-1/2}$ for the trapezoidal rule and Monte Carlo, respectively — that even for moderate dimensions s, the Monte Carlo method can outperform methods such as the trapezoidal rule.

Although the Monte Carlo error has the nice property that its convergence rate of $1/\sqrt{n}$ does not depend on the dimension, this rate is often considered to be quite slow. For example, to reduce the error by a factor of 10, one must increase the sample size n by 100 (on average). For this reason, a lot of work has been done on finding ways of improving the Monte Carlo error, and two different paths can be taken for that purpose. The first one is to try to find ways of reducing the variance σ^2 of f, or more precisely to try to find another function ϕ whose integral is also $I(f)$ but that has a smaller variance than f. Methods achieving this fall under the umbrella of *variance reduction techniques*, which will be discussed in Chap. 4. The second approach is to use an alternative sampling mechanism — often called *quasi-random* or *low-discrepancy* sampling — whose corresponding error has a better convergence rate. Using these alternative sampling mechanisms for numerical integration is usually referred to as "quasi–Monte Carlo" integration. For example, sampling methods based on *scrambled nets* [357, 359] have the property that, for sufficiently *smooth* functions, the corresponding integration error is in $O(n^{-3/2} \log^{s/2} n)$, which for a fixed dimension s is much better than the $O(1/\sqrt{n})$ associated with Monte Carlo integration. Chapters 5 and 6 discuss these alternative sampling mechanisms and the associated quasi–Monte Carlo integration methods. A Monte Carlo estimator to which no improvement technique has been applied is usually referred to as a "naive Monte Carlo" or "crude Monte Carlo" estimator.

Our assumption that the integration domain is the unit hypercube $[0,1)^s$ is usually not very restrictive since one can often perform a change of variables to satisfy this requirement. For instance, in our example with the normal density function, we could define $u = x/c$ and rewrite (1.1) as

$$I(f) = \int_0^1 \frac{c}{\sqrt{2\pi}} e^{-c^2u^2/2} du.$$

Applying the Monte Carlo method to this problem then amounts to generating n i.i.d. uniform points $u_1, \ldots, u_n$ in $[0,1)$ and constructing the estimator

$$Q_n = \frac{1}{n} \sum_{i=1}^{n} \frac{c}{\sqrt{2\pi}} e^{-c^2u_i^2/2},$$

which is exactly the same as before since $u \sim U(0,1)$ if and only if $x \sim U(0,c)$. More generally, the fact that we can reinterpret simulation problems as integration over the unit hypercube justifies our choice of focusing on this specific domain. The next section discusses how to do this reinterpretation.

1.2 Connection with stochastic simulation

When the Monte Carlo method is presented as a tool for multivariate integration, one question that often arises is: Are there any practical applications where such integrals have to be solved? The answer is a clear *yes* since many problems arising from the fields of physics, finance (see Chap. 7), and biology — just to name a few — can be formulated as integration problems.

In particular, and as mentioned before, a large class of problems that fit the integration formulation are those for which stochastic simulation is used to estimate a mathematical expectation. In that context, people use Monte Carlo simulation without necessarily using the integration formulation. Our point of view is that in such cases *Monte Carlo simulation* and *Monte Carlo integration* are just two different ways of viewing the problem and how it can be tackled by Monte Carlo methods. The former typically provides a more intuitive way of setting up the problem, while the latter can be more useful when studying theoretical properties of the estimators obtained, especially when variance reduction techniques or quasi–Monte Carlo are used.

It should be noted that simulation is a general tool that can be used to do more than just approximating mathematical expectations. With this in mind, we give in Fig. 1.6 a description of the integration versus simulation formulation.

1. Sample observations of a random vector $\mathbf{X}$ describing the simulation model and look at the distribution of $Y = h(\mathbf{X})$, which represents the output measure of interest or
2. sample the "source of randomness" $\mathbf{u}$ and look at the distribution of $f(\mathbf{u}) := h(g(\mathbf{u}))$, where g represents the function used to transform $\mathbf{u}$ into an observation of $\mathbf{X}$ (such functions are discussed in Chap. 2).

Fig. 1.6 The integration (2) versus the simulation (1) formulation.

Since this dual interpretation is a recurrent theme in this text, it must be well understood before proceeding to the following chapters. In Example 1.2 below, which is similar to other queueing examples that can be found in simulation textbooks such as [45, 243], we describe in detail how to perform the translation from simulation to integration.

Example 1.2. Consider a bank that operates from 10 am to 3 pm. We assume that there is only one teller, that the clients arrive according to a Poisson process at a rate of 1 per minute, and that each client stays with the teller for a random length of time that has an exponential distribution with mean 45 seconds. We assume these service times and all interarrival times are independent from each other. The goal is to estimate the expected number c_5 of clients that will wait more than 5 minutes for a teller at the bank during a given day of operation. (We suppose that all clients that arrived before 3 pm will eventually be served.)

To estimate by simulation the quantity c_5 described in Example 1.2, one would run, say, $n = 1000$ independent realizations of a given day at that bank — generating at random the arrival times and service times — count for each realization how many clients waited more than 5 minutes, and then take the average over the n runs.

To be more precise, we will describe with pseudocode how simulation can be used to estimate c_5. In general, a computer program that implements a simulation model requires the use of event lists, procedures to manage queues, statistical counters, etc. [45, 243]. These tools can be implemented from scratch, but there also exist several simulation software packages that have all of this built-in and that require very little programming from the user (see, for example, [243, 266] and the references therein).

Fortunately, our simple simulation model does not require any of these tools since we can use Lindley's equation [292], which gives us the following recurrence relation for the waiting times W_j based on the interarrival and service times:

$$W_j = \max(0, W_{j-1} + S_{j-1} - A_j), \qquad j \geq 1, \tag{1.6}$$

where

$$\begin{aligned} W_j &= \text{waiting time in the queue of the } j\text{th customer}, \\ A_j &= \text{interarrival time between the } (j-1)\text{th and } j\text{th customers}, \\ S_j &= \text{service time of the } j\text{th customer}, \end{aligned}$$

and $W_0 = S_0 = 0$. To understand where this relation comes from, imagine you enter the bank system described in Example 1.2. If the person that entered before you waited for 3 minutes before spending 40 seconds with the teller and you arrived 1 minute after that person, then your waiting time is $3-1 = 2$ minutes and 40 seconds. This is because you are not in the system during the first minute of waiting of the client in front of you, but then you enter and wait two minutes while that other person waits, and you wait an additional 40 seconds while that customer is being served. If, instead, the client in front of you only waits 15 seconds before being served, then by the time you enter the system, that client has left 5 seconds ago and therefore you do not wait.

Using the notation above, we can now say that the quantity we wish to estimate is

$$c_5 = \mathrm{E}\left(\sum_{j=1}^{N} \mathbf{1}_{W_j > 5}\right), \tag{1.7}$$

where we used the indicator function

$$\mathbf{1}_{W_j>5} = \begin{cases} 1 \text{ if } W_j > 5 \\ 0 \text{ otherwise}, \end{cases}$$

and N is the number of clients that arrived during the bank's hours of operation. Note that N itself is random since the number of clients that come to the bank on a given day depends on the interarrival times observed during that day. In our case, N is actually a Poisson random variable with mean

$5 \times 60 \times 1 = 300$ since we assumed we had a Poisson arrival process with rate one per minute over five hours.

From (1.7) and the description given so far, we can see how this problem fits the simulation framework and its associated notation, as given in Fig. 1.6. More precisely, we have that

$$\mathbf{X} = (A_1, S_1, A_2, S_2, \ldots)$$

$$\text{and } h(\mathbf{X}) = \sum_{j=1}^{N(\mathbf{X})} \mathbf{1}_{W_j(\mathbf{X})>5},$$

where $N(\mathbf{X})$ and $W_j(\mathbf{X})$ are functional representations of N and W_j used to highlight the dependence on $\mathbf{X}$.

Using Lindley's equation (1.6), we can then perform one simulation of this model as shown in Fig. 1.7, where $\beta_A = 1$ and $\beta_S = 0.75$ represent the mean interarrival time and the mean service time in minutes, respectively. In this pseudocode, we assume that the function **Exp**(β) returns an observation from the exponential distribution with mean β. That is, if $X \leftarrow$ **Exp**(β), then $P(X \leq x) = 1 - e^{-x/\beta}$ for $x > 0$. Equivalently, we say that **Exp**(β) returns a *random variate* from the exponential distribution with mean β.

```
OneSimBank(βA, βS)
    NbWait5 ← 0
    w ← 0
    a ← Exp(βA)
    time ← a
    while (time < 300) do
        s ← Exp(βS)
        a ← Exp(βA)
        time ← time + a
        w ← max(0, w + s − a)
        if ((time < 300) and (w > 5)) then
            NbWait5 ← NbWait5 + 1
    return NbWait5
```

Fig. 1.7 Pseudocode for Example 1.2. Times are in minutes.

To estimate the quantity c_5 given in (1.7) with $n = 1000$ independent simulations and, say, compute a 95% confidence interval for c_5, one would run the algorithm `Run1000Sim` described on the left-hand side of Fig. 1.8, where we assume that `ave(y)` and `var(y)` return the sample average and variance of the vector $\mathbf{y}$, respectively. On the right-hand side of this figure, we give an example of what an execution of this algorithm might look like.

Now, to see how estimating c_5 by simulation is equivalent to using Monte Carlo for numerical integration, we first need to say more about the function

```
Run1000Sim()                                    y(1) = 10
    for i = 1 to 1000 do                        y(2) = 16
        y(i) ← OneSimBank(1,0.75)               y(3) = 8
    hw = 1.96 × sqrt(var(y)/1000)               y(4) = 2
    print ("average is", ave(y))                y(5) = 95
    print ("95% CI half-width is", hw)          ...
                                                y(1000) = 70
                                                average is 40.875
                                                95% CI half-width is 2.163
```

Fig. 1.8 Pseudocode to estimate c_5 based on 1000 runs (left); example of output (right).

Exp(·). We assume here that its implementation has the representation given in Fig. 1.9, where, as discussed in Sect. 1.1, Rand01() returns i.i.d. observations from the $U(0,1)$ distribution.

```
Exp(β)
    u ← Rand01()
    return GenExpon(u, β)
```

Fig. 1.9 Pseudocode for generating exponential random variates with mean β.

The function GenExpon then uses this random number u and transforms it into an observation from the exponential distribution with mean β. More precisely, this function can be implemented using *inversion* (also called the *inverse-function method* by some authors). The idea of inversion is to generate observations from a given probability distribution by evaluating the inverse of the corresponding CDF at a value uniformly distributed between 0 and 1. Figure 1.10 illustrates the idea.

For example, since the exponential distribution with mean β has the CDF $F(x) = 1 - e^{-x/\beta}$, by setting $u = F(x) = 1 - e^{-x/\beta}$, we then find that

$$
\begin{aligned}
& 1 - u = e^{-x/\beta}, \\
\Leftrightarrow\ & -x/\beta = \ln(1-u), \\
\Leftrightarrow\ & \quad x = -\beta \ln(1-u),
\end{aligned}
$$

and thus $F^{-1}(u) = -\beta \ln(1-u)$. Therefore, we can implement the function GenExpon(u, β) as shown in Fig. 1.11.

If $U \sim U(0,1)$, then the value X returned by GenExpon(u, β) has the correct distribution since

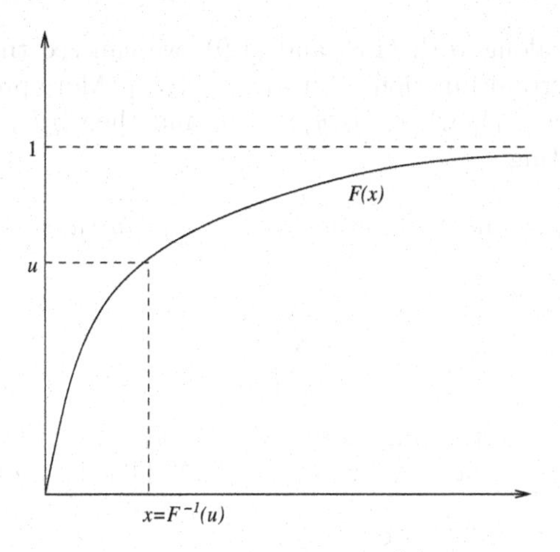

Fig. 1.10 Inversion. For a given u, find x such that $F(x) = u$.

```
GenExpon(u, β)
    return −β ln(1 − u)
```

Fig. 1.11 Pseudocode for generating an exponential random variate with mean β by inversion.

$$
\begin{aligned}
P(X \le x) &= P(-\beta \ln(1-U) \le x) = P(1 - U \ge e^{-x/\beta}) \\
&= P(U \le 1 - e^{-x/\beta}) = 1 - e^{-x/\beta}.
\end{aligned}
$$

Hence, within each simulation, each variate a_j and s_j can be viewed as a function of $\mathbf{u} = (u_1, u_2, \ldots)$ since, for $j = 1, 2, \ldots$, we can write

$$a_j = g_1(u_{2j-1}), \tag{1.8}$$

$$s_j = g_2(u_{2j}), \tag{1.9}$$

where $g_1(\cdot) =$ GenExpon$(\cdot, 1)$ and $g_2(\cdot) =$ GenExpon$(\cdot, 0.75)$. Similarly, N itself can be written as a function $\zeta(\cdot)$ of $\mathbf{u} = (u_1, u_2, \ldots)$ since

$$N = \sum_{j=1}^{\infty} \mathbf{1}_{a_1 + \ldots + a_j < 300},$$

and a_j is a function of u_{2j-1} for each $j \ge 1$. More precisely, based on (1.8), we can write

$$N = \zeta(u_1, u_2, \ldots) := \sum_{j=1}^{\infty} \mathbf{1}_{g_1(u_1) + g_1(u_3) + \ldots + g_1(u_{2j-1}) < 300}. \tag{1.10}$$

Also, using (1.6) along with (1.8) and (1.9), we can see that each w_j can be written as a certain function η_j of $u_1, \ldots, u_{2j-1}$. More precisely, we have $w_j = \eta_j(u_1, \ldots, u_{2j-1})$, where $\eta_1(u_1) = 0$, and then $\eta_j(\cdot)$ can be defined recursively by setting

$$\eta_j(u_1, \ldots, u_{2j-1}) = \max(0, \eta_{j-1}(u_1, \ldots, u_{2j-3}) + g_2(u_{2j-2}) - g_1(u_{2j-1})).$$

Hence the whole sum

$$C_5 := \sum_{j=1}^{N} \mathbf{1}_{w_j > 5} \tag{1.11}$$

can be written as a certain function f of $\mathbf{u} = (u_1, u_2, \ldots)$ by replacing each w_j by $\eta_j(u_1, \ldots, u_{2j-1})$ and N by $\zeta(u_1, u_2, \ldots)$. That is, we have

$$C_5 = f(\mathbf{u}) = \sum_{j=1}^{\zeta(u_1, u_2, \ldots)} \mathbf{1}_{\eta_j(u_1, u_2, \ldots, u_{2j-1}) > 5}. \tag{1.12}$$

When we run `OneSimBank(1,0.75)`, we end up evaluating this function f at a certain point $\mathbf{u} = (u_1, u_2, \ldots)$, where the u_j's are i.i.d. $U(0,1)$. Figure 1.12 illustrates the idea. In the case considered there, the point $(0.45, 0.14, 0.62, 0.97, 0.05, \ldots, 0.09, 0.07, 0.33, \ldots)$ produces a value of N equal to 288 and a value of

$$C_5 = \sum_{j=1}^{N} \mathbf{1}_{w_j > 5} = f(0.45, 0.14, 0.62, 0.97, 0.05, \ldots, 0.09, 0.07, 0.33, \ldots) = 36.$$

$u_1 = 0.45 \rightarrow a_1 = 35.9$	$\rightarrow w_1 = 0$	$C_5 = 0$	$A_1 = 35.9$
$u_2 = 0.14 \rightarrow s_1 = 6.8$			
$u_3 = 0.62 \rightarrow a_2 = 58.1$	$\rightarrow w_2 = 0$	$C_5 = 0$	$A_2 = 93.9$
$u_4 = 0.97 \rightarrow s_2 = 157.8$			
$u_5 = 0.05 \rightarrow a_3 = 3.1$	$\rightarrow w_3 = 154.7$	$C_5 = 0$	$A_3 = 97.0$
$\vdots$	$\vdots$	$\vdots$	
$u_{575} = 0.09 \rightarrow a_{288} = 5.7$	$\rightarrow w_{288} = 314.2$	$\mathbf{C_5 = 36}$	$A_{288} = 17980.5$
$u_{576} = 0.07 \rightarrow s_{288} = 3.3$			
$u_{577} = 0.33 \rightarrow a_{289} = 24.0$			$A_{289} = 18004.5$

Fig. 1.12 How to view `OneSimBank(1,0.75)` as a function evaluation. Times are in seconds. C_5 is updated each time a new waiting time w_j is computed. $A_j = a_1 + \ldots + a_j$ is the arrival time of the jth client.

The algorithm `Run1000Sim()` then returns an estimate

$$\frac{1}{1000} \sum_{i=1}^{1000} f(\mathbf{u}_i)$$

for c_5, where f is as defined in (1.12). More intuitively, each $f(\mathbf{u}_i)$ is an observation for the value C_5 given in (1.11) based on the input vector of uniform numbers $\mathbf{u}_i = (u_{i1}, u_{i2}, \ldots)$ required to generate the random observations $a_1, s_1, a_2 \ldots$ for the ith simulation. Hence we can say that using 1000 i.i.d. simulation runs to estimate c_5 is equivalent to using a sample of $n = 1000$ i.i.d. points to integrate the function f given in (1.12) using the Monte Carlo integration method. We also note that this estimator is unbiased since

$$\mathrm{E}(C_5) = \mathrm{E}\left(\sum_{j=1}^{N} \mathbf{1}_{w_j > 5} \right) = c_5.$$

From this example, we see that going from the simulation to the integration formulation simply amounts to rewriting the problem so that the input is the vector $\mathbf{u}$ of uniform numbers used to run the simulation. In that setting, we can think of f as the mechanism by which the simulation program takes a sequence of i.i.d. uniform numbers and transforms them into an observation of the quantity for which we want to estimate the expectation. The dimension s of the domain of f is the number of uniform numbers required to run the simulation. In the example above, we have that $s = \infty$ because there is no a priori upper bound on the number of uniform numbers required to run the simulation. However, for a given simulation run, only a finite number of coordinates is actually required to evaluate f. For instance, in Fig. 1.12, only the first 577 coordinates of $\mathbf{u} = (0.45, 0.14, \ldots, 0.09, 0.07, 0.33, \ldots)$ are used to get an observation for $C_5 = f(\mathbf{u})$.

More generally, for this problem, the required number of uniform numbers is equal to $2N + 1$, where N is the Poisson random variable corresponding to the number of clients who arrive during the bank's hours of operation. Indeed, we need to generate N service times and $N + 1$ interarrival times in order to determine N since we need to generate the arrival time of the first client that arrives after 3 pm in order to know how many clients arrived *before* 3 pm. This is because the arrival time of the last client entering before 3 pm is not a *stopping time*; i.e., when this client arrives, we do not have enough information to determine that this is indeed the last client. If the problem was instead to estimate the number of persons who wait more than 5 minutes *among the first 300 clients*, then s would be 599 since in that case we would only need to generate 300 interarrival times and 299 service times (we do not need a service time for the last client since we are only interested in his or her waiting time).

It is important to point out that the definition of the integrand f corresponding to a simulation problem depends on a number of choices that have to be made when designing the simulation model and its computer implementation. In particular, the definition of f is determined by which random variables need to be simulated (e.g., successive interarrival times as we did or increments of a Poisson process as in [128]), which method is used for non-uniform random variate generation, how the uniform random numbers are assigned to the random variables to be simulated, etc. An example illustrating these choices is given in the next section. These choices can make a significant difference in the definition of f, which in turn can affect the computation time required to evaluate it and, more importantly, have an impact on the effectiveness of variance reduction techniques and quasi–Monte Carlo methods meant to improve on naive Monte Carlo estimation. For instance, so far we only talked about the inversion method to generate observations from nonuniform distributions. But other methods are available, such as *acceptance-rejection* [332]. This method is quite popular, but it does not work too well with quasi–Monte Carlo methods because it has the effect of increasing the dimension of the underlying function f. For quasi-random sampling, inversion is usually preferred.

Since nearly all applications for which simulation is used require generation of random variates from a variety of distributions, and given the fact that how this step is performed has an important impact on how we go from the simulation formulation to the integration one, a discussion of different methods that are available for generating random variates will be given in Chap. 2.

1.3 Alternative formulation of the integration problem via f: an example

In this section, we give an example that shows how different choices in the design of the simulation model and its computer implementation can impact the corresponding integration formulation for a very simple estimation problem.

Example 1.3. Suppose we want to estimate by simulation the probability that a gamma random variable with shape parameter 2 and scale parameter 0.75 is greater than 2; i.e., we want $p = P(X > 2.5)$, where $X \sim \text{Gamma}(2, 0.75)$. To do so, we assume we have access to the following functions:

`GenExpon`(u, β):	if $u \sim U(0,1)$, it returns an exponential random variate with mean β using inversion;
`GenPoisson`(u, λ):	if $u \sim U(0,1)$, it returns a Poisson variate with mean λ using inversion;
`GenGamma`$(\mathbf{u}, \alpha)$:	if $\mathbf{u} \sim U([0,1)^\infty)$, it returns a Gamma$(\alpha, 1)$ variate using *acceptance-rejection*.

The reason why the input vector $\mathbf{u}$ of uniform numbers has unbounded dimension for GenGamma is that with acceptance-rejection methods random observations must be generated until some criterion is satisfied, and therefore there is no a priori bound on the number of uniform numbers required. This will be discussed in more detail in Chap. 2.

A first approach to estimating p by simulation would be to use the fact that if X_1 and X_2 are independent exponential random variables with mean $\beta = 0.75$, then $X_1 + X_2 \sim \text{Gamma}(2, 0.75)$. Based on this, we can generate two exponential random variates using GenExpon, add them up, and check whether they exceed 2.5 or not. In other words, here we are using the *convolution approach* — also to be discussed in Chap. 2 — to generate gamma variates. This is illustrated in the left panel of Fig. 1.13.

A second approach would be to use the fact that a Gamma(2, 0.75) random variable can be thought of as the arrival time of the second event for a Poisson process with arrival rate $\lambda = 1/0.75$. This is because a Poisson process with arrival rate λ is known to have corresponding interarrival times that are exponential with mean $1/\lambda$. If we denote by N the number of events (arrivals) that have occurred for such a process by time 2.5 and use the fact that N has a Poisson distribution with mean $2.5\lambda = 10/3$, then we have that $P(X > 2.5) = P(N < 2)$. This approach is illustrated in the middle panel of Fig. 1.13.

Finally, a third approach is to directly generate gamma variates and check whether they are larger than 2.5 or not. In that case, we can use the fact that if $X \sim \text{Gamma}(\alpha, 1)$, then $\beta X \sim \text{Gamma}(\alpha, \beta)$. This is illustrated in the right panel of Fig. 1.13. For all three approaches, p can be estimated by repeated calls to SimGamma$j$, each time using different random uniform numbers as the input.

SimGamma1(u_1, u_2)	SimGamma2(u_1)	SimGamma3($\mathbf{u}$)
$x_1 \leftarrow$ GenExpon(u_1,0.75)	$N \leftarrow$ GenPoisson(u_1,10/3)	$x \leftarrow$ GenGamma($\mathbf{u}$,2)
$x_2 \leftarrow$ GenExpon(u_2,0.75)	**if** ($N < 2$) **then**	$x \leftarrow 0.75x$
if ($x_1 + x_2 > 2.5$) **then**	**return** 1	**if** ($x > 2.5$) **then**
return 1	**else return** 0	**return 1**
else return 0		**else return** 0

Fig. 1.13 Pseudocode showing three different approaches to estimating the probability $p = P(\text{Gamma}(2, 0.75) > 2.5)$.

Using these three approaches, we can define three functions f_1 to f_3 — corresponding to SimGamma$j$ for $1 \le j \le 3$ — where each of them is such that its integral $I(f)$ equals the desired quantity p:

$$f_1(u_1, u_2) = \begin{cases} 1 \text{ if } \texttt{GenExpon}(u_1, 0.75) + \texttt{GenExpon}(u_2, 0.75) > 2.5, \\ 0 \text{ else;} \end{cases}$$

$$f_2(u_1) = \begin{cases} 1 \text{ if } \texttt{GenPoisson}(u_1, 10/3) < 2, \\ 0 \text{ else;} \end{cases}$$

$$f_3(u_1, u_2, \ldots) = \begin{cases} 1 \text{ if } 0.75 \times \texttt{GenGamma}((u_1, u_2, \ldots), 2) > 2.5, \\ 0 \text{ else.} \end{cases}$$

Estimators for p can then be obtained as

$$\hat{p}_j = \frac{1}{n} \sum_{i=1}^{n} f_j(\mathbf{u}_i).$$

Note that f_1 is bidimensional, while f_2 is one-dimensional, and f_3 is defined over $[0, 1)^\infty$. Figure 1.14 shows $f_1(u_1, u_2)$ (top) and $f_2(u)$ (bottom). Although these two functions look quite different, they both integrate to the desired quantity p, which is why in each case an approximation for their integral gives an (unbiased) estimator for p. Note that if we were to use inversion to generate gamma variates within the SimGamma3 approach, then the corresponding function would be almost the same as $f_2(u)$, but reflected around $u = 0.5$.

What can be learned from this example is that, for a given estimation problem for which simulation is used, there are many different equivalent integration formulations, and each of them yields a different estimator. Although all of these different estimators usually have the same expectation, their variance might be significantly different after applying variance reduction techniques and/or quasi-random sampling. The successful application of such techniques thus requires a clear understanding of this issue. This is one of the reasons why the integration versus simulation formulation is a recurrent theme in this book.

1.4 A primer on uniform random number generation

As we have seen already, the practical implementation of the Monte Carlo method requires the use of a random number generator. So far, this has been encapsulated within the generic function Rand01(), which is assumed to produce i.i.d. uniform numbers between 0 and 1. Algorithms that can be used to implement this type of function will be discussed in detail in Chap. 3. However, because these generators are so crucial to the Monte Carlo method, we want to say a few words on this important topic before going further.

Although intuitively one may think that the best way to generate random numbers is to use some kind of physical device, in practice (e.g., in programming languages and software) *pseudorandom number generators* are used instead. Those are deterministic programs that output numbers $u_1, u_2, \ldots$ that

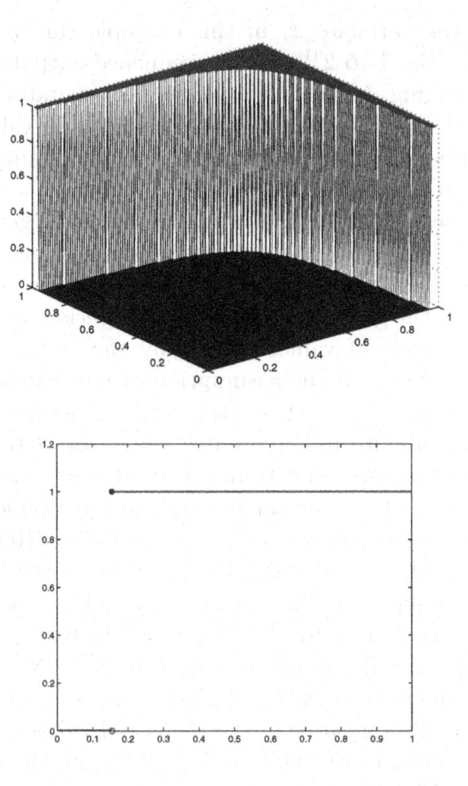

Fig. 1.14 Graphical representation of $f_1(u_1, u_2)$ (top) and $f_2(u)$ (bottom).

look like they are i.i.d. $U(0,1)$. They are preferred over physical devices because the latter are typically slower, do not allow the possibility to be "reset" so that the same sequence can be output again, and are also hard to analyze.

An example of a generator is the following function [123], which is a special case of a *linear congruential generator (LCG)* [276]:

$$\begin{aligned} x_i &= 950706376\ x_{i-1} \bmod m, \\ m &= 2^{31} - 1 = 2147483647, \\ u_i &= x_i/m, \\ x_0 &= 1 \text{ (seed)}. \end{aligned}$$

Hence, this generator outputs a sequence of numbers starting with

$$\begin{aligned} u_0 &= 1/2147483647 = 4.66\text{e} - 10, \\ u_1 &= 950706376/2147483647 = 0.44271, \\ u_2 &= (950706376^2 \bmod 2147483647)/2147483647 = 0.0601, \end{aligned}$$

and so on. Since the variables x_i in this example can only take values in the set of integers from 1 to $2^{31} - 2$, the sequence output by this generator eventually starts cycling. In fact, this particular generator can be shown to have a period length of $2^{31} - 2$, which is maximal for the value m used for the modulo but is quite short for many typical studies. Examples of generators with a longer period are discussed in Chap. 3.

An important question is: How do we know if a given generator is good or not? To answer this, several tests have been designed to assess the quality of random number generators. There are *theoretical tests*, which typically study structural aspects of the generator over its whole period, and *statistical tests*, which consider a sample of values output by the generator and use it to formally verify statistically if the assumption of true randomness should be rejected or not. We will discuss these tests in more detail in Chap. 3.

A word of caution about random number generators is that before starting a simulation study it is important to make sure the generator used has been tested appropriately and can be safely used. In our opinion, two examples of generators that can be safely used are L'Ecuyer's MRG32k3a, for which C code is given in [252], and Matsumoto and Nishimura's Mersenne-Twister [310]. The latter is implemented in MatLab®7 and has a period of $2^{19937} - 1$; the former has a period close to 2^{191}. Examples of bad generators found in commercial packages are discussed in [105, 109, 255, 269, 274]. Well-known examples are the infamous generator RANDU that was included in the IBM Scientific Subroutine Library used in the 1960s and 1970s [220], the LCG that was used in Excel®prior to its 2007 version [493], and the generator `ran1()` published in [379]. An important anomaly of that generator is discussed in [432].

Users are also advised not to attempt to change the seed of a generator without a proper understanding of its behavior. An interesting example of what can happen if this is not done properly can be found in [311] and at the link [492]. Also, a common approach is to use the computer's internal clock to choose a seed. This is typically done by users who do not like the idea that the same random numbers are used each time they call their program, something that in their view goes against the idea that the numbers are supposed to be random. One possible problem with this approach is that it is not guaranteed that the sequences used from two different seeds chosen in this way will not overlap. Another problem is that the seed returned by the clock itself might not be uniformly distributed, and thus to be safe one should actually test this uniformity (in addition to the generator's uniformity) before employing this method. Users who want different streams of numbers should instead use generators that can create different substreams of numbers that are *guaranteed* not to overlap. Such generators are discussed in [272].

1.5 Using Monte Carlo to approximate a distribution

In the examples we have seen so far, the Monte Carlo method has only been used to estimate expectations, and typically the first moment of the distribution is the focus of interest. However, the sample $h(\mathbf{x}_1), \ldots, h(\mathbf{x}_n)$ used to construct the estimator

$$\frac{1}{n}\sum_{i=1}^{n} h(\mathbf{x}_i)$$

for an expectation of the form $\mathrm{E}(h(\mathbf{X}))$ can clearly be used to extract more information on the distribution of $h(\mathbf{X})$ in addition to its mean.

In particular, the CDF of $h(\mathbf{X})$ can be approximated by the *empirical CDF*

$$\hat{F}_n(y) = \frac{1}{n}\sum_{i=1}^{n} \mathbf{1}_{y_i \leq y}, \tag{1.13}$$

where $y_i = h(\mathbf{x}_i)$, $i = 1, \ldots, n$. The empirical CDF $\hat{F}_n$ is discontinuous, but continuous variants can be obtained by using interpolation (see Prob. 1.15 and also [23]). Note that, for each y, $\hat{F}_n(y)$ is an unbiased estimator of $F(y) = P(h(\mathbf{X}) \leq y)$. Hence, by the strong law of large numbers, $\hat{F}_n$ converges in distribution to the CDF of $Y = h(\mathbf{X})$ as n goes to infinity.

Once we have an approximation for the CDF $F(\cdot)$ of the variable $Y = h(\mathbf{X})$ of interest, we can also get estimates for quantiles. That is, for $0 < p < 1$, we can estimate the $100p$*th quantile* of $Y = h(\mathbf{X})$, given by

$$q_p = F^{-1}(p) = \inf\{y : F(y) \geq p\}.$$

More precisely, based on $\hat{F}_n$, we can estimate $F^{-1}(p)$ by

$$\hat{q}_p = \inf\{y : \hat{F}_n \geq p\} = y_{(\lceil np \rceil)},$$

where $y_{(1)} \leq \ldots \leq y_{(n)}$ are the order statistics of the sample $y_1, \ldots, y_n$.

Alternative quantile estimators can be obtained based on variants of $\hat{F}_n$ (see Prob. 1.15). It is important to note that, in general, $\hat{q}_p$ is a biased estimator of q_p and that the method used to estimate the CDF influences the size of the bias. Example 1.4 illustrates this in a very simple setting. Although this bias goes to 0 with n under minimal conditions, in some circumstances it might be worthwhile to assess its magnitude using techniques such as *bootstrapping* (see Prob. 1.16).

Example 1.4. Suppose $n = 4$ and $y_1, \ldots, y_4$ is an i.i.d. sample from the $U(0,1)$ distribution. Then the estimator for the median is $\hat{q}_{0.5} = y_{(3)}$, for which the expectation is $\mathrm{E}(Y_{(3)}) = 3/5$ since $Y_{(3)}$ has a beta distribution with parameters $(3, n+1) = (3, 5)$. However, $F^{-1}(0.5) = 0.5$ for the $U(0,1)$ distribution since its CDF is $F(x) = x$ for $0 \leq x \leq 1$. Therefore, the estimator $y_{(3)}$ has a bias of $3/5 - 1/2 = 1/10$ in this case. More generally, for a sample of size n, we have

$$\hat{q}_{0.5} = \begin{cases} y_{((n+1)/2)} & \text{if } n \text{ is odd,} \\ y_{(n/2+1)} & \text{if } n \text{ is even.} \end{cases}$$

Therefore, for n odd, $\hat{q}_p$ has no bias, but for n even, the bias is

$$\frac{n/2+1}{n+1} = \frac{n+2}{2(n+1)} - \frac{1}{2} = \frac{1}{2(n+1)},$$

which goes to 0 with n. Note that if we were using linear interpolation to define $\hat{F}_n$, then the corresponding estimator for the median would be

$$\frac{1}{2}(y_{(n/2)} + y_{(n/2+1)})$$

with a bias of

$$\frac{1}{2}\left(\frac{(n/2)+(n/2)+1}{n+1} - 1\right) = 0.$$

Confidence intervals for quantiles can be obtained if we have a central limit theorem for the estimate $\hat{q}_p$, as discussed for example in [401] and also in [23, 156, 178]. The approximate confidence intervals for q_p thus obtained have the form

$$\hat{q}_p \pm \frac{\sqrt{p(1-p)}}{\sqrt{n}\psi(\hat{q}_p)} z_{\alpha/2}$$

at the $100(1-\alpha)\%$ level, where $\psi(\cdot)$ is the pdf of the random variable under study. Comparing this with the confidence interval

$$\hat{p}_n \pm \frac{\sqrt{\hat{p}_n(1-\hat{p}_n)}}{\sqrt{n}} z_{\alpha/2}$$

for $p = P(Y > y)$ based on $\hat{p}_n = 1 - \hat{F}_n(y)$, we see that the standard error is multiplied by a factor of $1/\psi(q_p)$ in the case of the quantile.

Another way of obtaining confidence intervals for quantiles is to find integers m_1 and m_2 such that $1 \le m_1 < m_2 \le n$ and $P(y_{(m_1)} < q_p < y_{(m_2)}) = 1-\alpha$ using the fact that

$$P(y_{(m_1)} < q_p < y_{(m_2)}) = \sum_{l=m_1}^{m_2-1} \binom{n}{l} p^l (1-p)^{n-l}$$

and then use $(y_{(m_1)}, y_{(m_2)})$ as a confidence interval.

1.6 Two more examples

We end this chapter by presenting two more examples taken from applications where simulation is a useful tool for estimating quantities of interest. In each case, we discuss the formulation of the problem using the function representation $f(\mathbf{u})$ and its relation to the choice of the simulation model. To facilitate this discussion, for each problem we present pseudocode in which the random source of input is represented by uniform numbers u_j, for instance obtained by prior calls to `Rand01()`.

Gillespie's method for chemical simulations

In biology and chemistry, systems that interact via a set of chemical reactions are often studied. More precisely, here we follow [142] and assume we have K different types of reactions and M different types of molecules inside a space whose volume is equal to V. Let $X_j(t)$ be the number of molecules of type j present in the system at time t. When a reaction takes place, it affects the system by modifying the number of molecules according to some vector $\boldsymbol{\nu}$ in $\mathbb{Z}^K$. For instance, if $K = 3$, then $\boldsymbol{\nu} = (-1, -1, 1)$ describes a reaction whereby one molecule of type 1 and one molecule of type 2 react and are transformed into one molecule of type 3. In what follows, $\boldsymbol{\nu}_k$ denotes this vector for the kth reaction type in the system.

Realistic models of such systems typically view these reactions as occurring randomly. The stochastic model proposed by Gillespie in [142] makes use of a *propensity function* $r_k(\mathbf{X})$ for each reaction k, where $\mathbf{X} = \mathbf{X}(t) = (X_1(t), \ldots, X_M(t))$ gives us the number of molecules of each type at time t. This function is such that

$$\begin{aligned} r_k(\mathbf{X})dt &= \text{probability, given } \mathbf{X} = \mathbf{X}(t)\text{, that a reaction of} \\ &\qquad \text{type } k \text{ will occur between } t \text{ and } t + dt \\ &= r_k N_k(\mathbf{X}), \end{aligned}$$

where r_k is a constant that depends on the physical properties of the reactants, the volume V, and the temperature of the system, while $N_k(\mathbf{X})$ is the number of possible subsets of molecules at time t that can be used as the reactants for the kth reaction. For example, if $\boldsymbol{\nu}_k = (-1, -1, 1)$ as above, then $N_k(\mathbf{X}) = X_1(t)X_2(t)$ is the number of different pairs that can be formed using one molecule of type 1 and one of type 2. If $\boldsymbol{\nu}_k = (-2, 0, 1)$, then $N_k(\mathbf{X}) = X_1(t)(X_1(t) - 1)/2$ since we then need two (unordered) molecules of type 1.

Given a certain system with an initial number $X_1(0), \ldots, X_M(0)$ of molecules and a number K of different reactions, we want to study its behavior as measured by the number of molecules of each type over a certain interval of time $[0, T]$.

This system can be simulated exactly simply by generating the time τ until the next reaction and the type κ of reaction. For this purpose, we use the fact that at time t the joint distribution of τ and κ conditioned on $\mathbf{X}$ is given by

$$\varphi_{\tau,\kappa}(\tau, k|\mathbf{X}, t) = r_\kappa(\mathbf{X}) \exp(-r_0(\mathbf{X})\tau), \qquad \tau \geq 0, \kappa = 1, \ldots, K,$$

where

$$r_0(\mathbf{X}) = \sum_{k=1}^{K} r_k(\mathbf{X}).$$

Moreover, this conditional joint density function can be rewritten as

$$\varphi_{\tau,\kappa}(\tau, \kappa|\mathbf{X}, t) = \varphi_\tau(\tau|\mathbf{X}, t)\varphi_\kappa(k|\tau, \mathbf{X}, t), \tag{1.14}$$

where

$$\varphi_\tau(\tau|\mathbf{X}, t) = r_0(\mathbf{X}) \exp(-r_0(\mathbf{X})\tau), \ \tau > 0,$$

is the marginal density function of τ and

$$\varphi_\kappa(k|\tau, \mathbf{X}, t) = \frac{r_k(\mathbf{X})}{r_0(\mathbf{X})}, \ k = 1, \ldots, K, \tag{1.15}$$

is the conditional density function of κ given τ and $\mathbf{X}$ at time t. Note that, given $\mathbf{X}$, κ is independent of τ and thus (1.15) is the marginal density function of κ given $\mathbf{X}$. Hence the marginal distribution of τ is exponential with mean $1/r_0(\mathbf{X})$, and κ has a discrete distribution with probabilities proportional to the individual propensity functions, evaluated at the current time t.

Using (1.14), we can proceed as in Fig. 1.15 to simulate this system for a certain period of time T. In that code, we assume that, for $\mathbf{r} = (r_1, \ldots, r_K)$ and $u \sim U(0, 1)$, the function DiscDist$(K, \mathbf{r}, u)$ returns a variate equal to k, with probability proportional to r_k, for $k = 1, \ldots, K$. As will be discussed in Chap. 2, one way to do this is to return the index k such that

$$\sum_{l=1}^{k-1} \tilde{r}_l \leq u < \sum_{l=1}^{k} \tilde{r}_l,$$

where

$$\tilde{r}_l = \frac{r_l}{\sum_{l=1}^{K} r_l}.$$

Also, we assume the function Number$(k, \mathbf{X})$ returns the number $N_k(\mathbf{X})$ of combinations of molecules that can react together for a reaction of type k, given the vector $\mathbf{X}$ containing the number of molecules of each type.

In the code given in Fig. 1.15, we have chosen to use the uniform numbers u_j in chronological order inside the simulation. That is, we use u_1 to generate the first reaction time, u_2 to generate that reaction's type, then u_3 to generate the second reaction time, and so on. This is similar to what we did in our bank example from Sect. 1.2.

```
Gillespie(T,X,r_1,...,r_K,u_1,u_2,...)
    t ← 0
    j ← 1 // keeps track of which uniform u_j we are using
    while t < T do
        r_0 ← 0
        for l = 1 to K
            r[l] ← r_l Number(l,X)
            r_0 ← r_0 + r[l]
        τ ← −r_0 ln(1 − u_j)
        t ← t + τ
        if t < T then
            κ ← DiscDist(K,r,u_{j+1})
            X ← X + ν_κ
            j ← j + 2
    return(X)
```

Fig. 1.15 Code for the simulation algorithm of Gillespie. The input $\mathbf{X}$ gives the initial number of molecules of each type, and $\mathbf{r} = (r[1], \ldots, r[K])$.

Now, suppose that for some molecule type j we want to write $X_j(T)$ as a function of the uniform numbers $u_1, u_2, \ldots$ that are used to generate the reaction times τ and types κ in this system. To do so, we first define

$$\begin{aligned} N &= \text{number of reactions that occurred in } [0,T] \\ &= \sum_{j=1}^{\infty} \mathbf{1}_{\tau_1+\ldots+\tau_j<T}; \\ N_k(T) &= \text{number of reactions of type } k \text{ that occurred in } [0,T] \\ &= \sum_{j=1}^{N} \mathbf{1}_{j\text{th reaction is of type } k}. \end{aligned}$$

Using this formulation, we might think that N is only a function of $u_1, u_3, \ldots,$ since τ_1 is a function of u_1, τ_2 is a function of u_3, and so on. But N also depends on the other uniform numbers — those used to generate the type of reaction — since the transformation

$$\tau_j = -r_0(\mathbf{X}) \ln(1 - u_{2j-1})$$

depends on the value $r_0(\mathbf{X}) = r_0(\mathbf{X}(T_{j-1}))$ of the sum of the propensity functions at the time $T_{j-1} = \tau_1 + \ldots + \tau_{j-1}$ when τ_j is generated, which in turn depends on which reactions were chosen at the previous times $T_1, \ldots, T_{j-2}$.

Similarly, the indicator

$$\mathbf{1}_{j\text{th reaction is of type } k}$$

is a function of u_{2j}, but it also depends on $r_0(\mathbf{X}(T_{j-1}))$ and each individual propensity function $r_l(\mathbf{X}(T_{j-1}))$, as these define the probability with which the jth reaction type is chosen.

Hence, for this problem, the formulation $f(\mathbf{u})$ that shows how the uniform numbers u_j are transformed into the output of interest is quite complex and not particularly enlightening. But we describe it anyway, and to do so it is useful to introduce the recursive function

$$\begin{aligned}\eta_j &:= \eta_j(u_2, u_4, \ldots, u_{2j}) \\ &= \text{value of } \mathbf{X}(t) \text{ between the } j\text{th and } (j+1)\text{th reactions} \\ &= \eta_{j-1}(u_2, u_4, \ldots, u_{2j-2}) + \boldsymbol{\nu}_{\mathtt{DiscDist}(K, \mathbf{r}(\eta_{j-1}(u_2, u_4, \ldots, u_{2j-2})), u_{2j})},\end{aligned}$$

where $\eta_0 = \mathbf{X}(0)$. We then have

$$\begin{aligned}T_j &:= T_j(u_1, u_2, \ldots, u_{2j-1}) = \text{time of the } j\text{th reaction} \\ &= T_{j-1}(u_1, \ldots, u_{2j-3}) - r_0(\eta_{j-1}(u_2, \ldots, u_{2j-2})) \ln(1 - u_{2j-1}),\end{aligned}$$

where $T_0 = 0$, and

$$N := N(u_1, u_2, \ldots) = \sum_{l=1}^{\infty} \mathbf{1}_{T_l(u_1, \ldots, u_{2l-1}) \leq T},$$

the value of which can be determined using only a finite number of variables u_j for reasons similar to those outlined in our discussion of Example 1.2. Finally, we have

$$N_k(T_j) := N_k(T_j; u_2, u_4, \ldots, u_{2j}) = \sum_{l=1}^{j} \mathbf{1}_{k = \mathtt{DiscDist}(K, \mathbf{r}(\eta_{l-1}(u_2, u_4, \ldots, u_{2l-2})), u_{2l})}$$

and then write

$$\mathbf{X}(T) = \mathbf{X}(0) + \sum_{k=1}^{K} N_k(T_{N(u_1, u_2, \ldots)}; u_2, u_4, \ldots, u_{2N}) \boldsymbol{\nu}_k,$$

for which the jth component is precisely $X_j(T)$. Hence $X_j(T)$ is a function of $u_1, u_2, \ldots, u_{2N+1}$, where N is the number of reactions that took place between 0 and T and is thus random. Hence the corresponding function has unbounded dimension.

Other approaches to simulate this system could lead to completely different functions f. For instance, an alternative simulation model described in [142] is to generate, after each reaction, a tentative time τ_k for the next reaction for each reaction type $k = 1, \ldots, K$. Then we let the next reaction time τ

be the minimum of these K reaction times $\tau_1, \ldots, \tau_K$, with the reaction type κ being the one corresponding to the time τ_κ that achieves the minimum. Hence a total of K uniform numbers are needed to generate each pair(τ, κ), instead of two as in the approach described in Fig. 1.15. If we are interested in replacing random sampling by quasi-random sampling, the first formulation is advantageous for $K > 2$, because it should require fewer random numbers, thus making use of point sets of lower dimension.

Equity-linked contract with a surrender option

Here we study a simplified version of a risk management problem for a life insurance company inspired by the problem discussed in [166]. Consider a life insurance policy with a maturity of 25 years issued to an individual of age x at time 0 and defined so that, in return for a premium P paid at time 0, the insured (or his or her estate) receives a payment

$$C(k) = \min\left(P, P\left(\frac{S(k)}{S(0)}\right)^\alpha\right)$$

at time k in case of death between ages $x+k-1$ and $x+k$, where $S(k)$ is the value of an index at time k and α is a constant in $(0,1)$. That is, the insured is guaranteed to at least get his or her money back, and if the index does well then he or she gets an appreciation of the amount P paid that is related to the return of the index. If the insured is still alive at age $x+25$, then $C(25)$ is paid at time 25. The insured also has the option of withdrawing the capital insured, but pays a penalty β in this case. That is, if the insured surrenders the contract between age $x+k-1$ and age $x+k$, where $k < 25$, then he or she receives $(1-\beta)C(k)$ at time k.

Suppose the company adopts the following (very naive) strategy. Upon receipt of the premium P, it invests all of it in the index, the behavior of which is described by $S(\cdot)$. Hence the company incurs a loss at time $k < 25$ if $S(k) < S(0)$ and the insured dies during that year since it then pays out P at time k while only holding $P \times S(k)/S(0) < P$. However, it realizes a gain if $S(k) > S(0)$ since in that case it pays $P \times (S(k)/S(0)^\alpha)$, which is smaller than the amount $P \times (S(k)/S(0))$ held.

Suppose the goal is to determine the probability that the value of the insurer's portfolio will become negative when holding a certain number m of these contracts, given the strategy above. To estimate this probability, assumptions must be made on mortality, surrender behavior, and the dynamics for the index. To keep things simple, here we assume that the decision to surrender is independent of the behavior of the index and that we have a multiple-decrement table providing both mortality and surrender rates at any age x. We denote by $q_x^{(d)}$ and $q_x^{(w)}$ the probability that between age x and $x+1$, an individual of age x will die or surrender his or her contract, respectively. For the index, we assume a lognormal model, where $\log(S(t)/S(0))$

follows a normal distribution with mean $(\mu - \sigma^2)t$ and variance $\sigma^2 t$, where μ and σ are the return rate and volatility of the index, respectively.

In Fig. 1.16, we give pseudocode to simulate a portfolio of 1000 contracts sold to individuals of age 40. The code returns 1 if the value of the fund at time k, denoted $V(k)$, becomes negative for some $k \in \{1, \ldots, 25\}$. We assume all payments are made at the end of the year. We also assume that individuals are independent, so that the number X_k of departures between age $x+k$ and $x+k+1$ — either due to death or surrender — has a binomial distribution with parameters (L_k, q_k), where

$$\begin{aligned} q_k &= q_{x+k}^{(d)} + q_{x+k}^{(w)}, \\ L_k &= \text{number of contracts still in place at time } k. \end{aligned}$$

We ask the reader to verify in Prob. 1.18 that, conditioned on X_k, the number of deaths D_k between age $x+k$ and age $x+k+1$ has a binomial distribution with parameters $(X_k, q_{x+k}^{(d)}/q_k)$.

```
EqLinked(25,μ,σ,P,α,β,u_1,...,u_75)
  L ← 1000 // number of contracts held
  V ← L × P // value of the portfolio
  k ← 1
  S ← 1 // normalized value of the index
  neg ← 0 // indicator of V < 0
  while k ≤ 25 and V > 0
      q ← q^(d)_{40+k-1} + q^(w)_{40+k-1}
      X ← Binom(L, q, u_{3k-2})
      D ← Binom(X, q^(d)_{40+k-1}/q, u_{3k-1})
      W ← X − D
      R ← exp(μ − σ^2/2 + σ× Norm01(u_{3k}))
      S ← S × R
      if S < 1 then
          C ← P
      else
          C ← P × S^α
      V ← V × R − D × C − W × (1 − β) × C
      L ← L − D − W
      k ← k + 1
      if V ≤ 0 then
          neg ← 1
  return(neg)
```

Fig. 1.16 Pseudocode for the risk management problem. We assume the function Norm01(u) returns a variate from the standard normal distribution if $u \sim U(0,1)$ (see Prob. 1.10).

Define

$$\tau = \min\{k : V(k) \leq 0\}$$

to be the first year where the fund's value became zero or less. To show how the indicator function

$$\mathbf{1}_{\tau \leq 25} \tag{1.16}$$

can be written as a function of $\mathbf{u} = (u_1, \ldots, u_{75})$, we use the following intermediate functions:

$$\begin{aligned}
X_k &:= X_k(u_1, u_4, \ldots, u_{3k-2}) = \text{number of departures in year } k \\
&= \texttt{Binom}(L_{k-1}, q_k, u_{3k-2}), \\
L_k &:= L_k(u_1, u_4, \ldots, u_{3k-2}) = L_{k-1} - X_k \\
&= \text{number of contracts still in place at time } k, \text{ where } L_0 = m, \\
D_k &:= D_k(u_1, u_2, u_4 \ldots, u_{3k-2}, u_{3k-1}) = \text{number of deaths in year } k \\
&= \texttt{Binom}(X_k, q_{x+k}^{(d)}/q_k, u_{3k-1}), \\
W_k &:= W_k(u_1, u_2, u_4, \ldots, u_{3k-2}, u_{3k-1}) = X_k - D_k, \\
R_k &:= R_k(u_3, u_6, \ldots, u_{3k}) = S(k)/S(0) \\
&= \text{cumulative return on the index at time } k \\
&= \exp(k(\mu - \sigma^2/2) + \sigma Z_k), \\
Z_k &:= Z_k(u_3, u_6, \ldots, u_{3k}) = \texttt{Norm01}(u_3) + \ldots + \texttt{Norm01}(u_{3k}), \\
C_k &:= C_k(u_3, u_6, \ldots, u_{3k}) = \text{death capital paid at time } k \\
&= P \times \min(1, R_k).
\end{aligned}$$

We can then write the value of the fund at time k as

$$\begin{aligned}
V(k) := V(k; u_1, \ldots, u_{3k}) &= V(k-1) \times \exp[(\mu - \sigma^2/2) + \sigma \texttt{Norm01}(u_{3k})] \\
&-C_k(u_3, u_6, \ldots, u_{3k}) \times (D_k(u_1, u_2, u_4, \ldots, u_{3k-2}, u_{3k-1}) \\
&+(1-\beta) \times W_k(u_1, u_2, u_4, \ldots, u_{3k-2}, u_{3k-1})).
\end{aligned}$$

Finally, we write the indicator function (1.16) as

$$\begin{aligned}
f(u_1, \ldots, u_{75}) = &\mathbf{1}_{V(1;u_1,u_2,u_3) \leq 0} + \mathbf{1}_{V(1;u_1,u_2,u_3)>0, V(2;u_1,\ldots,u_6) \leq 0} + \cdots \\
&+\mathbf{1}_{V(1;u_1,u_2,u_3)>0,\ldots,V(24,u_1,\ldots,u_{72})>0,V(25,u_1,\ldots,u_{75}) \leq 0}.
\end{aligned}$$

Let us discuss how alternative simulation models for this problem would affect the definition of f. First, since the behavior of the index is independent from the other random variables in this problem — the number of deaths and surrenders each year — we could simulate it first, for instance using the first 25 uniform numbers $u_1, \ldots, u_{25}$. In addition, rather than generating the values of the index sequentially in time, we could have generated them in any desired order by using the *Brownian bridge* formulation. This will be discussed in more detail in Chap. 6. Next, rather than simulating the number of deaths and surrenders each year by using the binomial distribution, we could also have chosen to generate each year, for each contract that is still held in the portfolio, whether the individual holding the contract will die, surrender, or stay in the portfolio. That would require one uniform number per contract each year, which for a large portfolio represents a big increase in the dimension of the problem, in addition to making the number of uniform numbers, and thus the dimension, random. A less naive approach would be to simulate at time 0, for each individual, the pair (k, j) indicating in which year k they leave the portfolio and for what reason $j \in \{\text{death, surrender, end of contract}\}$, something that can be done using at most two uniform random numbers per individual. However, this would still require more uniform numbers than in the approach described in Fig. 1.16.

Problems

1.1. Consider the function $f(u) = \sqrt{1-u^2}$ defined over $[0,1)$. (a) Evaluate $I(f) = \int_0^1 f(u)du$. (b) If $U \sim U(0,1)$, what is $\sigma^2 = \mathrm{Var}(f(U))$? (c) Use the Monte Carlo method to estimate Q_n based on $n = 10, 1000$, and 100,000 points. In each case, compute a 95% confidence interval for $I(f)$ based on your estimates for $I(f)$ and σ. Comment on the behavior of the size of the half-width of your confidence interval as n grows, and compare it with the exact size of the half-width.

1.2. Consider the function $f(u_1, u_2, u_3) = u_1 + \sin(2\pi u_2) + u_3^2$ defined over $[0,1)^3$. (a) Evaluate $I(f)$. (b) Estimate $I(f)$ using (i) the Monte Carlo method with $n = 1000$ points and (ii) the multivariate trapezoidal rule with $N = 9$. Repeat with $n = 8000$ and $N = 19$. Compare the error $|Q_n - I(f)|$ obtained for each of the two methods.

1.3. Consider the functions (i) $f_1(\mathbf{u}) = (\prod_{j=1}^s e^{u_j})/c_1$ and (ii) $f_2(\mathbf{u}) = \sum_{j=1}^s e^{u_j}/c_2$. (a) Find the constants c_1 and c_2 such that f_1 and f_2 both have an integral of 1. (b) Using the constants found in (a), compare for both (i) and (ii) the error obtained by the Monte Carlo method, the multivariate trapezoidal rule, and Simpson's rule for $s = 5, 10$ and n =59,049 and then for $s = 15$ and n =14,348,907. (c) If you cannot afford more than 10^6 function evaluations, what is the largest dimension s for which you can still use Simpson's rule (or the trapezoidal rule)?

1.4. Show that for any function $f(\mathbf{u})$ of the form $f(\mathbf{u}) = \sum_{j=1}^s g(u_j)$, where $g : [0,1) \to \mathbb{R}$, and any given integer $N \geq 1$, the integration error obtained for this function with the trapezoidal rule based on $n = (N+1)^s$ points is constant as s increases.

1.5. We mentioned in Sect. 1.2 that the function GenExpon(u, β) could be implemented using inversion. This means that the value x returned by GenExpon(u, β) is such that $F(x) = u$, where $F(x) = 1 - e^{-x/\beta}$ is the CDF of an exponential random variable with mean β. Using this, implement GenExpon(u, β) and create an i.i.d. sample of size $n = 1000$ from the exponential distribution with $\beta = 2$. Construct a relative frequency plot based on this sample and compare it with the pdf of an Exp(2). Verify that the sample average and sample variance are "close" to their theoretical values of 2 and 4, respectively.

1.6. *Hit-and-Miss.* The hit-and-miss method [45, 165, 391] is an alternative to Monte Carlo integration. It works as follows. Suppose you want to compute

$$I(f) = \int_A f(\mathbf{u})d\mathbf{u},$$

where $f(\mathbf{u}) < \infty$ for all $\mathbf{u} \in A$ and $A \subseteq \mathbb{R}^s$ is such that you can generate random variables $\mathbf{u}_i$ that are uniformly distributed over A. The idea is to find a constant M such that $f(\mathbf{u}) \leq M$ for all $\mathbf{u} \in A$ and then generate an i.i.d. sample $(\mathbf{u}_1, w_1), \ldots, (\mathbf{u}_n, w_n)$ uniformly distributed over $A \times [0, M]$. Then, let

$$y_i = \begin{cases} 0 & \text{if } f(\mathbf{u}_i) \leq w_i \\ 1 & \text{otherwise,} \end{cases}$$

and estimate $I(f)$ by

$$H_n = \frac{1}{n} \sum_{i=1}^{n} y_i \times \text{Vol}(A) \times M.$$

(a) Show that H_n is an unbiased estimator of $I(f)$. (b) Devise a hit-and-miss algorithm to estimate the integral $I(f)$ of $f(u) = \sqrt{1-u^2}$ over $u \in [0, 1)$. (b) Compute a 95% confidence interval for $I(f)$ based on $n = 1000$ with your hit-and-miss algorithm from (a). (c) Compare the half-width of the interval obtained in (b) with that of a 95% confidence interval based on Monte Carlo integration and $n = 1000$, as computed in Prob. 1.1. (d) Compare the theoretical variance of your hit-and-miss estimator with that of the Monte Carlo estimator based on the same value of n. Is this comparison consistent with your answer to part (c)?

1.7. *Verifying matrix multiplication.* In theoretical computer science, the term "Monte Carlo algorithm" typically refers to a probabilistic algorithm that tests a certain property of a mathematical system and returns a correct answer with probability at least p for any instance considered. The algorithm is then said to be p-correct. A well-known example is the *Miller-Rabin* algorithm for testing primality [323, 381]. Here we consider *Freivald's algorithm* [42, 131], which can be used to verify matrix multiplication and works as follows. Suppose you have three $d \times d$ matrices, A, B, and C, and want to test whether $C = AB$ or not. The idea is to randomly generate a binary vector X in $\{0,1\}^d$ and return yes if $(XA)B = XC$ and no otherwise. This algorithm can be shown to be p-correct with $p = 0.5$. Furthermore, when it returns *false*, we can be sure that it is correct. The problem is that when it returns *true* we cannot determine if it is correct or if it is making a mistake in the case where $AB \neq C$. (a) Show that if you run this algorithm five times, the probability of obtaining a correct answer is at least $31/32$ for any choice of matrices A, B, and C. (b) Implement this algorithm with

$$A = \begin{bmatrix} 1\,2\,3 \\ 4\,5\,6 \\ 7\,8\,9 \end{bmatrix}, B = \begin{bmatrix} 3\,1\,2 \\ 4\,6\,5 \\ 8\,7\,9 \end{bmatrix},$$

and the following three cases for C:

$$C_1 = \begin{bmatrix} 35 & 34 & 39 \\ 80 & 76 & 87 \\ 125 & 118 & 135 \end{bmatrix}, C_2 = \begin{bmatrix} 35 & 34 & 38 \\ 80 & 77 & 87 \\ 125 & 118 & 135 \end{bmatrix}, C_3 = \begin{bmatrix} 35 & 34 & 38 \\ 80 & 76 & 87 \\ 125 & 118 & 135 \end{bmatrix}.$$

Run it (at most) five times to make a decision (each time with a different X), and then repeat the process 1000 times. For each of C_1, C_2, and C_3, how many times do you get the correct answer out of 1000 trials? Comment on your result in light of your answer to (a). (c) Repeat part (b) with ten trials instead of five to make a decision. What is the value p such that you can say that your algorithm is at least p-correct for any choice of matrices A, B, and C?

1.8. (a) Show that each of the functions f_1 to f_3 defined in Example 1.3 has an integral $I(f)$ equal to p. (b) Compute the variance $\mathrm{Var}(f_i(\boldsymbol{U}))$ for $1 \leq i \leq 3$.

1.9. (a) Estimate the probability p defined in Example 1.3 using the function f_1 based on $n = 5000$ function evaluations. Compare this with the true value of p. (b) Extend the idea used in (a) to estimate $\tilde{p} = P(\mathrm{Gamma}(20, 0.75) > 25)$ and compare it with an approximate value for $\tilde{p}$, for instance by using the function `gammainc` in Matlab.

1.10. We mentioned in Sect. 1.1 that there was no closed-form formula for the CDF $\Phi(x)$ of a Normal(0,1) random variable. However, there exist good approximations for $\Phi(x)$, such as Hasting's approximation, which is presented in [1, p. 932]. Also, several mathematical software packages have functions that compute such approximations (for example, `normcdf` in Matlab). (a) Compare the value returned by the Monte Carlo estimate $Q_n + 0.5$ with $n = 1000$ and as described in (1.2) for (i) $\Phi(1.282)$, (ii) $\Phi(1.645)$, and (iii) $\Phi(1.96)$, with an approximation such as those mentioned above. (b) For a given value of c, what is the theoretical variance of Q_n? (c) For each of the three values of c used in (a) (i.e., 1.28, 1.645, and 1.96), compare the theoretical variance of Q_n with its estimated variance (based on $n = 1000$). (d) Consider the following alternative approach for estimating $\Phi(c)$, where we assume that the function `Norm01()` returns i.i.d. standard normal variates (this is of course a very unlikely way to proceed in practice since if we have a way to generate normal variates via the `Norm01` function, we could also probably find a better approximation for $\Phi(c)$):

```
SimPhi(c)
    sum ← 0
    for i = 1 to n
        x ← Norm01()
        if (x < c)
            sum ← sum + 1
    R_n ← sum/n
    return (R_n)
```

What is the theoretical variance of R_n? (e) Use the estimator R_n to estimate (i) $\Phi(1.28)$, (ii) $\Phi(1.645)$, and (iii) $\Phi(1.96)$, and compare the estimated variance of R_n with its theoretical variance. To implement `Norm01()`, you can use inversion and an approximation for Φ^{-1} such as the one given in [216, pp. 95–96], or you can use the function `randn` in Matlab.

1.11. Consider the queueing example discussed in Sect. 1.2. (a) Compute an approximate 95% confidence interval for the *fraction* of clients that will wait more than 10 minutes on a given day using (i) $n = 10$, (ii) $n = 100$, (iii) $n = 1000$, and (iv) $n = 10{,}000$ simulations. Compute the *relative half-width* of the confidence interval in each case, and discuss its behavior as n increases. (b) We explained in Sect. 1.2 that $s = \infty$ for this problem, although for a given simulation the number S of uniform numbers required to evaluate f is finite. Compute the expected value of S. (c) For $n = 1000$, what is the average value of S obtained with your simulation program?

1.12. Assume the following model for one share of IBM stock. At time $t > 0$, the value of the stock $S(t)$ follows a lognormal distribution; i.e., $\ln S(t)$ has a normal distribution with mean $\ln(S(0)) + (r - \sigma^2/2)t$ and variance $\sigma^2 t$, where σ is the *volatility* of the stock and r is the risk-free interest rate. In what follows, assume $S(0) = 100$, $r = 0.05$, and $\sigma = 0.2$. (a) Write an expression involving the CDF Φ of a Normal(0,1) to describe the probability p_K that $S(T)$ is larger than some fixed quantity $K > 0$. (b) Using pseudocode, describe how you could use n simulations of the price $S(T)$ to estimate p_K. (c) For $T = 1$ and $K = 110$, compare an approximate value for p_K (based on approximations for $\Phi(x)$ such as those mentioned in Prob. 1.10) with the one obtained using simulation as in (b), with $n = 1000$. Does the 95% confidence interval based on these n simulations contain the approximate value? (d)

Describe two different functions f_1 and f_2 defined over $[0,1)^s$ for some s (s does not need to be the same for f_1 and f_2) whose integral is equal to $E(S(T))$ for $T = 1$.

1.13. (a) Repeat part (c) of the previous problem using a *different* stream of pseudorandom numbers. (Make sure you do this correctly. One possibility is to perform two computations (e.g., using a loop) within one call to the program. Changing the seed/state arbitrarily can lead to overlapping streams, as discussed on p. 24.) Are the two confidence intervals obtained with the two different streams comparable? (b) Perform the same comparison as in (a), but with (i) $n = 10$ and $n = 10{,}000$. Comment on the differences observed with respect to the value of n.

1.14. Repeat Prob. 1.2, only with Monte Carlo, but with $n = 10$ and $n = 1000$, and repeat the process $m = 100$ times (computing a 95% confidence interval each of the m times). (a) Compute the sample variance of your estimator Q_n based on these $m = 100$ samples, and compare it with the true variance of Q_n. (b) Out of the m times, how many times did your confidence interval contain the true value for $I(f)$?

1.15. Consider the empirical CDF $\hat{F}_n$ described in (1.13). (a) Propose a modification to $\hat{F}_n$ based on linear interpolation, thereby obtaining a continuous empirical CDF $\tilde{F}_n$. (b) Derive an expression for the estimate $\tilde{q}_p$ of the pth quantile of F based on $\tilde{F}_n$.

1.16. Suppose we have a sample $X_1, \ldots, X_n$ of i.i.d. observations from a certain distribution and an estimator $\theta = \theta(X_1, \ldots, X_n)$ defined over the sample. For instance, θ might be the sample variance. It is sometimes of interest to investigate the properties of the distribution of θ. One way to do this is to use the *bootstrap technique*, introduced by Efron in [97] and surveyed in detail in [98, 100]. This technique is based on the following approach:

1. For $i = 1, \ldots, B$:
   a. Randomly and uniformly choose n indices $l_{i,1}, \ldots, l_{i,n}$ from $\{1, \ldots, n\}$ with replacement.
   b. Compute
   $$\hat{\theta}_i = \theta(x_{l_{i,1}}, \ldots, x_{l_{i,n}}).$$
2. Use the obtained sample $\hat{\theta}_1, \ldots, \hat{\theta}_B$ to infer on the desired property. For instance, if the goal is to estimate $\mathrm{Var}(\theta)$, then we can use the estimator
   $$\hat{\sigma}_\theta^2 = \frac{1}{B-1} \sum_{i=1}^{B} (\hat{\theta}_i - \bar{\theta})^2,$$
   where
   $$\bar{\theta} = \frac{1}{B} \sum_{i=1}^{B} \hat{\theta}_i.$$

To get a confidence interval for $\mathrm{E}(\hat{\theta}_i)$, we can either use a percentile approach and construct the $100(1-\alpha)\%$ confidence interval

$$\left(\hat{\theta}_{(B\alpha/2)}, \hat{\theta}_{(B(1-\alpha/2))}\right)$$

or a central limit theorem approach with

$$\left(\bar{\theta} - z_{\alpha/2}\frac{\hat{\sigma}_\theta}{\sqrt{B}}, \bar{\theta} + z_{\alpha/2}\frac{\hat{\sigma}_\theta}{\sqrt{B}}\right).$$

(a) Use bootstrapping to estimate the variance of the relative half-width of the confidence interval discussed in Prob. 1.11 (with $n = 1000$) using $B = 100$ draws. Based on this, compute a 95% confidence interval for this relative half-width. (b) Explain how you would perform Step 2 of the bootstrapping algorithm described above in order to estimate the bias of an estimator θ.

1.17. In the τ-*leap approach* suggested by Gillespie in [143], the chemical system described in Sect. 1.6 is simulated approximately by using a discretization in time steps of size τ within which the propensity functions are assumed to remain constant throughout $[k\tau, (k+1)\tau)$ for $k = 0, 1, \ldots, T/\tau - 1$. (a) For this approximate model, what is the distribution of the number of reactions of type k occurring between $k\tau$ and $(k+1)\tau$? (b) Based on your answer to (a), propose an algorithm for simulating (approximately) the chemical system described in Sect. 1.6 based on the τ-leap approach.

1.18. Show that for the equity-linked problem whose pseudocode is given in Fig. 1.16, conditioned on X_k, the number of deaths between age $x + k$ and age $x + k + 1$ has a binomial distribution with parameters $(X_k, q_{x+k}^{(d)}/q_k)$.

1.19. Explain how you would proceed to generate the pair (k, j) giving the time k of "departure" from the portfolio and reason j for one individual at time 0, as discussed at the end of Sect. 1.6.

Chapter 2
Sampling from Known Distributions

In this chapter, we give an overview of different methods that can be used to generate random variates from a given distribution. Even if inversion should be the preferred choice for quasi–Monte Carlo users, it is important to be aware of other methods that are available for that purpose. First of all, inversion is sometimes slower and more difficult to apply than other methods. In such cases, Monte Carlo users may prefer these other methods. Also, when working with predefined functions (e.g., `randn` in Matlab) to generate observations from a given distribution, it is quite possible that the underlying method is not based on inversion. In addition, there are applications for which the common approach used by people working in that area is to use something other than inversion (e.g., in computer graphics, for ray generation). In such cases, even if ultimately the quasi–Monte Carlo user will try to use inversion instead of these other methods in order to modify code or algorithms appropriately, it is important to understand what the other method does. Finally, in some cases inversion may not be directly applicable, and an alternative method needs to be used.

We assume the reader is familiar with common distributions such as those already encountered in Chap. 1 — exponential, gamma, binomial, and normal — and will not describe specifically how to handle each one of these in this chapter. Instead, we wish to describe general techniques that can be used for a variety of models. More precisely, we describe four general approaches that can be used for generating random variates from a given (univariate) distribution and then talk about the multivariate case. Much more extensive coverage of specific distributions and algorithms can be found in [45, 75, 196, 243, 391]. In particular, Luc Devroye's book (which is out of print) can be downloaded from his Web page [485].

Before we do this, we want to briefly discuss a few distributions that are often encountered in simulation models.

C. Lemieux, *Monte Carlo and Quasi–Monte Carlo Sampling*,
Springer Series in Statistics 692, DOI: 10.1007/978-0-387-78165-5_2,

2.1 Common distributions arising in stochastic models

Our goal in this section is simply to talk about a few distributions that are commonly used in stochastic models. Our discussion is by no means extensive, as we restrict ourselves to distributions arising in the different examples used throughout the book.

Normal and Lognormal Distribution

The normal distribution arises very often in financial simulation models. We already saw an example in Sect. 1.6 when discussing equity-linked contracts. One reason why it arises so often is that the *Brownian motion* is often used as a building block to model asset prices, and the increments of a Brownian motion are normally distributed. Because of the importance of this process, we give a formal definition before going further. The reader is referred to [212, 350, 388] for more information.

Definition 2.1. A *standard Brownian motion* is a continuous-time stochastic process $\{B(t), t \geq 0\}$ with the following properties:

1. $B(0) = 0$.
2. The increments over disjoint intervals are independent. That is, for $r < s < t < u$, $B(u) - B(t)$ and $B(s) - B(r)$ are independent.
3. The increments are stationary. That is, for any $r, s, t > 0$, $B(r+t) - B(r)$ and $B(s+t) - B(s)$ have the same probability function, which is normal with mean $\mu = 0$ and variance t.

If $\{B(t), t \geq 0\}$ is a standard Brownian motion, then for $\sigma > 0$ and $\mu \in \mathbb{R}$, the process $\{\sigma B(t) + \mu t, t \geq 0\}$ is a Brownian motion with *drift* μ and *diffusion coefficient* σ.

The simplest financial model that uses a Brownian motion is the lognormal model encountered in Chap. 1, which amounts to having the asset price $S(t)$ at time t given by

$$S(t) = S(0) \exp\left((\mu - \sigma^2/2)t + \sigma B(t)\right),$$

where μ and σ are the instantaneous return rate and volatility of the asset price, respectively. Since $B(t) \sim N(0, t)$, we have that $S(t)$ has a lognormal distribution with parameters $((\mu - \sigma^2/2)t, \sigma^2 t)$.

In financial simulations, the multinormal distribution is also often encountered either when modeling a vector of financial assets — in which case they are driven by Brownian motions that are correlated — or when looking at a given asset value at different times.

Exponential, Gamma, Weibull, and Poisson distributions

The exponential distribution is frequently encountered in simulation models, partly because Poisson processes are often used to model stochastic processes that count the occurrence of a certain event — for example, client arrivals in a queue, molecular reactions in a chemical system, claims arrivals for an insurance company — and in this case the interarrival time between two events is known to have an exponential distribution.

The gamma distribution shows up in financial models that include jumps, as we discuss in Sect. 7.2 of our chapter on financial applications. It also arises as the distribution of the kth event from a Poisson process and more generally as a sum of exponential random variables. The Weibull distribution arises as the minimum of a sample of i.i.d. exponential random variables. All three distributions can also be used to model failure times.

The Poisson distribution is used to count the number of events in a Poisson process. An example was discussed in Prob. 1.17. Users may sometimes want to draw from it directly rather than generating exponential interarrival times until a certain time limit is reached. Inversion can be used to do that, and specific aspects of this task are discussed in [129].

Beta distribution

The beta distribution often arises when studying order statistics. More precisely, it comes up when we look at a sample of n i.i.d. $U(0,1)$ random variables $u_1, \ldots, u_n$, because then the ith smallest observation $u_{(i)}$ has a beta distribution with parameters $(i, n+1-i)$.

Copula-based models

Models based on copulas have become increasingly popular over the last ten years or so, for instance in biostatistics and risk management [104, 130]. Formally, a copula is a joint distribution C defined over $[0,1]^k$ and such that each marginal distribution is a $U(0,1)$. A theorem by Sklar [404] says that for any joint CDF $F(x_1, \ldots, x_k)$ with given marginal CDFs $H_1(x_1), \ldots, H_k(x_k)$, there exists a copula such that we can write

$$F(x_1, \ldots, x_k) = C(H_1(x_1), \ldots, H_k(x_k)). \tag{2.1}$$

By writing the joint CDF $F(x_1, \ldots, x_k)$ in this way, we specify the distribution in two steps. We start by choosing the marginal distributions and then introduce the dependence relation between the variables X_j via the copula function C. This formulation also naturally suggests the use of inversion to generate $(x_1, \ldots, x_k)$. We will come back to copulas in Sect. 2.6.

2.2 Inversion

This method goes back to the beginnings of Monte Carlo. It was proposed by von Neumann in a letter to Stan Ulam discussing their "random numbers work" [95]. We discussed on p. 16 of Chap. 1 how to use inversion for the exponential distribution. More generally, for a continuous distribution with CDF $F(\cdot)$, it can be applied as in Fig. 2.1.

1. $U \leftarrow$ `Rand01()`.
2. Return $X = F^{-1}(U)$.

Fig. 2.1 Steps to apply inversion for continuous distributions.

This looks very simple, but the applicability and effectiveness of this method rests on how easy it is to compute the inverse CDF F^{-1}. For the exponential, Weibull (see Prob. 2.2), and other distributions, the inverse function can be determined rather easily. But for the normal, gamma, beta, and other distributions, in particular those that do not have closed-form expressions for the corresponding CDF, inversion cannot be applied directly, and an approximation for F^{-1} must first be determined. For instance, Kennedy and Gentle discuss rational fraction approximations for the inverse CDF of a normal distribution [216, pp. 95–96]. In that setting, $F^{-1}(u)$ can be approximated by a function of the form [349]

$$F^{-1}(u) \approx t + \frac{p_0 + p_1 t + p_2 t^2 + p_3 t^3 + p_4 t^4}{q_0 + q_1 t + q_2 t^2 + q_3 t^3 + q_4 t^4}$$

for $u > 0.5$ and constants q_i, p_i, where $t = (\ln(1/u^2))^{1/2}$. The case $u < 0.5$ is handled by using the symmetry of the normal pdf, which implies that $F^{-1}(u) = -F^{-1}(1-u)$. Another well-known approximation for the inverse CDF of a normal, which is particularly popular in finance [145, p. 68], is the one proposed by Moro [324]. For other distributions, approximations have been implemented in various software packages and libraries, for example in Matlab's statistical toolbox.

For a distribution that is not continuous, inversion is applied as shown in Fig. 2.2. We give in Fig. 2.3 an example where a simple discrete distribution with $P(X = x)$ equal to 0.22, 0.16, 0.33, and 0.29 for $x = 0, 1, 2, 3$, respectively, is inverted. If U falls in the interval $[0, 0.22)$, we return $X = 0$; in $[0, 22, 0.38)$, we return $X = 1$; in $[0.38, 0.71)$, we return $X = 2$; and in $[0.71, 1]$, we return $X = 3$. This clearly causes X to have the correct distribution.

Several known discrete distributions are such that $\inf\{y : F(y) \geq u\}$ can be determined explicitly. For instance, if X has a geometric distribution with parameter p, then $P(X = x) = p(1-p)^x$, where $x \in \{0, 1, \ldots\}$. Therefore,

1. $U \leftarrow$ `Rand01()`.
2. Return $X = \inf\{y : F(y) \geq U\}$.

Fig. 2.2 Steps to apply inversion for noncontinuous distributions.

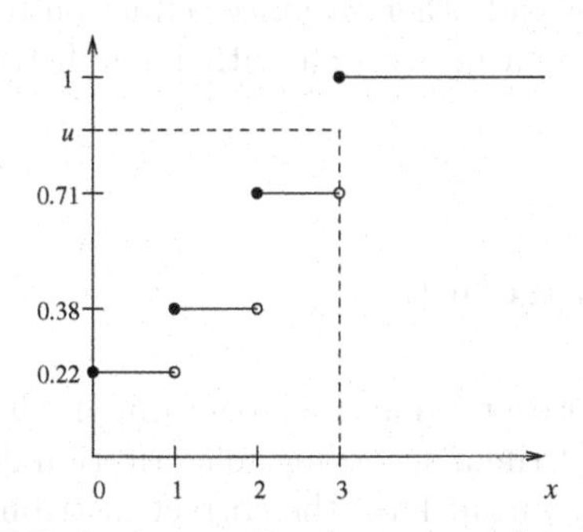

Fig. 2.3 Inverting the CDF of a discrete distribution over $\{0, 1, 2, 3\}$. The u shown is such that inversion returns $x = 3$.

$$F(x) = \sum_{y=0}^{x} p(1-p)^y = (1 - (1-p)^x),$$

and thus

$$\begin{aligned}
\inf\{y : F(y) \geq u\} &= \inf\{y : (1 - (1-p)^y) \geq u\} \\
&= \inf\{y : 1 - u \geq (1-p)^y\} \\
&= \inf\{y : (1-u)^{1/y} \geq 1 - p\} \\
&= \inf\{y : (1/y)\ln(1-u) \geq \ln(1-p)\} \\
&= \lceil \ln(1-p)/\ln(1-u) \rceil.
\end{aligned}$$

Just as in the continuous case, though, for some distributions we might not be able to derive an explicit expression for $\inf\{y : F(y) \geq u\}$. When this happens, using inversion turns out to be a searching problem, where for a given U the goal is to quickly find the index i such that

$$\sum_{j=0}^{i-1} p_j < U \leq \sum_{j=0}^{i} p_j, \tag{2.2}$$

where $p_j = P(X = x_j)$, and we assumed the domain of X was $\{x_0, x_1, \ldots\}$, where $x_j \leq x_{j+1}$ for all $j \geq 0$. (We also assumed that the sum $\sum_{j=0}^{-1} p_j = 0$.) As required, the index i satisfying (2.2) is the smallest one such that $F(x_i) \geq U$. Of course, one can perform a simple linear search starting from $i = 0$ in order to identify the correct index, but more efficient methods can (and should) be used. For instance, we can use a binary search rather than a linear one, or a "bucket scheme" meant to improve on binary search [45].

Even if inversion is sometimes slower than other methods, the fact that it uses one uniform number per random variate and transforms this number in a monotone way makes it the preferred choice when used in combination with quasi–Monte Carlo and other variance reduction techniques. As we will see below, it also works naturally well with joint distributions specified by copula functions.

2.3 Acceptance-rejection

Here the idea is to generate random variates from an alternative distribution and then accept or reject them according to a criterion designed so that overall the variates that are output have the correct distribution. More precisely, to generate random variates with a pdf $\varphi(x)$, we first find a function $t(x)$ that is majoring $\varphi(x)$ over its domain (i.e., $t(x) \geq \varphi(x)$ for all x) and whose integral is finite. Note that $t(x)$ itself usually is not a density function since

$$T = \int t(x)dx \geq \int \varphi(x)dx = 1, \tag{2.3}$$

but $r(x) := t(x)/T$ is a density function. The function $t(x)$ should be chosen so that it is easy to generate observations from $r(x)$. The algorithm described in Fig. 2.4 can then be used.

1. Generate Y having density $r(x)$.
2. Generate $U \sim U(0, 1)$, independent of Y.
3. If $U \leq \varphi(Y)/t(Y)$, then return $X = Y$; otherwise go back to step 1.

Fig. 2.4 Steps for acceptance-rejection.

To understand why acceptance-rejection works, we follow the proof given in [243, App. 8A]. We first notice that each time we go through the three steps above, a pair (Y, U) is generated. To be accepted, a pair must be such that $U \leq \varphi(Y)/t(Y)$. Hence, an observation X output by this algorithm has the same distribution as $(Y|U \leq \varphi(Y)/t(Y))$; i.e., the conditional distribution of Y given that Y is accepted. Therefore,

$$P(X \leq x) = P(Y \leq x|U \leq \varphi(Y)/t(Y)) = \frac{P(Y \leq x, U \leq \varphi(Y)/t(Y))}{P(U \leq \varphi(Y)/t(Y))}.$$

Now,

$$P\left(Y \le x, U \le \frac{\varphi(Y)}{t(Y)}\right) = \int_{-\infty}^{x} P\left(U \le \frac{\varphi(y)}{t(y)}\right) r(y)dy = \int_{-\infty}^{x} \frac{\varphi(y)}{t(y)} r(y)dy$$
$$= \frac{1}{T}\int_{-\infty}^{x} \varphi(y)dy = \frac{F(x)}{T},$$

where $F(x)$ is the CDF corresponding to $\varphi(x)$, and T is as defined in (2.3). In addition, we have

$$P\left(U \le \frac{\varphi(Y)}{t(Y)}\right) = \int_{-\infty}^{\infty} \frac{\varphi(y)}{t(y)} r(y)dy = \frac{1}{T}.$$

Hence $P(X \le x) = F(x)$, as required.

Figure 2.5 illustrates the acceptance-rejection method in the case where $\varphi(x) = 12x^2(1-x)$ for $0 \le x \le 1$, which corresponds to the Beta distribution with parameters $\alpha = 3$ and $\beta = 2$. Since the maximum of $\varphi(x)$ occurs at $x = 2/3$, where $\varphi(x) = 16/9$, this means we can take $t(x) = 16/9$, for $x \in [0,1]$, corresponding to a uniform density $r(x)$ over $[0,1]$. In Fig. 2.5, we show $\varphi(x)$, $t(x)$, and 200 points corresponding to trials $(Y, Ut(Y))$. When the second coordinate $Ut(Y)$ is below $\varphi(Y)$, the point is accepted; otherwise it is rejected. For this particular sample, 111 points were accepted and 89 were rejected for a proportion $111/200 = 0.555$ of acceptance, not too far from the theoretical one of $1/T = 9/16 = 0.5625$.

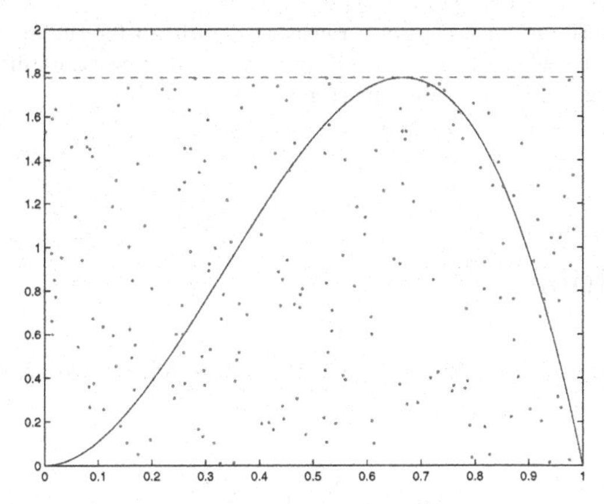

Fig. 2.5 Acceptance-rejection method for $\varphi(x) = 12x^2(1-x)$ (solid line); $t(x) = 16/9$ is the dotted line.

For practical applications, one should obviously try to use a majoring function $t(x)$ that more closely follows the pdf under consideration. By doing so, the probability $1/T$ of accepting Y increases, which causes the expected number of trials to decrease. To illustrate this, Fig. 2.6 gives an example of an acceptance-rejection algorithm for the Gamma$(k, 1)$ distribution [391, 431]. The majoring function in this case is based on a Laplace distribution and is such that

$$\frac{\varphi(x)}{t(x)} = \left|\frac{(\theta-1)x}{\theta(k-1)}\right|^{k-1} \exp\left(-x + \frac{|x-(k-1)| + (k-1)(\theta+1)}{\theta}\right). \quad (2.4)$$

The Laplace distribution with location parameter $k-1$ and scale parameter θ — also called *double exponential* — is described by the pdf [391]

$$r(x) = \frac{1}{2\theta} \exp\left(\frac{|x-(k-1)|}{\theta}\right). \quad (2.5)$$

The alternative name double exponential comes from the fact that, when $k = 1$, for $x > 0$ the pdf (2.5) is just a scaled exponential pdf, which is reflected around the y-axis to get the $x < 0$ part. The pdf (2.5) is simple enough that we can easily use inversion to perform Step 1 of the algorithm described in Fig. 2.6; see Prob. 2.10.

1. Generate a Laplace variate Y with location parameter $k-1$ and scale $\theta = 1 + \sqrt{4k-3}/2$.
2. If $Y < 0$, then return to Step 1.
3. $U \leftarrow$ `Rand01()`.
4. If $U \leq \varphi(Y)/t(Y)$, then return Y; otherwise go back to Step 1.

Fig. 2.6 Steps describing an acceptance-rejection algorithm for the gamma distribution with parameters $(k, 1)$, where $\varphi(\cdot)/t(\cdot)$ is given in (2.4). At least two uniform numbers are used every time we go through these four steps.

2.4 Composition

This method can be used when the CDF from which we want to generate observations can be written as a sum,

$$F(x) = \sum_{i=1}^{\infty} p_i F_i(x), \quad (2.6)$$

where $p_i \geq 0$, $\sum_{i=1}^{\infty} p_i = 1$, and each $F_i(\cdot)$ is a CDF. Hence a random variable with a CDF of the form (2.6) is such that with probability p_i it has a distribution determined by $F_i(\cdot)$. We can then use the algorithm shown in Fig. 2.7 to generate variates from a CDF of the form (2.6).

Of course, each of the two steps themselves require that some generating method be used, for instance inversion based on two independent uniform numbers U_1 and U_2 (one for generating I, the other for X). Note also that,

1. Generate I according to $P(I = i) = p_i$.
2. Return an observation X having CDF $F_I(\cdot)$ and independent from I.

Fig. 2.7 Steps describing how to use composition to generate random variates.

unlike inversion, we need at least two uniform numbers to generate one variate.

The composition method arises naturally for mixture distributions, but it can also be useful for tackling complicated density functions by breaking them down into different components, in which case p_i corresponds to the area under the curve of the ith component. We illustrate this idea in Example 2.2.

Example 2.2. Consider the beta density function $\varphi(x) = 12x^2(1 - x)$ for $0 \leq x \leq 1$. Here we can form a piecewise linear function as illustrated in Fig. 2.8. This function passes through the maximum of $\varphi(x)$ occurring at $(2/3, 16/9)$; the inflection point $(1/3, 8/9)$, where the second derivative of $\varphi(x)$ becomes negative; the endpoint (1,0); and the point $(1/9, 0)$ obtained by drawing a line from the inflection point $(1/3, 8/9)$ that has the same slope as $\varphi(x)$ at that point. (This slope is given by 4.) The remainder of the area under the curve of $\varphi(x)$ can then be split into three areas. The area under the curve of the piecewise linear function can be shown to be 68/81, which means that about 84% of the draws based on the composition method will require generating observations from a distribution with a piecewise linear pdf, something that is relatively easy to achieve (see Prob. 2.6). Problem 2.5 at the end of the chapter asks you to find the corresponding values of p_i and $F_i(x)$, $i = 1, \ldots, 4$, for Fig. 2.8.

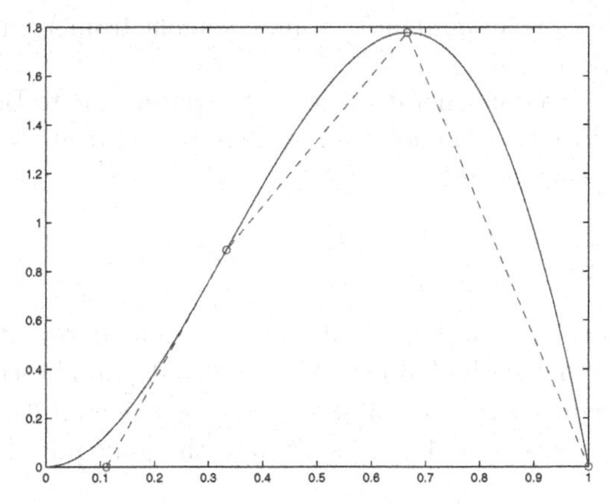

Fig. 2.8 Composition applied to the beta pdf $12x^2(1 - x)$. The area under the curve is partitioned into four pieces.

2.5 Convolution and other useful identities

The convolution method is useful for random variables that can be written as a sum of i.i.d. random variables, typically coming from a simpler distribution. More precisely, we assume $X = Y_1 + \ldots + Y_n$, where the Y_i are i.i.d. random variables. Well-known examples are as follows:

1. $X \sim \text{Gamma}(n, \beta)$: $Y_i \sim \text{Exp}(\beta)$.
2. $X \sim \chi^2(n)$: $Y_i = Z_i^2$, where $Z_i \sim N(0,1)$.
3. $X \sim \text{Binomial}(n, p)$: $Y_i \sim \text{Bernoulli}(p)$.
4. $X \sim \text{Negative Binomial}(n, p)$: $Y_i \sim \text{Geometric}(p)$.

The main disadvantage of this method is that it requires that n random variates be generated in order to get a single observation from X.

More generally, relationships between different distributions can be used for random variate generation. For instance, Fox [126] uses the fact that, for a sample of n i.i.d. uniform variates in $[0, 1]$, the ith order statistic has a beta distribution with parameters $(i, n + 1 - i)$. Based on this, he suggests the method shown in Fig. 2.9 for generating a random variate $X \sim \text{Beta}(a, b)$, where a and b are positive integers.

1. Generate $a + b - 1$ i.i.d. uniform numbers in $(0, 1)$.
2. Return the ath smallest observation.

Fig. 2.9 Steps for generating a beta variate with parameters (a, b) using ranked data.

Another way of generating a beta variate is to use the fact that if Y_1 is a Gamma$(a, 1)$ and Y_2 is a Gamma$(b, 1)$, independent from Y_1, then $Y_1/(Y_1 + Y_2)$ is a Beta(a, b).

Finally, a clever way of generating normal variates, due to Box and Muller [36], exploits the idea that the joint pdf of two independent standard normal variables x and y is given by

$$\varphi_{X,Y}(x, y) = \frac{1}{2\pi} e^{-(x^2+y^2)/2}, -\infty < x, y < \infty.$$

We can then perform a change of variables using polar coordinates — which is why a variation of this method, due to Marsaglia [301] and based on rejection, is called the *polar method* — as follows: $r = \sqrt{x^2 + y^2}$ and $\theta = \arctan(y/x)$. Hence we have $x = r\cos\theta$ and $y = r\sin\theta$, and the joint pdf of r and θ is

$$\varphi_{R,\Theta}(r, \theta) = \frac{|J|}{2\pi} e^{-r^2/2}, r > 0, 0 \le \theta \le 2\pi,$$

where $|J|$ is the Jacobian of the transformation given by

$$\begin{vmatrix} \cos\theta & -r\sin\theta \\ \sin\theta & r\cos\theta \end{vmatrix} = r\cos^2\theta + r\sin^2\theta = r.$$

Hence $\varphi_{R,\Theta}(r,\theta) = (r/2\pi)e^{-r^2/2}$ with corresponding CDF

$$F_{R,\Theta}(r,\theta) = (\theta/2\pi)(1 - e^{-r^2/2}), r > 0, 0 \le \theta \le 2\pi.$$

Thus r and θ are independent, and we can generate them by inversion as

$$r = \sqrt{-\ln 2(1-U_1)},$$
$$\theta = 2\pi U_2.$$

Transforming these back into x and y gives us the Box-Muller method described in Fig. 2.10. This method is quite popular for generating normal variates, but users should know that the sample produced when the source of randomness is a simple LCG has abnormal properties, as is illustrated nicely in [314].

$U_1 \leftarrow$ `Rand01()`
$U_2 \leftarrow$ `Rand01()`
$X_1 \leftarrow \sqrt{-2\ln(1-U_1)}\cos(2\pi U_2)$
$X_2 \leftarrow \sqrt{-2\ln(1-U_1)}\sin(2\pi U_2)$
return (X_1, X_2)

Fig. 2.10 Pseudocode for the Box-Muller method. It returns two independent standard normal variates.

2.6 Multivariate case

Here we consider the problem of generating vectors $(x_1, \ldots, x_k)$ of observations with a joint CDF $F(x_1, \ldots, x_k)$. First, a general approach that can be used is what we could call *nested conditioning* [243], where we generate each variate $x_1, \ldots, x_k$ successively, starting with x_1, for which we need the marginal distribution $F_{X_1}(x)$ given by

$$F_{X_1}(x) = \int_{-\infty}^{x}\int_{-\infty}^{\infty}\cdots\int_{-\infty}^{\infty}\varphi(x_1,\ldots,x_k)dx_k\ldots dx_2dx_1,$$

where $\varphi(x_1,\ldots,x_k)$ is the joint pdf associated with the CDF F. Once we have x_1, then we generate x_2 *conditionally* on x_1. That is, we generate an observation x_2 from $F_{X_2|X_1}(x|x_1)$ given by

$$F_{X_2|X_1}(x|x_1) = \int_{-\infty}^{x}\int_{-\infty}^{\infty}\dots\int_{-\infty}^{\infty}\frac{\varphi(x_1,\dots,x_k)}{\varphi_1(x_1)}dx_k\dots dx_3dx_2,$$

where $\varphi_1(x_1)$ is the marginal pdf of X_1. We continue like this until the last variate x_k, generated from the conditional distribution

$$F_{X_k|X_1,\dots,X_{k-1}}(x|x_1,\dots,x_{k-1}).$$

Of course, for this method to be applicable, we need to be able to determine the marginal and conditional distributions and have a way of generating variates from each of them. Also, the efficiency of the method depends heavily on the order we chose for generating the variates x_i. That is, among the $k!$ possible choices, some might lead to a much faster generation of the vector $(x_1,\dots,x_k)$ [391].

Here is a simple example to illustrate this method.

Example 2.3. Suppose we want to generate a vector (x_1, x_2) having the joint pdf

$$\varphi(x_1,x_2) = \begin{cases} 2 & \text{if } 0 \le x_2 \le x_1 \le 1 \\ 0 & \text{else.} \end{cases} \tag{2.7}$$

We have that the marginal pdf of X_1 is

$$\varphi_1(x_1) = \int_0^{x_1} 2dx_2 = 2x_1, \qquad 0 \le x_1 \le 1,$$

and thus the marginal CDF of X_1 is

$$F_{X_1}(x_1) = x_1^2, \qquad 0 \le x_1 \le 1.$$

We must then get the conditional pdf of X_2 given X_1,

$$\varphi_{X_2|X_1}(x_2|x_1) = \frac{1}{x_1}, \qquad 0 \le x_2 \le x_1 \le 1,$$

so that the conditional CDF of X_2 given $X_1 = x_1$ is

$$F_{X_2|X_1}(x_2|x_1) = \frac{x_2}{x_1}, \qquad 0 \le x_2 \le x_1.$$

Overall, the algorithm shown in Fig. 2.11 can be used to generate (x_1, x_2).

Second, an important case to discuss is the multinormal distribution. That is, suppose we want to generate a vector $(x_1,\dots,x_k)$ that follows a multinormal distribution with mean $\boldsymbol{\mu} = (\mu_1,\dots,\mu_k)^{\mathrm{T}}$ and covariance matrix Σ. In that case, we can use the fact that if $\mathbf{Z} = (Z_1,\dots,Z_k)^{\mathrm{T}}$ is a vector of i.i.d. standard normal random variables, then $A\mathbf{Z}$ has a multinormal distribution with mean zero and covariance matrix AA^{T}. Hence, by using a matrix C such that $CC^{\mathrm{T}} = \Sigma$, we can use the identity

```
U_1 ← Rand01()
x_1 ← √U_1
U_2 ← Rand01()
x_2 ← x_1 U_2
return(x_1, x_2)
```

Fig. 2.11 Pseudocode for using nested conditioning for the simple bivariate distribution (2.7).

$$\mathbf{X} = \boldsymbol{\mu} + C\mathbf{Z},$$

where $\mathbf{X} = (x_1, \ldots, x_k)^{\mathrm{T}}$. To get a matrix C such that $CC^{\mathrm{T}} = \Sigma$, we can use the lower-triangular matrix obtained from the Cholesky decomposition of Σ. As we will see in Chap. 6, other choices might be more suitable when using quasi–Monte Carlo sampling.

The third case we discuss is the use of copulas to model a joint distribution. The general approach to generate a vector $(x_1, \ldots, x_k)$ of variates having the joint CDF $F(x_1, \ldots, x_k)$ given by (2.1) is shown in Fig. 2.12.

Generate $(u_1, \ldots, u_k)$ according to C.
return $x_j = H_j^{-1}(u_j)$, $j = 1, \ldots, k$.

Fig. 2.12 Steps describing the general approach for generating random variates modeled using a copula C and having marginal CDF $H_1, \ldots, H_k$.

We illustrate with the following two examples how models described by copulas tend to lend themselves nicely to the use of inversion. More examples are given in [130, 462], for instance.

Example 2.4. Consider a bivariate *Gaussian copula*. In this case, we have $C(u, v) = \Phi_{2,\rho}(\Phi^{-1}(u), \Phi^{-1}(v))$, where Φ^{-1} denotes the inverse standard normal CDF, and $\Phi_{2,\rho}$ represents the CDF of a bivariate normal with correlation coefficient ρ, for which the covariance matrix is

$$\Sigma = \begin{pmatrix} 1 & \rho \\ \rho & 1 \end{pmatrix}.$$

Here, we can generate (U_1, U_2) so that they follow C by first generating a vector (Z_1, Z_2) from the bivariate normal with correlation ρ and then set $U_1 = \Phi(Z_1)$ and $U_2 = \Phi(Z_2)$. This works since then we have

$$
\begin{aligned}
P(U_1 \le u_1, U_2 \le u_2) &= P(\Phi(Z_1) \le u_1, \Phi(Z_2) \le u_2) \\
&= P(Z_1 \le \Phi^{-1}(u_1), Z_2 \le \Phi^{-1}(u_2)) \\
&= \Phi_{2,\rho}(\Phi^{-1}(u_1), \Phi^{-1}(u_2)) = C(u_1, u_2).
\end{aligned}
$$

Note that the second equality in the display above holds because the inverse transform Φ^{-1} is a continuous and monotonically increasing function. Once we have (U_1, U_2) with the desired dependence structure — as prescribed by the copula — then we get X_1 and X_2 by applying the chosen marginal distribution to U_1 and U_2. That is, we let $X_1 = H_1^{-1}(U_1)$ and $X_2 = H_2^{-1}(U_2)$. This clearly produces a pair (X_1, X_2) with the correct distribution since

$$
\begin{aligned}
P(X_1 \le x_1, X_2 \le x_2) &= P(H_1^{-1}(U_1) \le x_1, H_2^{-1}(U_2) \le x_2) \\
&= P(U_1 \le H_1(x_1), U_2 \le H_2(x_2)) \\
&= C(H_1(x_1), H_2(x_2)).
\end{aligned}
$$

Example 2.5. A well-known family of copulas are the *Archimedean copulas*, which can be expressed as

$$
C(u_1, \ldots, u_d) = \phi^{-1}(\phi(u_1) + \ldots + \phi(u_k)),
$$

where ϕ is a convex, decreasing function with domain $(0, 1]$ and range $[0, \infty)$ such that $\phi(1) = 0$, and is called the *generator* of the copula. A member of this family is *Frank's bivariate copula*, where

$$
C(u_1, u_2) = \frac{1}{\alpha} \ln \left(1 + \frac{(\exp(\alpha u_1) - 1)(\exp(\alpha v_1) - 1)}{\exp(\alpha) - 1} \right).
$$

For this special case, correlated uniform numbers (U_1, U_2) following this bivariate CDF can be generated as in Fig. 2.13, where $\tilde{\alpha} = e^{\alpha}$ [136].

```
FrankBivCopula(α̃)
    V1 ← Rand01()
    V2 ← Rand01()
    T ← α̃^V1 + (α̃ − α̃^V1)V2
    U1 ← V1
    U2 ← log_α̃[T/(T + (1 − α̃)U2)]
    return (U1, U2)
```

Fig. 2.13 Pseudocode showing how to generate (U_1, U_2) according to Frank's bivariate copula.

Problems

2.1. Show that if $\{B(t), t \geq 0\}$ is a standard Brownian motion, then we have that $\text{Cov}(B(s), B(t)) = \min(s, t)$ for $t, s \geq 0$.

2.2. A Weibull random variable has a pdf given by

$$\varphi(x) = \frac{k}{\lambda}\left(\frac{x}{\lambda}\right)^{k-1} e^{-(x/\lambda)^k},$$

where $k > 0$ is the shape parameter and $\lambda > 0$ is the scale parameter. Describe an algorithm that uses inversion to generate random variates having a Weibull distribution with generic parameters (k, λ).

2.3. Suppose you want to generate observations from a truncated distribution. That is, for some real numbers $a < b$ and some pdf $\varphi(x)$ (with associated CDF $F(\cdot)$), $\infty < x < \infty$, you want to generate random variates having the truncated pdf

$$\tilde{\varphi}(x) = \begin{cases} \frac{\varphi(x)}{F(b)-F(a)} & a \leq x \leq b \\ 0 & \text{else.} \end{cases}$$

Assume the inverse CDF $F^{-1}(\cdot)$ can be computed. Describe an algorithm to generate variates from the truncated pdf above.

2.4. Describe an algorithm to generate observations from the continuous empirical distribution $\tilde{F}_n$ defined in Prob. 1.15.

2.5. Compute the values of p_i and $F_i(x)$ for the composition method applied to the beta pdf $\varphi(x) = 12x^2(1-x)$ discussed in Example 2.2.

2.6. Consider the pdf that corresponds to the piecewise linear function shown in Fig. 2.8, which, as discussed in Example 2.2, accounts for about 85% of the draws when using the composition method. (a) Give an expression for that pdf. (b) Give an algorithm to generate variates from this pdf using inversion.

2.7. For the beta pdf $\varphi(x) = 12x^2(1-x)$, $0 \leq x \leq 1$, implement the acceptance-rejection approach described on p. 47, and for a sample of 100,000 beta variates compute the average number of uniform variates required to output one beta variate.

2.8. An example of an acceptance-rejection algorithm to generate random variates is given in [11, p. 25]. In this case, the goal is to generate three-dimensional random unit vectors. To do so by acceptance-rejection, the idea is to generate a random point uniformly in $[-1, 1)^3$, accept it if it is within the unit sphere centered at (0, 0, 0) (and then rescale it so that its length is one), and reject it otherwise. (a) Prove that this method correctly generates a random unit vector. (b) What is the expected number of trials required in order to generate one vector? (c) Use a two-dimensional version of that method to perform the *Buffon's needle* experiment, which can be used to estimate π as follows [42]. Throw n needles of length 0.5 on a floor with planks of width 1 and infinite length; estimate π by the fraction n/k, where k is the number of times the needle fell across a crack in the floor. To simplify things, assume we want to estimate $1/\pi$ and thus can use the approximation k/n. Use $n = 1000$, and verify whether a 95% confidence interval based on this sample contains $1/\pi$ or not.

2.9. Consider a random variable X having the following probability distribution:

$$\begin{aligned} P(X = 0) &= 0.05, \\ P(X = 1) &= 0.10, \\ P(X = 2) &= 0.15, \\ P(x < X \leq y) &= c(y - x) \text{ for } 0 < x < y < 1 \\ &\qquad \text{and } 1 < x < y < 2. \end{aligned}$$

(a) Find the value of c such that the distribution above is a valid probability distribution. (b) Give an algorithm using inversion to generate random variates having the distribution above. Make sure the transformation you use is monotone.

2.10. Consider the Laplace distribution whose pdf is given in (2.5). (a) Describe one way of applying composition to generate Laplace random variates. (b) Describe how to use inversion to generate Laplace random variates.

2.11. Consider the bivariate distribution under study in the pseudocode given in Fig. 2.11. Suppose the goal is to estimate $\mu = \mathrm{E}(X_1+X_2)$ by drawing n i.i.d. pairs of observations $(x_{i,1}, x_{i,2})$ for $i = 1, \ldots, n$. (a) Compute the variance of the estimator obtained based on the approach described in Fig. 2.11. (b) Give pseudocode for the approach that consists in first generating X_2 instead of X_1. (c) Compare the variance of the estimator for μ obtained using the approach in (b) with the one from (a).

2.12. Consider a multivariate normal vector $\mathbf{X}$ with covariance matrix Σ having entries of the form $\sigma_{ij} = \sigma_i \sigma_j \rho_{ij}$, where σ_i^2 is the variance of X_i, for $i = 1, \ldots, d$, and ρ_{ij} is the correlation between X_i and X_j for $1 \leq i, j \leq d$. Give a formula for the entries of the $d \times d$ lower-triangular matrix C obtained by Cholesky decomposition of Σ.

2.13. Find the generator ϕ corresponding to the *Gumbel-Hougaard copula* [130]

$$C(u, v) = \exp\left\{-\left[(-\ln u)^\alpha + (-\ln v)^\alpha\right]^{1/\alpha}\right\}.$$

2.14. Show that the pair (U_1, U_2) output by the algorithm described in Fig. 2.13 has the desired distribution.

Chapter 3
Pseudorandom Number Generators

As seen in the previous chapters, the use of the Monte Carlo method relies on the availability of *uniform random numbers* in order to perform random sampling. Although theoretical results for this method are based on the assumption that truly uniform random numbers are used, in practice, and as mentioned in Sect. 1.4, pseudorandom numbers are used. That is, we use sequences of numbers that *look* like they are random but that are in fact produced by a deterministic algorithm called a *pseudorandom number generator* (PRNG).

The concept of randomness is hard to define and can lead to philosophical considerations that we will not attempt to discuss here. Unfortunately, the "aura of mystery" that surrounds this concept sometimes leads people to think that they can invent some bizarre function to generate random numbers or "tweak" an existing generator so that "it behaves more randomly". But as Knuth wisely said [220]: "Random numbers should not be chosen with a method chosen at random. Some theory should be used." A useful discussion of "what is a random sequence?" can also be found in Knuth's book [220, Sect. 3.5].

If we agree that randomness is a concept that is difficult to define, then it becomes even less clear what we mean by "sequences of numbers that look like they are random". A pragmatic explanation is to say we want those pseudorandom numbers to be such that results from computations based on them should lead to conclusions *similar* to those that would have been obtained with true random numbers. The approach that has been taken in the literature on random number generators in order to verify if this (vague and general) property holds for a given generator is to devise various "tests" assessing their quality. Several such tests will be discussed in this chapter.

As we mentioned in Sect. 1.4, in the past there have been bad generators proposed in the literature and/or used in various software, "bad" meaning that such generators are likely to provide invalid results for several applications. Hence it is important for anyone using pseudorandom numbers to have at least some basic knowledge about PRNGs and what makes them good or

C. Lemieux, *Monte Carlo and Quasi–Monte Carlo Sampling*,
Springer Series in Statistics 692, DOI: 10.1007/978-0-387-78165-5_3,

bad. This chapter is aimed at providing such knowledge. We do not cover all the generators that have been proposed or present all tests that can be used to assess their quality. But we think the information provided in this chapter will at least allow the reader to correctly use PRNGs and have enough background information to be able to read more complete references on this topic such as [120, 221, 248, 257, 339, 441] if needed. Also, we pay special attention to aspects of random number generators that are related to the construction of low-discrepancy point sets for quasi–Monte Carlo.

This chapter is organized as follows. First, we review basic concepts and definitions pertaining to PRNGs. We then discuss generators based on linear recurrences, which include several widely used families. A brief discussion of add-with-carry and subtract-with-borrow generators comes next, as well as a short description of nonlinear generators. We conclude with a discussion of tests that can be used to assess the quality of PRNGs. The material presented here is largely based on [248, 257].

3.1 Basic concepts and definitions

Before we start, let us first take a step back and explain why "true" random number generators are not used. Although in principle such a generator could be implemented — for example, based on principles of quantum mechanics — in practice, random number generators based on physical devices are not the ideal thing. First, measurement errors and other technical details may introduce some kind of bias or deviation from true randomness that would be hard to assess. Second, such generators may be too slow for many applications where millions of numbers are required. Third, it is sometimes useful to be able to generate more than once a sequence of "random numbers" either for debugging purposes or to use certain variance reduction techniques, such as "common random numbers", as discussed in Chap. 4. This property of generators is usually referred to as "repeatability".

Instead of using some kind of physical device, approaches based on the use of computers started to be studied and proposed around 1950. For instance, in 1955 the RAND Corporation published a table with one million random digits produced by an electronic roulette wheel [64]. Alternatively, John von Neumann proposed at the end of the 1940s the "mid-square" method to generate random numbers, whereby random digits are extracted by squaring the previous number and outputting its middle digits [332]. For example, if the current number is 3456, we square it and obtain 11,943,936, from which we extract the four middle digits 9439 and repeat the process. Although this method was quickly found not to be very useful in practice, it contains the major ingredients used to construct the generators that are used nowadays in that it uses a *deterministic algorithm* based on some kind of *recurrence* and implemented on a *computer* to generate numbers that attempt to *look random*.

More precisely, a PRNG can be described as a structure of the form (S, T, τ, ξ, x_0) [248], where

$$\begin{aligned} S &= \text{state space}, \\ T &= \text{output space}, \\ \tau : S \rightarrow S &= \text{transition function}, \\ \xi : S \rightarrow T &= \text{output function}, \\ x_0 &= \text{seed}. \end{aligned}$$

The sequence $u_0, u_1, \ldots$ produced by the PRNG is then defined as $u_i = \xi(x_i)$, for $i \geq 0$, where $x_i = \tau(x_{i-1})$ for $i \geq 1$. In other words, the function τ is used to go from one state x_{i-1} to the next x_i, and then each of these states is transformed into a number in the output space T using the function ξ. Unless otherwise stated, we assume that τ is a bijection and ξ is one-to-one. Also, all the generators that we will be looking at in this chapter have an output space T given by $[0, 1)$.

Example 3.1. Let $S = \mathbb{Z}_{11}$, the ring of integers modulo 11, $T = [0, 1)$, $\tau(x) = 6x \bmod 11$, $\xi(x) = x/11$, and $x_0 = 1$. The first 12 numbers of the resulting sequence are then

$$\begin{aligned} u_0 &= x_0/11 = 1/11, \\ u_1 &= x_1/11 = \tau(1)/11 = 6/11, \\ u_2 &= x_2/11 = \tau(x_1)/11 = \tau(6)/11 = (36 \bmod 11)/11 = 3/11, \\ u_3 &= x_3/11 = \ldots = 18 \bmod 11/11 = 7/11, \\ u_4 &= x_4/11 = \ldots = 42 \bmod 11/11 = 9/11, \\ u_5 &= x_5/11 = \ldots = 54 \bmod 11/11 = 10/11, \\ u_6 &= x_6/11 = \ldots = 60 \bmod 11/11 = 5/11, \\ u_7 &= x_7/11 = \ldots = 30 \bmod 11/11 = 8/11, \\ u_8 &= x_8/11 = \ldots = 48 \bmod 11/11 = 4/11, \\ u_9 &= x_9/11 = 2/11, \\ u_{10} &= x_{10}/11 = 1/11, \\ u_{11} &= x_{11}/11 = 6/11. \end{aligned}$$

Of course, this extremely small generator should not be used in practice. As shown above, it produces only ten different numbers between 0 and 1, and the sequence repeats these ten numbers in the same order forever. In other words, the *period* of this generator is 10, period meaning the smallest integer ρ such that $u_{i+\rho} = u_i$ for all $i \geq 0$. In this example, the small period is due to the fact that we chose a very small state space S with only 11 elements, and, by definition, a PRNG has a period of at most $|S|$ since every time $x_i = x_0$ we have that $\tau(x_{i+j}) = \tau(x_j)$ for all $j \geq 0$. In Example 3.1, any seed x_0 different from 0 produces a sequence with a period of length 10, while taking $x_0 = 0$ produces the sequence $0, 0, 0 \ldots$ of period 1. In this case, we say the generator has two possible *cycles*.

From this discussion, it is obvious that one of the important properties that a good generator should have is a very long period. How long? The period should be orders of magnitude larger than the total number N of values to be output by the generator. By orders of magnitude, a rule of thumb might be to say that the period should be at least N^2, or maybe even N^3. These magic numbers come from systematic testing of generators [262, 271], where it has been shown that if $N/\sqrt{\rho}$ is large enough — that is, N is a large enough fraction of the square root of the period ρ — then tests that look at the first N numbers output by some types of generators can detect a departure from true randomness. From this point of view, generators with a period of about 2^{31} should not be used since the square root of such periods is less than one million. Note that generators with periods of that size are still in use. For instance, one of the three generators available in Matlab 7.3.0 is a *multiplicative congruential generator* with a period of $2^{31} - 2$ [499].

In addition to the period length, there are many other quantitative properties that can be used to assess the quality of a generator by making use of various theoretical and statistical tests that have been developed for that purpose. We will come back to this in Sect. 3.5.

The important qualitative properties that a generator should have are as follows [248]: efficiency (both in terms of space and time), repeatability (as discussed on p. 58), portability (that is, the sequence output by the generator does not depend on the programming language, compiler, or machine used), ease of implementation, and jumping ahead capabilities, which are useful when the sequence output by the generator is subdivided into substreams so that one can "jump" to the next substream without having to generate all the intermediate values [272].

3.2 Generators based on linear recurrences

In this section, we discuss a few basic generators whose transition function is described by a linear recurrence over a state space of the form $\mathbb{Z}_m$ for some positive integer m.

3.2.1 Recurrences over $\mathbb{Z}_m$ for $m \geq 2$

We will start with a very simple construction called a *linear congruential generator*, which was introduced by Lehmer in 1949 [276].

Definition 3.2. A *linear congruential generator* (LCG) is a PRNG for which $S = \mathbb{Z}_m$ for some positive integer m, called the *modulus*, $\tau(x) = (ax+c) \bmod m$, where $a \in \mathbb{Z}\backslash\{0\}$ is called the *multiplier*, $c \in \mathbb{Z}\backslash\{0\}$ is the *increment*, and $\xi(x) = x/m$.

Hence, an LCG is completely determined by the modulus m, the multiplier a, and the increment c. The toy PRNG mentioned in Example 3.1 had $m = 11$, $a = 6$, and $c = 0$. When $c = 0$, the maximal period of an LCG is $m - 1$ and is obtained when m is a prime and a is a *primitive element modulo m* [291]. That is, a must be a generator of the cyclic group $(\mathbb{Z}_m^*, \cdot)$, where $\mathbb{Z}_m^*$ represents $\mathbb{Z}_m$ without the element 0, and $\cdot$ denotes multiplication modulo m. In what follows, we drop the modulo m notation for convenience, as we assume all operations are carried out in the ring $\mathbb{Z}_m$. To see why the maximal period of $m-1$ is reached when the multiplier a is a primitive element modulo m, consider the sequence output by the LCG in this case. It has the form

$$\mathcal{X} = (x_0/m, a \cdot x_0/m, (a^2 \cdot x_0)/m, \ldots, (a^{m-1} \cdot x_0)/m, (a^m \cdot x_0)/m, \ldots).$$

Since a is a primitive element modulo m, we have that $a^i \neq a^j$ for all $0 \leq i \neq j \leq m-2$ and $a^{m-1} = 1$. Since $(\mathbb{Z}_m^*, \cdot)$ is a cyclic group, $x_0 a^i \neq x_0 a^j$ for all $0 \leq i \neq j \leq m-2$, so the first $m-1$ elements of $\mathcal{X}$ are all distinct, and the mth one, $(a^{m-1} \cdot x_0)/m$, is equal to x_0/m. This means the sequence starts repeating itself at that point.

An LCG with $c = 0$ and m prime is usually called a *multiplicative linear congruential generator* (MLCG) (or sometimes just *multiplicative congruential generator*). Note that taking a nonzero increment c when m is prime only has the benefit of allowing a period of m instead of $m-1$ for the LCG. A nonzero increment is more useful when m is not prime. For example, a popular choice is to take m equal to a large power of two because arithmetic operations in $\mathbb{Z}_m$ then become easy to perform. That is, one can take $m = 2^e$, where e is the word size of the computer (e.g., $e = 32$), and then the modulo m operations are done automatically as arithmetic operations overflow. However, when m is a power of two, if $c = 0$, then the maximal period of the generator is $m/4$ and is reached when $a \bmod 8 = 5$ and x_0 is odd. If c is a nonzero odd integer and $a \bmod 8 = 5$, then the maximal period of m can be reached with m a power of two [220]. The infamous generator RANDU, known for severe defects due to abnormal correlations, is an LCG with a power-of-two modulus defined by the recurrence $x_i = 65539 x_{i-1} \bmod 2^{31}$ [220].

In practice, since $\log_2 m$ cannot exceed the word length of the computer used, LCGs cannot have a very long period and therefore should not be used.

The reason why we talk about them here is that they provide a nice first example of PRNG that can be understood easily. Also, this construction can be used to construct *recurrence-based point sets* for quasi–Monte Carlo, as discussed in Chap. 5. In addition, they can be used as the component of a *combined generator*, as given in Def. 3.4.

One way of constructing a PRNG with a longer period than an LCG is to use a recurrence of higher order for the transition function. This leads to the more general notion of a *multiple recursive generator* (MRG) [159, 220].

Definition 3.3. Let $k \geq 1$ and m be prime. A *multiple recursive generator* is a PRNG for which $S = \mathbb{Z}_m^k$, and the state $\mathbf{y}_i = (x_i, \ldots, x_{i-k+1})$ at step i evolves through the recurrence

$$x_i = \tau(\mathbf{y}_{i-1}) = (a_1 x_{i-1} + \ldots + a_k x_{i-k}) \bmod m, \qquad i \geq k, \qquad (3.1)$$

where $a_j \in \mathbb{Z}$ for $j = 1, \ldots, k$, $a_k \neq 0$, and the output is $\xi(\mathbf{y}_i) = x_i/m$.

The case where $k = 1$ corresponds to the MLCG, which, as we saw, is a special case of Def. 3.2. Another special case of an MRG is the additive *lagged-Fibonacci generator*, where the transition function is given by

$$x_i = (x_{i-r} + x_{i-k}) \bmod m.$$

For instance, Mitchell and Moore in 1958 proposed a generator based on the recurrence

$$x_i = (x_{i-24} + x_{i-55}) \bmod 2^{24}.$$

Other types of lagged-Fibonacci generators are obtained by replacing $(\mathbb{Z}_m, +)$ by another pair of operation and state space.

The maximal period that can be reached by an MRG is $m^k - 1$ and is attained when the characteristic polynomial $P(z) = z^k - a_1 z^{k-1} - \ldots - a_k$ of the recurrence is a *primitive polynomial* (over the Galois field $\mathbb{F}_m$) [291]. This means $P(z)$ must be such that the smallest integer r for which

$$z^r \equiv 1 \bmod P(z)$$

is $r = m^k - 1$. That is, the powers of z (modulo $P(z)$) from 0 to $m^k - 1$ generate the set of nonzero polynomials over $\mathbb{F}_m$ of degree less than k. Methods for testing primitivity are given in [221]. In particular, a necessary condition for $P(z)$ to be primitive is that a_k and at least one other coefficient a_r with $1 \leq r < k$ must be nonzero. For this reason, MRGs based on trinomials are often used, which then give a recurrence of the form

$$x_i = (a_r x_{i-r} + a_k x_{i-k}) \bmod m$$

that can be implemented efficiently [252].

Another way of constructing a PRNG with a long period is to combine several generators. More precisely, the idea is to run J generators in parallel and then combine their respective states in some way to get an output for the combined generator. Here we describe how to combine MRGs, an idea that has led to several successful PRNGs currently used in practice.

Definition 3.4. For $j = 1, \ldots, J$, let

$$x_{j,i} = (a_{j,1}x_{j,i-1} + \ldots + a_{j,k_j}x_{j,i-k_j}) \bmod m_j, i \geq k_j \tag{3.2}$$

be the recurrence defining the transition function of the jth generator. Let $\delta_1, \ldots, \delta_J$ be arbitrary integers and define the outputs

$$z_i = \delta_1 x_{1,i} + \ldots + \delta_J x_{J,i} \bmod m_1, \qquad u_i = z_i/m_1,$$

and

$$w_i = \left(\frac{\delta_1 x_{1,i}}{m_1} + \ldots + \frac{\delta_J x_{J,i}}{m_J}\right) \bmod 1. \tag{3.3}$$

Then both z_i and w_i can be used as output for a combined MRG.

Let ρ_j be the period of the MRG defined by the recurrence (3.2). It can be proved that under some conditions both sequences $u_0, u_1, \ldots$ and $w_0, w_1, \ldots$ output by (3.3) have a period length ρ equal to the least common multiple of $\rho_1, \ldots, \rho_J$, and the sequence (3.3) is equivalent to an MRG with a composite modulus and coefficients a_j that can be computed explicitly, as explained in [252]. This connection is useful when investigating the theoretical properties of generators like this.

To illustrate the concept of combined generators, we can use the generator MRG32k3a [252] mentioned in Chap. 1, which is a combined MRG with two components and for which

$$\begin{aligned}
x_{1,i} &= (1403580x_{1,i-2} - 810728x_{1,i-3}) \bmod (2^{32} - 209), \\
x_{2,i} &= (527612x_{2,i-1} - 1370589x_{2,i-3}) \bmod (2^{32} - 22853), \\
z_i &= (x_{1,i} - x_{2,i}) \bmod (2^{32} - 209), \\
u_i &= z_i/(2^{32} - 209).
\end{aligned}$$

The parameters of this generator were found through extensive searches based on theoretical and statistical tests.

Prior to this, Wichmann and Hill [474, 475] proposed a combined generator based on three components and defined by

$$\begin{aligned}
x_{1,i} &= 171x_{1,i-1} \bmod 30269, \\
x_{2,i} &= 172x_{2,i-1} \bmod 30307, \\
x_{3,i} &= 170x_{3,i-1} \bmod 30323, \\
w_i &= \left(\frac{x_{1,i}}{30360} + \frac{x_{2,i}}{30307} + \frac{x_{3,i}}{30323}\right) \bmod 1.
\end{aligned}$$

This generator is apparently used in Excel 2003 and Excel 2007 [493].

3.2.2 Recurrences modulo 2

Because of the binary nature of computers, it certainly makes sense to try using PRNG constructions that are defined directly in terms of binary operations. More formally, one can use recurrences over $\mathbb{F}_2$, the Galois field with two elements, which we identify as 0 and 1. A first simple construction based on this idea was proposed by Tausworthe in 1965 [434]as follows.

Definition 3.5. A *linear feedback shift register* (LFSR) (or *Tausworthe generator*) has a transition function based on the recurrence

$$x_i = (a_1 x_{i-1} + \ldots + a_k x_{i-k}) \bmod 2 \tag{3.4}$$

and output value

$$u_i = \sum_{j=1}^{L} x_{i\nu+j-1} 2^{-j}, \tag{3.5}$$

where the step size ν and word length L are positive integers. (L is usually taken to be equal to the word size of the machine; i.e., $L = 32$ or $L = 64$.)

If the recurrence (3.4) has a maximal period ρ of $2^k - 1$ and $\gcd(\rho, \nu) = 1$, then the sequence $u_0, u_1, \ldots,$ also has period ρ [434]. Note that (3.4) is just a special case of (3.1) with $m = 2$, which is why the maximal period is $2^k - 1$, and it is reached if the characteristic polynomial of the recurrence $P(z) = z^k - a_1 z^{k-1} - \ldots - a_k$ is a primitive polynomial over $\mathbb{F}_2$.

This construction has been generalized by replacing the "bits" x_i by *vectors* $\mathbf{x}_i$ of L bits. That is, we can replace (3.4) by a recurrence of the form

$$\mathbf{x}_i = a_1 \mathbf{x}_{i-1} + \ldots + a_k \mathbf{x}_{i-k}, \tag{3.6}$$

where $\mathbf{x}_i = (x_{i,1}, \ldots, x_{i,L})$ and all operations are performed modulo 2. In other words, $\mathbf{x}_i$ is obtained by performing a bitwise exclusive-or operation* on the vectors $\mathbf{x}_{i-j}$ for which $a_j = 1$. The state $\mathbf{y}_i$ is then given by the vector $(\mathbf{x}_i, \ldots, \mathbf{x}_{i-k+1})$ of kL bits, and the output is obtained as

$$u_i = \sum_{j=1}^{L} x_{i,j} 2^{-j}. \tag{3.7}$$

This type of generator is called a *generalized feedback shift register* (GFSR) [289]. It can be shown that the maximal period that can be reached by this type of generator is still $2^k - 1$. Recall that, in principle, the period can be as large as $|S|$, which in this case is 2^{kL} since the state vector $\mathbf{y}_i$ contains kL bits. In order to get closer to this upper bound, the recurrence defining the transition function of a GFSR needs to be generalized further, leading to

* The exclusive-or operation $\oplus$ is defined by the rule $0 \oplus 0 = 1 \oplus 1 = 0, 0 \oplus 1 = 1 \oplus 0 = 1$.

a class called *twisted generalized feedback shift register* (TGFSR) [309]. This general class includes the well-known *Mersenne-Twister* [310]. To describe this class, it is useful to first rewrite the GFSR using the matrix notation [248]

$$\mathbf{x}_i = A\mathbf{x}_{i-1},$$

where the $\mathbf{x}_i$ are vectors of kL bits and A is a $kL \times kL$ matrix of the form

$$A = \begin{pmatrix} 0 & I_L & \dots 0 \\ \vdots & \vdots & \ddots \vdots \\ 0 & 0 & \dots I_L \\ a_k I & a_{k-1} I_L & \dots a_1 I_L \end{pmatrix},$$

where I_L is the $L \times L$ identity matrix. The twisted GFSR amounts to replacing the matrices $a_j I_L$ on the last row of A by more general matrices. In addition, the output function (3.7) can be generalized by using *tempering transformations*. The well-known Mersenne-Twister MT19937 described in [310] is a TGFSR to which such tempering has been applied. It is shown to have a period of $2^{kL-r} - 1 = 2^{19937} - 1$, where $k = 624$, $L = 32$, and $r = 31$ is a parameter that is used in the definition of the recurrence that determines the transition function. Several implementations of this generator can be found on the Internet [487]. As we mentioned in Sect. 1.4, it is offered as one of three possible generators in Matlab 7.3.0 and is the default generator in Matlab 7.4.

It turns out that all these constructions can be defined using the following general setup [267] based on matrix notation, which are referred to as $\mathbb{F}_2$-*linear generators* in the recent survey [269].

Definition 3.6. An $\mathbb{F}_2$-*linear generator* has a state space $S = \mathbb{F}_2^k$ for some positive integer k, and for $\mathbf{x}_i \in S$,

$$\mathbf{x}_i = \tau(\mathbf{x}_{i-1}) = A\mathbf{x}_{i-1},$$

where A is a $k \times k$ matrix with entries in $\mathbb{F}_2$. The output is defined as

$$u_i = \xi(\mathbf{x}_i) = \sum_{l=1}^{L} y_{i,l-1} 2^{-l}, \tag{3.8}$$

where $\mathbf{y}_i = (y_{i,0}, \dots, y_{i,L-1})^{\mathrm{T}}$, $\mathbf{y}_i = B\mathbf{x}_i$, and B is an $L \times k$ matrix with entries in $\mathbb{F}_2$.

In the definition above, the matrix A is the *transition matrix* and B is the *output matrix*, which typically includes tempering transformations. The maximal period length of $2^k - 1$ for this type of generator is attained if $P(z) = \det(A - zI_k)$ is a primitive polynomial over $\mathbb{F}_2$ [248, 339]. Several generators based on this construction are proposed in [267], with periods ranging between $2^{64} - 1$ and $2^{128} - 1$.

These $\mathbb{F}_2$-linear generators can be combined using similar ideas as for MRGs [249, 268, 461]. More precisely, one can choose J generators respectively based on matrices $A_1, B_1, \ldots, A_J, B_J$. Then, at step i, compute the state $\mathbf{x}_{i,j}$ for each generator as $\mathbf{x}_{i,j} = A_j \mathbf{x}_{i-1,j}$ and then define $\mathbf{y}_i = B_1 \mathbf{x}_{i,1} \oplus \ldots \oplus B_J \mathbf{x}_{i,J}$, where the $\oplus$ operation is a bitwise exclusive-or performed on the L-bit vector operands. The output can then be defined as

$$u_i = \sum_{l=1}^{L} y_{i,l} 2^{-l}.$$

Examples of good combined Tausworthe generators with the relevant code are given in [254]. Examples of good combined TGFSRs with tempering and very long periods (up to about 2^{1250}) are given in [268], and more recent constructions can be found in [372].

3.3 Add-with-carry and subtract-with-borrow generators

This class of generators was proposed by Marsaglia and Zaman [304]. They have similarities with MRGs but do not exactly fit Def. 3.3 due to their "add" or "carry" features. More precisely, the add-with-carry (AWC) generator is defined by the recurrence

$$\begin{aligned} x_i &= x_{i-r} + x_{i-k} + c_i \bmod m, \\ c_{i+1} &= \mathbf{1}_{x_{i-r}+x_{i-k}+c_i \geq m}, \end{aligned} \tag{3.9}$$

where b and $k > r$ are positive integers and c_i is called the *carry*. Since there is no multiplication involved and the value of c_{i+1} indicates whether m must be subtracted or not when performing the modulo m operation in (3.9), this generator is very fast. The recurrence defining the subtract-with-borrow (SWB) generator is given by

$$\begin{aligned} x_i &= x_{i-r} - x_{i-k} - c_i \bmod m, \\ c_{i+1} &= \mathbf{1}_{x_{i-r}-x_{i-k}-c_i < 0}, \end{aligned}$$

where $k > r$ and c_i is called the *borrow*. A variant can be obtained by exchanging r and k in these recurrences. For both the AWC and SWB, the carry/borrow can be thought of as a way of adding noise to an otherwise simple lagged-Fibonacci generator. In addition, the output is produced using ideas similar to those used for LFSRs. That is, rather than defining $u_i = x_i/m$, the output of the AWC and SWB can be defined more generally as

$$u_i = \sum_{j=0}^{L-1} x_{Li+j} m^{j-L}.$$

Note that this is different from (3.5) in that the successive digits

$$x_{iL}, x_{iL+1}, \ldots, x_{iL+L-1}$$

are defining u_i from the least significant to the most significant digit (and also ν in (3.5) is taken equal to L here).

These generators are attractive because they are fast and can have a very long period. For instance, one of the generators proposed in [304] is an SWB with $m = 2^{32} - 5$, $k = 43$, $r = 22$, and a period of $m^{43} - m^{22} \approx 2^{1376}$. They can also be combined and generalized in different ways [158].

It turns out that these two generators have been shown by Tezuka and L'Ecuyer to be very closely related to an MLCG with modulus $\tilde{m} = m^k + m^r \pm 1$ for the AWC and $\tilde{m} = m^k - m^r \pm 1$ for the SWB [447]. More precisely, for $\tilde{m}$ prime, their output is equal (up to the first L digits) to that of an MLCG with the modulus $\tilde{m}$ given above and the multiplier $a = m^{(\tilde{m}-2)L} \bmod \tilde{m}$. Therefore, the numbers u_i produced by an AWC or SWB are within m^{-L} from those of the approximating MLCG. Unfortunately, this fact implies that these generators have bad theoretical properties related to their lattice structure and should therefore be avoided, as discussed in [65, 158, 448]. We will briefly come back to this point in Sect. 3.5.1.

3.4 Nonlinear generators

The generators we have seen so far were all based on linear recurrences for the transition function and linear output functions. For some applications — such as cryptography — linear generators are not suitable because their structure is too simple and makes it easy to predict the next number in the output sequence. *Nonlinear generators* are based on transition functions and/or output functions that are not linear and therefore have a structure that is much more complicated than for linear generators. This makes them better suited for applications where the *unpredictability* of the sequence is important. However, these generators are often quite slow and therefore are usually not suitable for applications where speed is important. An interesting idea to get the best of both worlds is to combine a small nonlinear generator with a large linear generator, as done in [261]. Further work in that direction seems a promising research area.

We will not discuss nonlinear generators in detail here and instead illustrate the idea with a simple example. We refer the reader to the recent survey [344] for information on these generators.

Definition 3.7. An *explicit inversive congruential generator* [102] is described by a transition function

$$x_i = (ai + c) \bmod m$$

and an output function

$$u_i = \frac{x_i^{-1}}{m},$$

where x_i^{-1} is the inverse of x_i modulo m (that is, x_i^{-1} is such that $x_i^{-1} x_i = 1 \bmod m$).

For instance, if $m = 11$, $a = 6$, and $c = 1$, then we have $x_0 = 1, x_1 = 7, x_2 = 13 \bmod 11 = 2, x_3 = 19 \bmod 11 = 8, x_4 = 25 \bmod 11 = 3, x_5 = 31 \bmod 11 = 9, \ldots$, and so on. Therefore, $u_0 = 1/11, u_1 = 8/11$ (since $7 \times 8 \bmod 11 = 1$), $u_2 = 6/11$, $u_3 = 7/11$, $u_4 = 4/11$, $u_5 = 5/11$, etc.

It can be shown that, for m prime, the inverse x_i^{-1} can be computed as $x_i^{-1} = (ai + c)^{m-2} \bmod m$ and the period of this generator is m. Also, here the choice of parameters (a and c in this case) is not as crucial as it is for the linear generators described in the previous section. This type of generator was tested empirically alongside other well-known generators in [274].

3.5 Theoretical and statistical testing

For all the different families of generators that we have seen in the previous sections, parameters must be chosen to define a specific generator. For example, for an MRG, we need to choose a prime m, an order k, and coefficients $a_1, \ldots, a_k$ in $\mathbb{Z}_m$. How do we do this? In practice, a typical approach is to perform a search (exhaustive or not) in which several sets of parameters are tested by analyzing the quality of the resulting generator. In the first stage, the tests performed are often of a theoretical nature. That is, they look at certain properties of the generators *over the whole period* and that can be analyzed in a precise, quantitative way. An obvious one is the period length. Depending on the type of generator, we can define other criteria, as discussed below. Once a few "good" generators have been found, they can then be tested further using *statistical tests*, which analyze samples produced by the generator and try to detect obvious discrepancies from "true randomness".

When talking about theoretical tests for a generator defined over a state space S, it is useful to consider the following set:

$$\Psi_s = \{(u_0, u_1, \ldots, u_{s-1}) : \mathbf{x}_0 \in S\}. \tag{3.10}$$

That is, we look at all possible initial states (seeds) $\mathbf{x}_0$ and for each of them form an s-dimensional point by taking the first s successive numbers

$u_0, u_1, \ldots, u_{s-1}$ output by the generator with this seed. Hence Ψ_s contains $|S|$ points.

For instance, for the toy MLCG based on $m = 11$ and $a = 6$, for $s = 2$ we have

$$\begin{aligned}\Psi_2 = \{&(0,0),(1/11,6/11),(2/11,1/11),(3/11,5/11),(4/11,2/11),\\&(5/11,8/11),(6/11,3/11),(7/11,9/11),(8/11,4/11),\\&(9/11,10/11),(10/11,5/11)\}.\end{aligned}$$

Note that for an MRG with maximal period $m^k - 1$, the set Ψ_s can be written using the alternative definition

$$\Psi_s = \{(u_i, \ldots, u_{i+s-1}), i = 0, \ldots, m^k - 2\} \cup \{\mathbf{0}\},$$

assuming the seed used to initialize the sequence $u_0, u_1, \ldots$ is not zero. That is, here we build Ψ_s by forming overlapping s-tuples from the sequence output by the generator until the cycle starts to repeat itself. The s-dimensional point $\mathbf{0}$ — corresponding to the zero seed — is then added to the $|S| - 1 = m^k - 1$ points obtained. Compared to (3.10), this simply lists the points in a different order. Again using the toy MLCG with $m = 11$ and $a = 6$, this alternative definition amounts to writing

$$\begin{aligned}\Psi_2 = \{&(1/11,6/11),(6/11,3/11),(3/11,7/11),(7/11,9/11),(9/11,10/11),\\&(10/11,5/11),(5/11,8/11),(8/11,4/11),(4/11,2/11),(2/11,1/11)\}\\&\cup\{(0,0)\}.\end{aligned}$$

The reason why the set Ψ_s is useful in understanding the theoretical properties of a generator is as follows [248]. Suppose the initial seed $\mathbf{x}_0$ of the generator is randomly chosen. If one uses the generator in an application where s random numbers are needed for each run, then we can think of the vector $\mathbf{u}$ containing these s numbers as being randomly chosen from the set Ψ_s. Ideally (if we had a true random number generator), $\mathbf{u}$ should be uniformly distributed over $[0,1)^s$. However, since Ψ_s is finite, the actual distribution is only *approximately* uniform, and the quality of the approximation depends on the structure of Ψ_s. If Ψ_s contains a very large number of points — which amounts to asking S to be large — that are well spread out over $[0,1)^s$, then the approximation should be reasonably good. If Ψ_s does not contain too many points or if they are not very well spread out, then the approximation will not be very good. As explained below, most theoretical tests thus look at Ψ_s for different values of s and try to measure its uniformity.

3.5.1 Theoretical tests for MRGs

For MRGs, the set Ψ_s has a *lattice structure* [30, 220, 302, 384]. That is, it can be written as $\Psi_s = L_s \cap [0,1)^s$, where L_s is a *lattice* defined by

$$L_s = \left\{ \mathbf{x} = \sum_{j=1}^{s} z_j \mathbf{v}_j : \mathbf{z} = (z_1, \ldots, z_s) \in \mathbb{Z}^s \right\}, \tag{3.11}$$

for some vectors $\mathbf{v}_1, \ldots, \mathbf{v}_s \in \mathbb{R}^s$ that depend on the coefficients a_j of the recurrence and the modulus m [259]. These vectors are said to form a *basis* for L_s because L_s is obtained by considering all possible integer linear combinations of the vectors $\mathbf{v}_j$. Note that the choice of basis is not unique [54]. Figure 3.1 shows an example of the set Ψ_2 for an MLCG with $m = 251$ and $a = 33$.

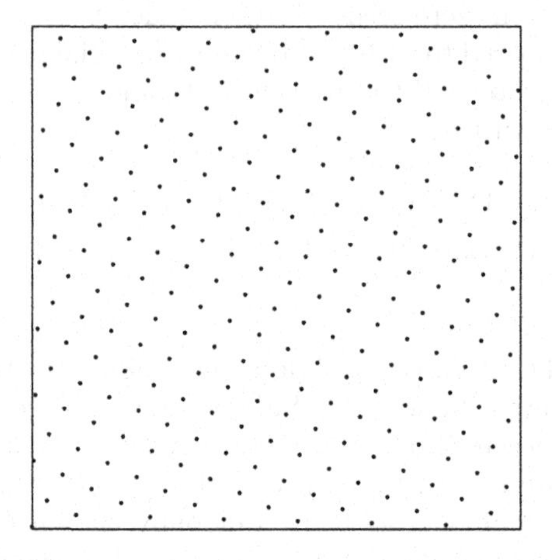

Fig. 3.1 Lattice structure of Ψ_2 for the MLCG based on $m = 251$ and $a = 33$.

For an MLCG, note that

$$\Psi_s = \left\{ \left(\frac{x_0}{m}, \frac{a x_0}{m}, \frac{a^2 x_0}{m}, \ldots, \frac{a^{s-1} x_0}{m} \right) : 1 \le x_0 \le m-1 \right\} \cup \{\mathbf{0}\},$$

where all operations are performed modulo m. Hence, in this case, a possible choice for the basis $\mathbf{v}_1, \ldots, \mathbf{v}_s$ defining the lattice L is to take

$$\begin{aligned}
\mathbf{v}_1 &= (1, a/m, a^2/m, \ldots, a^{s-1}/m), \\
\mathbf{v}_2 &= (0, 1, 0, \ldots, 0), \\
&\vdots \\
\mathbf{v}_s &= (0, \ldots, 0, 1).
\end{aligned}$$

The coefficient z_1 in (3.11) is then used to determine one of the m points in Ψ_s, and the other coefficients $z_2, \ldots, z_s$ simply determine a unit cube in $\mathbb{Z}^s$.

Theoretical tests for MRGs usually consider the lattice structure of Ψ_s for some value of s and try to measure its uniformity. For example, in the *spectral test* [68], one measures the largest distance d_s between adjacent parallel hyperplanes that together cover the points in Ψ_s. The smaller this distance is, the better the uniformity of Ψ_s. Figure 3.2 shows two successive hyperplanes separated by d_2 for two small MLCGs. On the left-hand side, $d_2 = 0.128$, while on the right-hand side it is equal to 0.196. (We will explain shortly how to compute these numbers.) Thus, from the point of view of the spectral test, the set of the left-hand side is better because the corresponding value of d_2 is smaller. The spectral test can also be generalized in a way that makes it useful for studying generators that do not necessarily have a lattice structure (e.g., nonlinear generators) [174], but we will not discuss these generalizations here.

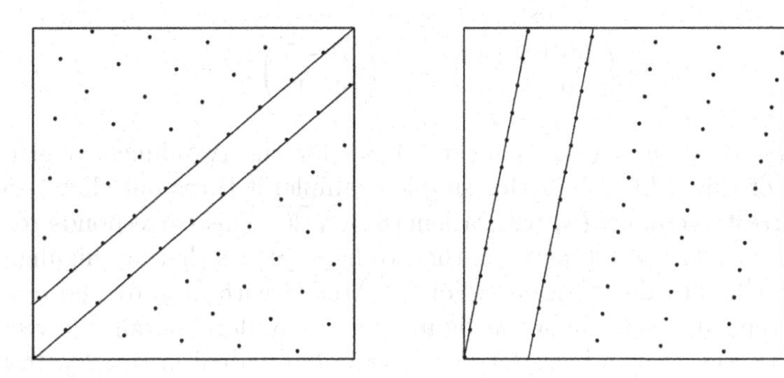

Fig. 3.2 Hyperplanes at a distance d_2 for the MLCG based on $n = 61$ and $a = 11$ (left) or $a = 5$ (right).

Another possibility is to count the number of hyperplanes that intersect $[0, 1)^s$ for the family of hyperplanes that are the farthest apart [85, 302]. For instance, in Fig. 3.2, on the left-hand side there are ten lines for the family of lines that are the farthest apart, while there are only five such lines for the MLCG shown on the right-hand side. It turns out that one of the weaknesses of the generator RANDU that was mentioned on p. 24 of Chap. 1 is that the set Ψ_3 for the generator falls on only 15 parallel hyperplanes.

To give an idea of how to compute d_s, we will use simple examples to illustrate what is at stake here. Methods for doing this actually compute $\ell_s = d_s^{-1}$, which turns out to be the length of the shortest vector — using the L_2 norm — in the *dual lattice* L_s^* of L_s, defined by

$$L_s^* = \{\mathbf{h} \in \mathbb{R}^s : \mathbf{h} \cdot \mathbf{v} \in \mathbb{Z} \text{ for all } \mathbf{v} \in L\},$$

where the operation $\cdot$ in $\mathbf{h} \cdot \mathbf{z}$ is the product $h_1 z_1 + \ldots + h_s z_s$. So, for instance, because $(1/m, a/m, \ldots, a^{s-1}/m)$ is in L_s, a necessary condition for $\mathbf{h}$ to be in L_s^* is that we must have

$$\frac{h_1}{m} + \frac{h_2 a}{m} + \ldots + \frac{h_s a^{s-1}}{m} \in \mathbb{Z},$$

which holds if and only if

$$h_1 + h_2 a + \ldots + h_s a^{s-1} = 0 \bmod m.$$

Another way of understanding how this dual lattice is related to the original lattice L_s (3.11) on which the points of Ψ_s lie is to observe that a basis for L_s^* can be obtained by taking the columns of the inverse of the matrix that has the vectors $\mathbf{v}_1, \ldots, \mathbf{v}_s$ used in (3.11) on its rows [54]. For instance, if $m = 61$ and $a = 5$, then for $s = 2$ we can use the basis $\mathbf{v}_1 = (1/61, 5/61)$ and $\mathbf{v}_2 = (0, 1)$ for L_2. Since

$$\begin{pmatrix} 1/61 & 5/61 \\ 0 & 1 \end{pmatrix}^{-1} = \begin{pmatrix} 61 & -5 \\ 0 & 1 \end{pmatrix},$$

the vectors $(61, 0)$ and $(-5, 1)$ form a basis for the two-dimensional dual lattice L_2^* of this MLCG. For this simple example, it turns out that $(-5, 1)$ is the shortest vector in L_2^*, with a length of $\sqrt{26}$. This corresponds to the distance of $1/\sqrt{26} = 0.196$ between the two hyperplanes that are highlighted on the right-hand side of Fig. 3.2. For the MLCG with $m = 61$ and $a = 11$ shown on the left-hand side of this figure, we can similarly obtain the vectors $(61, 0)$ and $(-11, 1)$ as a basis for the corresponding dual lattice L_2^*. In this case, the vector $(61, 0) + 5(-11, 1) = (6, 5)$ is the shortest vector for L_2^*, with a length $\sqrt{61}$, whose inverse $1/\sqrt{61} = 0.128$ equals the distance between the hyperplanes highlighted on the left-hand side of Fig. 3.2.

For these two simple examples, it was easy to find the shortest vector in L_s^*. In general, sophisticated methods such as those described in [106, 108, 118, 221, 259] need to be used in order to determine this quantity. The problem can be formulated using integer programming with a quadratic objective function because the goal is to find $(z_1, \ldots, z_s)$ in $\mathbb{Z}^s$ such that $\|z_1 \mathbf{w}_1 + \ldots + z_s \mathbf{w}_s\|^2$ is minimized, where $\mathbf{w}_1, \ldots, \mathbf{w}_s$ is a basis for L_s^*. The choice of the basis turns out to be quite important for methods that attempt to solve this problem.

Note that for the general problem of finding the shortest vector in a lattice, there is no known polynomial-time algorithm that can find an exact solution. Actually, this problem is hard enough that some people study cryptographic systems based on the difficulty of finding such vectors, just like RSA-type cryptographic systems are based on the fact that factoring large integers is a difficult problem for classical computers. In this context, norms other than the Euclidean norm might be used. Alternative norms can also be useful for testing MRGs. For instance, the length of the shortest vector in the dual lattice *measured using the* L_1 *norm* (i.e., using the norm $\|x\|_1 = |x_1| + \ldots + |x_s|$) is equal to one plus the *number* of hyperplanes on which the points of Ψ_s lie. As before, this is for the family of hyperplanes that are the farthest apart [85].

Interestingly, the original description and name of the spectral test were not based on the geometrical interpretation above, but instead on looking at the quantity

$$S(\mathbf{h}) = \frac{1}{|S|} \sum_{\mathbf{u} \in \Psi_s} e^{\mathbf{h} \cdot \mathbf{u}}, \tag{3.12}$$

where $|S|$ is the cardinality of the state space S (and thus of Ψ_s), and the operation $\cdot$ in $\mathbf{h} \cdot \mathbf{u}$ is the product $h_1 u_1 + \ldots + h_s u_s$. For a truly uniform vector $\mathbf{u}$, we have that

$$\mathrm{E}(e^{\mathbf{h} \cdot \mathbf{u}}) = \begin{cases} 1 & \text{if } h_j = 0 \bmod m \text{ for all } j, \\ 0 & \text{else.} \end{cases}$$

From this point of view, $S(\mathbf{h})$ represents an approximation for the expectation $\mathrm{E}(e^{\mathbf{h} \cdot \mathbf{u}})$, which is obtained by averaging over the points in Ψ_s. Coveyou and MacPherson argue in [68] that "wave functions" $S(\mathbf{h})$ with a "smaller" $\mathbf{h}$ (i.e., low-frequency waves) are the most important, and for that reason one should know what is the worst (smallest) vector $\mathbf{h}$ for which the corresponding wave function $S(\mathbf{h})$ fails to correctly approximate the true value $\mathrm{E}(e^{\mathbf{h} \cdot \mathbf{u}})$. Using the fact that if $\mathbf{h} \in L_s^*$, then $S(\mathbf{h}) = 1$, we see that this is precisely what l_s measures.

Now, assuming that d_s can be computed (or at least approximated), the next step is to decide for which values of s we should compute d_s. Typically, when a generator is designed, the broad area for which it will be used may be known, but not the specific applications. Therefore, generators should be designed so that they do well for a variety of applications. From this point of view, choosing a single value of s for which d_s will be computed is not realistic. Instead, what is typically done is that d_s is computed for several values of s. For example, in one of the first papers where the spectral test was used to systematically search for good MLCGs [123], d_s was computed for $s = 2, \ldots, 6$.

The next thing to do is to determine how these values d_s obtained for different s should be compared when assessing the generator. When s increases, the notion of distance changes too and therefore one should attempt to scale the different d_s values so that they can be compared more fairly. One possibility is to try to use theoretical lower bounds d_s^* for d_s and then scale each d_s as d_s^*/d_s, which will be a value between 0 and 1. These lower bounds can be computed exactly for $s \le 8$, and otherwise certain bounds can be found, as discussed in [61, 253]. These lower bounds represent the shortest possible distance between hyperplanes that can be achieved for s-dimensional lattices whose basis vectors are in $\mathbb{R}^s$ and thus cannot necessarily be realized among the set of all possible MRGs. That is, even the best possible MRGs might have $d_s^*/d_s < 1$ as they are restricted to rational vectors for their basis.

Once the values d_s are normalized like this, one can define a figure of merit such as

$$M_T = \min_{2 \le s \le T} d_s^*/d_s,$$

which returns the smallest (worst) normalized d_s for all s considered. For instance, in [119, 123], exhaustive searches to find all multipliers satisfying $M_6 \ge 0.8$ were done for $m = 2^{31} - 1$ and $m = 2^{32}$, respectively. Just to give an idea, for the modulus $m = 2^{31} - 1$, out of the 534 million multipliers yielding a maximal period, only 414 satisfied the bound $M_6 \ge 0.8$.

In addition, one can compute d_s for sets of the form

$$\Psi_I = \{(u_{i_1}, u_{i_2}, \ldots, u_{i_s}) : x_0 \in S\}, \tag{3.13}$$

where $I = \{i_1, \ldots, i_s\}$ and $1 \le i_1 < i_2 < \ldots < i_s$ [250]. Using *lacunary indices* $i_1, \ldots, i_s$ like this can help detect problems that would not be uncovered by restricting the assessment only to successive indices, as is done with Ψ_s since in that case $I = \{1, 2, \ldots, s\}$. For instance, L'Ecuyer shows in [250] that one of the SWB generators recommended in [204] is such that the set $\Psi_{\{1,11,25\}}$ lies within a distance of 2^{-24} from a pair of planes that are $1/\sqrt{3}$ apart, which is very large. This means that if this generator is used for a problem whose dimension (i.e., the number of uniform numbers used per run) is at least 25, then severe three-dimensional correlations might create abnormal results.

Using this broader type of subset leads to the general criterion

$$M_{\mathcal{I}} = \min_{I \subseteq \mathcal{I}} d_{|I|}^*/d_I$$

for testing MRGs, where $\mathcal{I}$ is a set of subsets I, d_I is the quantity computed by the spectral test for Ψ_I (i.e., the maximal distance between hyperplanes), and $d_{|I|}^*$ is a lower bound on d_I. Criteria like this have been proposed and used in [264] to find LCGs that can be used for quasi–Monte Carlo integration. There, the set $\mathcal{I}$ was of the form

$$\mathcal{I} = \{\{1, 2, \ldots, t_1\}, \{1, t_2\}, \{1, s_3, t_3\}, \{1, r_4, s_4, t_4\}, 2 \le t_1 \le d_1, 2 \le t_2 \le d_2, \\ 2 \le s_3 < t_3 \le d_3, 2 \le r_4 < s_4 < t_4 \le d_4\}$$

for integers $d_1, d_2, d_3, d_4 \geq 2$ (for example, $d_1 = 32, d_2 = 24, d_3 = 12, d_4 = 8$ are used in [264]). More recent work in this area can be found in [106, 108].

To conclude our discussion of the spectral test, we would like to mention that despite the fact that d_s is difficult to compute, there are useful bounds that can be used to make an initial assessment about the quality of the lattice structure of a generator [159, 250].

Theorem 3.8. *(i) For an MRG of order k and based on the coefficients $a_1, \ldots, a_k$, we have that*

$$d_s \geq \left(1 + \sum_{i=1}^{k} a_i^2\right)^{-1/2}.$$

(ii) For an MLCG with multiplier a, if the modulus m can be expressed as

$$m = \sum_{j=1}^{t} c_{i_j} a^{i_j}$$

for some integers c_{i_j}, for $j = 1, \ldots, t$, then for $I = \{i_1, \ldots, i_t\}$ we have that

$$d_I \geq \left(\sum_{j=1}^{t} c_{i_j}^2\right)^{-1/2}.$$

Result (ii) above is the reason why the AWC and SWB generators do not do well in the spectral test. Recall that these generators can be closely approximated by an MLCG with m of the form $a^k \pm a^r \pm 1$, which means that $d_I \geq 1/\sqrt{3}$ for $I = \{1, r-1, k-1\}$.

3.5.2 Theoretical tests for PRNGs based on recurrences modulo 2

Here we are still trying to measure the uniformity of sets of the form Ψ_s, but the tools are different because the structure of Ψ_s is different. However, it is interesting to note that, for $\mathbb{F}_2$-linear generators, Ψ_s also has a lattice structure, but in a different mathematical sense. Although the tests we are about to present can be explained in this lattice setting, we prefer to use a geometrical interpretation to describe them, and we refer the reader to [66, 135, 256, 268, 435, 437, 441] for more information on the lattice structure of these generators, which will also be discussed in Chap. 5.

We discuss two quantities that can be used to measure the uniformity of Ψ_s for generators based on recurrences modulo 2. They are both related to the concept of $(q_1, \ldots, q_s)$-*equidistribution*, which we now define:

Definition 3.9. Let $q_1, \ldots, q_s$ be nonnegative integers, and let $q = q_1 + \ldots + q_s$. A set Ψ_s of 2^k points in $[0,1)^s$ is $(q_1, \ldots, q_s)$*-equidistributed* (in base 2) if every cell of the form

$$\prod_{j=1}^{s} \left[\frac{r_j}{2^{q_j}}, \frac{r_j + 1}{2^{q_j}} \right), \tag{3.14}$$

for $0 \le r_j < 2^{q_j}, j = 1, \ldots, s$, contains 2^{k-q} points from Ψ_s.

In other words, here we partition the unit cube in 2^q congruent boxes of size 2^{-q_j} in dimension j and verify that each box contains the same number of points. Obviously, this condition can only be satisfied if there are at least as many points as there are boxes, which means we must have $q \le k$. The boxes (3.14) are often referred to as *elementary intervals* [335].

In Fig. 3.3, we show the point set Ψ_2 obtained from an LFSR with $k = 6$ and illustrate its $(1,3)$-equidistribution and $(3,1)$-equidistribution on the left-hand side and right-hand side, respectively. Details about the LFSR used to produce this figure are given at the end of this subsection.

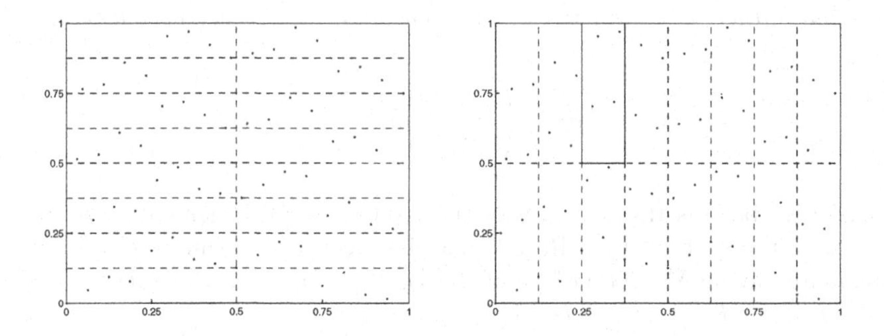

Fig. 3.3 $(1,3)$-equidistribution (left) and $(3,1)$-equidistribution (right) of a set Ψ_2 with 64 points. In both cases, each of the 16 boxes contains four points.

The first criterion that we present is called the *resolution* in [249].

Definition 3.10. The *resolution* of Ψ_s is the largest integer ℓ_s such that Ψ_s is $(\ell_s, \ldots, \ell_s)$-equidistributed.

The geometric interpretation for the resolution is that ℓ_s is the largest integer such that we can partition $[0,1)^s$ into congruent cubic boxes of volume $2^{-s\ell_s}$ — this is done by partitioning each axis into 2^{ℓ_s} intervals of size $2^{-\ell_s}$ — and get an equal number of points from Ψ_s in each box. Alternatively, a generator that has a resolution of ℓ_s in s dimensions is said to be (s, ℓ_s)-*equidistributed*, or s*-distributed to* ℓ_s *bits of accuracy* [135, 220, 450]. For instance, the Mersenne-Twister MT19937 is said to be "623-distributed" up to 32 bits of accuracy. This means that the resolution of the corresponding 623-dimensional point set Ψ_{623} has resolution $\ell_{623} = 32$.

By definition, $\ell_s \leq \ell_s^* := \min(\lfloor k/s \rfloor, L)$, where L is the number of bits used in the representation of the numbers output by the generator, as given in (3.8), and k is such that $|S| = 2^k$. Figure 3.4 shows that the point set Ψ_2 with $k = 6$ from Fig. 3.3 has a resolution ℓ_2 of 2. That is, each of the $2^{2\times2} = 16$ squares shown in this figure contains $2^{k-4} = 2^2 = 4$ points from Ψ_2. But $\ell_2 \neq 3$ since for the 64 squares of size $1/8 \times 1/8$, half of them contain two points, while the other half contain none. For the Mersenne-Twister MT19937, since $k = 19937$ and $L = 32$ in this case, the 623-dimensional resolution ℓ_{623} of 32 is maximal because $\lfloor 19937/623 \rfloor = 32$.

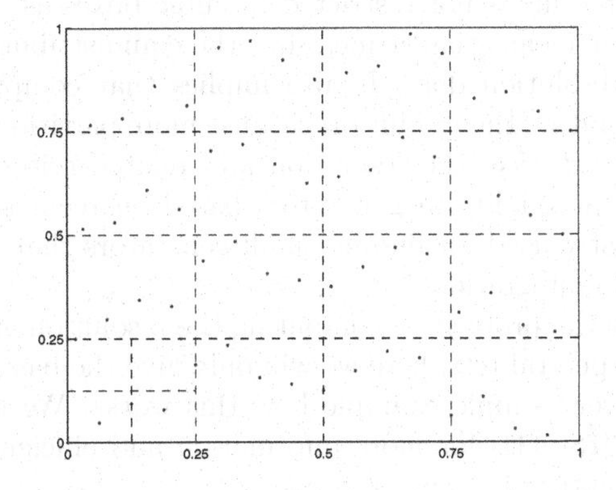

Fig. 3.4 Ψ_2 has a resolution $\ell_2 = 2$.

If a generator is such that $\ell_s = \ell_s^*$ for $s = 1, \ldots, k$, then it is said to be *maximally equidistributed* [135, 248], or *asymptotically random* [450]. The Mersenne-Twister MT19937 is not maximally equidistributed since for instance $\ell_{6241} < \ell_{6241}^* = \lfloor 19937/6241 \rfloor = 3$ (see [310, Table II]). Note that since $k = 19937$ for the Mersenne-Twister, the resolution ℓ_s for all $s = 1, \ldots, 19937$ would need to reach its maximal upper bound in order for this generator to be maximally equidistributed. Generators that are maximally equidistributed can be found in [254, 267, 268].

Similarly to what was discussed for the spectral test, more complex criteria based on the resolution can be defined, such as

$$\Delta_{\mathcal{I}} = \min_{I \in \mathcal{I}} (\ell_I / \ell_{|I|}^*),$$

where ℓ_I is the resolution of the set Ψ_I defined in (3.13), $\ell_{|I|}^* = \min(k/|I|, L)$ is the maximum resolution for a set of 2^k points — defined over L bits — in dimension $|I|$, and $\mathcal{I}$ is a set of subsets I [268].

We now present a second criterion for generators based on recurrences modulo 2. The terminology used here comes from [82].

Definition 3.11. The *t-value* of Ψ_s is the smallest integer t such that Ψ_s is $(q_1, \ldots, q_s)$-equidistributed for all $(q_1, \ldots, q_s)$ satisfying $q \leq k - t$, where $q = q_1 + \ldots + q_s$.

The origin of this criterion goes back to Sobol' [415], who labeled it as τ to measure the quality of his so-called *LP_τ-sequence*, which is now usually referred to as the *Sobol' sequence.* The notation with the letter t was introduced by Niederreiter in [335] and is widely used in the study of quasi–Monte Carlo methods. The smaller t is, the better the equidistribution.

If we compare it with the resolution, we observe that the equidistribution measured by the t-value is not restricted to cubic boxes as was the case for the resolution. This means that it measures the equidistribution to a greater extent than the resolution does. It also implies that computing t is more difficult than computing the resolution because more partitions of boxes must be considered. In practice, the resolution and related criteria are typically used to evaluate the quality of generators based on recurrences modulo 2. The t-value is mostly used for finding small generators that can be used for quasi–Monte Carlo integration.

We now turn to the problem of computing the resolution and t-value. Just as we did for the spectral test, here we will only give the basic principles and illustrate with a very simple example how this works. We refer the reader to [66, 135, 249, 378, 441] for more information and efficient algorithms to compute these quantities.

The first thing to note is that the dyadic elementary intervals defined in (3.14) play a key role in these two quality criteria. It is useful to label these intervals using the integers $(r_1, \ldots, r_s)$ introduced in (3.14), which determine the position of the elementary interval in the unit cube $[0,1)^s$. For instance, if $s = 2$ and $q_1 = 3, q_2 = 1$, then $(r_1, r_2) = (2, 1)$ refers to the rectangle with corners at (0.25, 0.5), (0.25, 1), (0.375, 0.5), and (0.375, 1), which is shown with complete (nondashed) lines on the right-hand side of Fig. 3.3. The next thing to understand is that to verify the $(q_1, \ldots, q_s)$-equidistribution of Ψ_s, for each point $\mathbf{u} = (u_0, \ldots, u_{s-1})$ in Ψ_s, we need to look at the first q_1 bits of u_0, the first q_2 bits of u_1, and so on, finishing off with the first q_s bits of u_{s-1}. These s bit strings identify a label $(r_1, \ldots, r_s)$ that indicates in which elementary interval $\mathbf{u}$ is. The third thing to notice is that each point $\mathbf{u}$ in Ψ_s is obtained by choosing one of the 2^k possible initial states $\mathbf{x}_0$ to initialize the generator. Hence, we are looking at a system of the form

$$\mathbf{C}\mathbf{x}_0 = \mathbf{y}, \tag{3.15}$$

where $\mathbf{x}_0$ runs over the set of k-bit vectors that can be used as initial states, $\mathbf{y} = (y_{0,1}, \ldots, y_{0,q_1}, \ldots, y_{s-1,1}, \ldots, y_{s-1,q_s})$ is a q-bit vector containing the first q_1 bits of u_0, the first q_2 bits of u_1, and so on, and it identifies an elementary interval. The $q \times k$ matrix $C := C(A, B, q_1, \ldots, q_s)$ depends on the generator and represents the linear transformation used to turn $\mathbf{x}_0$ into $\mathbf{y}$. Based on the general setup given in Def. 3.6, it is possible to verify that

$$C(A, B, q_1, \ldots, q_s) = \begin{pmatrix} B_{q_1} \\ B_{q_2} A \\ \vdots \\ B_{q_s} A^{s-1} \end{pmatrix}, \tag{3.16}$$

where the notation B_r represents the $r \times k$ matrix formed by the first r rows of the output matrix B of the generator. That is, (3.15) and (3.16) tell us that the bits $y_{0,1}, \ldots, y_{0,q_1}$ are obtained by applying the output matrix B_{q_1} to $\mathbf{x}_0$, the q_2 bits $y_{1,1}, \ldots, y_{1,q_2}$ are obtained by first applying A to $\mathbf{x}_0$ and then applying B_{q_2}, and so on until the q_s bits $y_{s-1,1}, \ldots, y_{s-1,q_s}$, obtained by applying A a total of $s-1$ times to $\mathbf{x}_0$, and then applying B_{q_s}.

Now, in this setup, being $(q_1, \ldots, q_s)$-equidistributed means that when $\mathbf{x}_0$ runs over all 2^k possible k-bit vectors, each possible q-bit vector (there are 2^q of them) occurs the same number of times when the linear transformation C is applied to all possible $\mathbf{x}_0$'s. As noted before, this "number of times" is necessarily given by 2^{k-q}, which implies that we must have $q \leq k$ for this property to make sense. Note that the matrix $\mathbf{C}$ is a $q \times k$ binary matrix. Therefore, the property that we are looking for is that we want this matrix to have rank q. Based on this fact, if, for example, we want to know whether or not Ψ_s has maximal resolution ℓ_s^*, then we just need to construct the matrix $C(A, B, \ell_s^*, \ldots, \ell_s^*)$ required to test the $(\ell_s^*, \ldots, \ell_s^*)$-equidistribution of Ψ_s and verify that it has rank $s\ell_s^*$. If instead we want to verify whether or not the t-value equals a certain value T, then we need to verify *for each* matrix $C(A, B, q_1, \ldots, q_s)$ corresponding to vectors $(q_1, \ldots, q_s)$ such that $q \leq k-T$ if the matrix has the desired rank q. The fact that there are several such vectors is the reason why computing t is more time-consuming than computing the resolution.

To conclude this discussion, we look at a very simple example and compute the resolution and t-value for $s = 2$. The generator we use is the LFSR whose corresponding two-dimensional sample set Ψ_2 is shown in Figs. 3.3 and 3.4. This generator is based on the recurrence

$$x_i = (x_{i-5} + x_{i-6}) \bmod 2$$

and output function

$$u_i = \sum_{j=1}^{6} x_{4i+j-1} 2^{-j}.$$

The corresponding matrix A is given by

$$A = \begin{pmatrix} 0\,1\,0\,0\,0\,0 \\ 0\,0\,1\,0\,0\,0 \\ 0\,0\,0\,1\,0\,0 \\ 0\,0\,0\,0\,1\,0 \\ 0\,0\,0\,0\,0\,1 \\ 1\,1\,0\,0\,0\,0 \end{pmatrix}^4 = \begin{pmatrix} 0\,0\,0\,0\,1\,0 \\ 0\,0\,0\,0\,0\,1 \\ 1\,1\,0\,0\,0\,0 \\ 0\,1\,1\,0\,0\,0 \\ 0\,0\,1\,1\,0\,0 \\ 0\,0\,0\,1\,1\,0 \end{pmatrix},$$

and $B = I_6$, the 6×6 identity matrix. We will look at $(q_1 + q_2) \times k$ matrices $C(A, B, q_1, q_2)$ obtained by taking the first q_1 rows of I_6 and the first q_2 rows of A. First consider the matrix $C(A, B, 3, 3)$. If its rank is $3 + 3 = 6$, then it means that $\ell_2 = 3$. It turns out that its rank is 5. However, the rank of the matrix $C(A, B, 2, 2)$ is 4, and thus $\ell_2 = 2$, as we can see in Fig. 3.4.

Moving on to the determination of the t-value, we already know that $t > 0$ since $\ell_2 < 3$, and thus Ψ_2 is not $(3, 3)$-equidistributed. So now we want to check if $t \leq 2$. We know the $(2, 2)$-equidistribution holds, but we need to verify the $(0, 4)$-, $(4, 0)$-, $(3, 1)$-, and $(1, 3)$-equidistributions. That is, we need to verify that $C(A, B, q_1, q_2)$ has rank 4 for (q_1, q_2) in $\{(0, 4), (4, 0), (3, 1), (1, 3)\}$, which they all do. So we know $t \leq 2$. Similarly, to determine if $t = 1$, we need to check that $C(A, B, q_1, q_2)$ has rank 5 for (q_1, q_2) in $\{(5, 0), (0, 5), (1, 4), (4, 1), (2, 3), (3, 2)\}$. They all do except $C(A, B, 2, 3)$, and so $t = 2$. The failure to be $(2, 3)$-equidistributed can be seen in Fig. 3.4, where it is clear that if we slice each of the cubic boxes horizontally, one of the $1/4 \times 1/8$ rectangles thus obtained contains two points, while the other contains none.

3.5.3 Statistical tests

Once a generator with good theoretical properties has been identified — for instance, a combined MRG with a long period and good results with respect to the spectral test — the next step is to test its local properties with the help of various statistical tests.

In general, statistical tests for random number generators test the hypothesis

H_0 : the sequence $u_0, u_1, \ldots$ output by the generator is a sequence of i.i.d. $U(0, 1)$ random variables.

They do so by forming a *test statistic* of the form

$$Z = \zeta(u_0, u_1, \ldots, u_{n-1})$$

based on the n first numbers $u_0, \ldots, u_{n-1}$ output by the generator, and whose distribution under H_0 is known or can be approximated. We can then formally test H_0 by computing the associated p-value. For instance, if we fix a level of type I error α (the probability of rejecting H_0 given that H_0 is true), then we reject H_0 if the p-value is smaller than α. Alternatively, without formally fixing α, we can compute the p-value and become "suspicious" when it is considered "small".

Of course, the deterministic nature of PRNGs implies that H_0 is necessarily false for them. From this point of view, it seems like applying such tests is a waste of time because we know there exists at least one test for which H_0 will be rejected. The reason why these tests are still useful is that although we know H_0 is false, if it is difficult to gather statistical evidence showing that H_0 is false, then we can have more confidence in the underlying generator than if it is very easy to find a test for which H_0 is rejected.

Now, the next question is: Which tests should be performed? The setup above, where each function ζ gives rise to a different test, provides us with an unmanageably large number of choices. A reasonable approach is to choose functions ζ that share similarities with the applications for which the generator is likely to be used. Unfortunately, we usually do not have this kind of knowledge when generators are designed. As a compromise, we can look for functions ζ that measure a more "intuitive" notion of uniformity and/or that seem more "natural".

There are several packages for testing randomness that include a wide variety of tests like that. Examples are the *Diehard* package of George Marsaglia [503], the *TestU01* package of Pierre L'Ecuyer and Richard Simard [497], and a package developed by NIST (National Institute of Standards and Technology) [486]. Here we mention a few tests that are commonly used and refer the reader to [120, 221, 251, 262, 269, 391, 471] and the references therein for more examples. Just to give an idea, NIST recommends 16 tests be used when assessing a generator, while Knuth recommends 13 tests [221]. In the library TestU01, the smallest battery of tests offered computes a total of 144 test statistics and p-values for each generator tested.

We will not attempt to create our own list of tests that should be performed on a generator to make sure it is *safe*. Our "recommendation" for users who need random numbers is to either use a well-tested generator like MRG32k3a or else, if one wants to use a known generator (perhaps one that is implemented in the programming language/software used), at least make sure it is not on a "blacklist" somewhere for having failed too many tests (see, for instance, Table 1 in [269]). If it is not a known generator, then at least apply some battery of tests (such as *SmallCrush* from TestU01) to make sure there is no gross defect. The information given below should be sufficient for understanding the contents of such batteries of tests and how they operate.

We start by describing a very common test called the *serial test* [220], and then we will describe a general setup that includes several tests used in practice.

The serial test is simply a Pearson chi-square goodness-of-fit test such as those typically done when testing if a sample of observations follows a given distribution. In our case, the sample is obtained by forming r vectors of s points obtained by $n = rs$ successive calls to the generator. That is, we look at

$$\mathbf{u}_i = (u_{si}, u_{si+1}, \dots, u_{si+s-1}), \qquad i = 0, \dots, r-1.$$

We then consider a group of $k = d^s$ cubic cells in $[0,1)^s$ obtained by partitioning each interval $[0,1)$ into d subintervals of length $1/d$. The statistic Z for the serial test is formed by counting the number of points in the sample $\mathbf{u}_0, \dots, \mathbf{u}_{r-1}$ that fall in each cell. More precisely, let N_j be the number of points that fall in cell j for $j = 1, \dots, k$ (assuming a given labeling has been chosen for the cells). The vector $(N_1, \dots, N_k)$ then has a multinomial distribution with parameters $(k, p_1, \dots, p_k)$, where $p_j = 1/k$ for each $j = 1, \dots, k$. From standard results in statistics, under H_0 the distribution of the quantity

$$X^2 = \sum_{j=1}^{k} \frac{(N_j - r/k)^2}{r/k}$$

approaches a chi-square distribution with $k-1$ degrees of freedom as r goes to infinity. Typically, a rule of thumb is to say that if the expected number of points per cell r/k is at least 5, then the approximation by a chi-square should be reasonably good. We can then compute the value taken by X^2 for a given sample — call it x — and determine $p = P(X^2 > x|H_0)$ or $p = P(X^2 < x|H_0)$. For instance, suppose $s = 2$, $d = 5$, r =10,000, and that we get $x = 5$. For a chi-square random variable X^2 with 24 degrees of freedom, we have that $p = P(X^2 < 5) = 1.26 \times 10^{-5}$ is very small, which suggests that H_0 should be rejected. In other words, the sample considered is "too uniform". A two-sided test with $\alpha > 2.52 \times 10^{-5}$ would reject H_0 in this case.

It turns out that several other tests can be derived from the vector $(N_1, \dots, N_k)$ described above. More generally, we can use the following setup to describe several statistical tests commonly used for random number generators [262]. Consider

$$Z = \sum_{j=1}^{k} \beta(N_j),$$

where β is a real-valued function. For example, we have

$$\begin{aligned} \text{Serial test}: Z = X^2 &= \textstyle\sum_{j=1}^{k} \frac{(N_j - r/k)^2}{r/k}, \\ \text{Negative entropy}: Z = -H &= \textstyle\sum_{j=1}^{k} (N_j/r)\log_2(N_j/r), \\ \text{Collisions}: \ Z = C &= \textstyle\sum_{j=1}^{k} (N_j - 1)\mathbf{1}_{N_j > 1}. \end{aligned}$$

For these three examples, a larger value of Z means the points are less uniformly distributed in $[0,1)^s$. A very small value of Z means the points are very (maybe too much) uniformly distributed in $[0,1)^s$. The negative entropy can be related to a *loglikelihood ratio test* for the multinomial distribution [251]. The collisions test counts the number of collisions that occur within the sample, where by collision we mean that a point falls in a cell already occupied by at least one other point.

Once a function β is chosen, the next step is to determine the distribution of Z under H_0 or at least get an approximation for it. We already saw that, under H_0 and the assumption that $r/k \geq 5$, the distribution of X^2 was approximately chi-square. An important factor that determines the distribution of Z is whether we are working in a *dense case* setting or a *sparse case* setting. The sparse case means roughly that r/k is small, so that we are very likely to observe zero values for some of the variables N_j. The dense case means r/k is quite large. For instance, the setting we described for the serial test was the dense case because we (implicitly) assumed k was fixed and looked at the distribution of X^2 as $r \to \infty$. More generally, we have Theorem 3.12.

Theorem 3.12. *[262] (Dense case) Under H_0 and when k is fixed and $r \to \infty$, under some mild conditions we have*

$$\frac{Z - \mathrm{E}(Z) + (k-1)\sigma_c}{\sigma_c} \Rightarrow \chi^2(k-1),$$

where $\sigma_c^2 = \mathrm{Var}(Z)/(2(k-1))$, $\chi^2(k-1)$ denotes the chi-square distribution with $k-1$ degrees of freedom.

The *mild conditions* mentioned in the statement of this theorem are satisfied by X^2 and $-H$ but not C. Knuth shows how to compute the exact distribution of C in [220]. The case $Z = X^2$ discussed previously fits the setup of Theorem 3.12, with $\mathrm{E}(Z) = k-1$ and $\sigma_c = 1$. For $Z = -H$, the connection with the loglikelihood ratio test can be used to show that $\mathrm{E}(Z) = \log_2(k) - (k-1)/2n \ln 2$ and $\mathrm{Var}(Z) = (k-1)/(2n^2(\ln 2)^2)$ [251].

The sparse case differs from the dense case in that as r goes to infinity we also make the number of cells k go to infinity in such a way that the average number of points per cell r/k tends toward a constant δ. More precisely, we have Theorem 3.13.

Theorem 3.13. *[262] (Sparse case) Under H_0 and when $k \to \infty$, $r \to \infty$, and $r/k \to \delta$, where $0 < \delta < \infty$, under mild conditions*

$$\frac{Z - \mathrm{E}(Z)}{\sqrt{\mathrm{Var}(Z)}} \Rightarrow N(0,1). \tag{3.17}$$

In the sparse case, the *mild conditions* are satisfied by X^2, $-H$, and C. General expressions for $\mathrm{E}(Z)$ and $\mathrm{Var}(Z)$ are given in [273]. Once those are evaluated, we can compute p-values; i.e., $p = P(Z > z|H_0)$, where z is the value of Z obtained for a given sample. If p is too small, then H_0 should be rejected. It should be noted that since the statistic C is integer-valued, the approximation (3.17) by the normal distribution is good only if the expectation of C is large enough. If it is too small (e.g., smaller than 50 or so [262]), then a Poisson approximation should be used instead. For instance, in [255], the collision test is performed on several widely used generators with $s = 2$, $d = r/16$, and r equal to different powers of two ranging between 2^{15} and 2^{20}. The distribution of Z in this case is Poisson with a mean $\lambda = r^2/(2k)$, which in the setting of [255] gives $\lambda = 128$.

Another family of tests that can be defined using the setup above is as follows [271]. Define I_i as the number (label) of the cell where $\mathbf{u}_i$ has fallen. Then sort these variables in increasing order, thereby obtaining $I_{(0)} \leq I_{(1)} \leq \ldots \leq I_{(r-1)}$. Compute the spacings $S_j = I_{(j)} - I_{(j-1)}$ for $j = 1, \ldots, r-1$, and let $Z := B$ be the number of *collisions* between these spacings, that is, the number of j in $\{1, \ldots, r-2\}$ such that $S_{(j)} = S_{(j+1)}$, where $S_{(1)} \leq \ldots \leq S_{(r-1)}$ are the order statistics of the spacings $S_1, \ldots, S_{r-1}$. This test is called the *birthday spacings test* in [303], where it was introduced, because we can view each point $\mathbf{u}_i$ as a "person" with a "birthday" I_i in a year with k days.

For instance, suppose we have a sample of $r = 8$ and $k = 4$ cells. Assume the eight points fall in cells 4, 4, 2, 1, 1, 3, 1, 4. Then $I_{(0)} = I_{(1)} = I_{(2)} = 1$, $I_{(3)} = 2$, $I_{(4)} = 3$, and $I_{(5)} = I_{(6)} = I_{(7)} = 4$, so that $S_1 = 0, S_2 = 0, S_3 = 1, S_4 = 1, S_5 = 1, S_6 = 0, S_7 = 0$. Hence $S_{(1)} = \ldots S_{(4)} = 0$ and $S_{(5)} = \ldots = S_{(7)} = 1$, which means $B = 5$.

It can be shown that if r is large and $\lambda = r^3/4k$ is small, then under H_0, B follows approximately the Poisson distribution with mean λ [271]. One can then compute $P(B \geq z|H_0)$ or $P(B < z|H_0)$, where z is the value of Z for a given sample, and reject H_0 if the p-value is too small. To give an idea of what is a "large" r and a "small" λ, in [271] values of r of about $\rho^{1/3}$ are used (where ρ is the period of the generator under study) and d chosen so that $r/k = r/d^s$ is about 1. See Prob. 3.15 for more specific examples of parameters.

So far, we have assumed that the sample $\mathbf{u}_0, \ldots, \mathbf{u}_{r-1}$ was formed by using nonoverlapping numbers produced by the generator. For this reason, under H_0, these r points are assumed to be independent. Alternatively, one can use overlapping points. That is, the points in the sample are then defined as $\mathbf{u}_i = (u_i, u_{i+1}, \ldots, u_{i+s-1})$ for $i = 0, \ldots, r-1$. One advantage of these *overlapping tests* over their nonoverlapping counterpart is that they can detect departures from H_0 almost as well, although they require $n = r + s - 1$ numbers to be output by the generator rather than $n = rs$. On the other hand, finding the distribution of the corresponding test statistic in the overlapping case is usually much more difficult [471].

Finally, in addition to computing p-values in order to give us an idea of how likely it is to have observed a value z for the statistic Z, another possibility is to perform a *second-level* test. That is, for a given test statistic Z, generate a sample $Z_1, \ldots, Z_m$ of Z and then perform a statistical test that compares the empirical distribution thus obtained with the distribution of Z under H_0. For instance, Knuth [221] suggests generating m replications $B_1, \ldots, B_m$ of the birthday spacings test and then performing a Pearson chi-square goodness-of-fit test for these B_i to see if they are close enough to a Poisson distribution, as would be the case under H_0.

Problems

3.1. For each of the two MLCGs (i) $m = 61$ and $a = 17$ and (ii) $m = 61$ and $a = 3$, (a) find all the cycles of the generator; (b) determine a set Σ of seeds such that for each cycle there is exactly one seed in Σ that generates that cycle, and (c) plot Ψ_2.

3.2. Consider an MLCG with a prime modulus $m > 1000$ and a multiplier a that is a primitive element modulo m. Suppose that from a given output u_i you want to jump ahead to u_{i+1000} without having to generate all 999 intermediate values. How would you proceed?

3.3. A well-known result in number theory [291] says that if m is prime, then there are $\phi(m-1)$ elements in $\{1,\ldots,m-1\}$ that are primitive elements modulo m, where $\phi(p)$ is the Euler function, which gives the number of elements i in $\{1,\ldots,p-1\}$ such that $\gcd(i,p) = 1$. Also, once a primitive element modulo m has been identified — call it a — the other ones take the form $a^r \bmod m$, where r runs over all integers in $\{1,\ldots,m-2\}$ such that $\gcd(r, m-1) = 1$. Use this result to find all primitive elements modulo 31.

3.4. Show that for $a_r, a_k \in \{1,\ldots,m-1\}$ and m prime, the recurrence

$$x_i = (a_r x_{i-r} - a_k x_{i-k}) \bmod m$$

is equivalent to the recurrence

$$x_i = (a_r x_{i-r} + \tilde{a}_k x_{i-k}) \bmod m,$$

where $\tilde{a}_k = (m-1)a_k \bmod m$.

3.5. Give expressions for the matrices A (transition matrix) and B (output matrix) in Def. 3.6 that correspond to an LFSR generator.

3.6. Plot Ψ_2 for the LFSR generator defined by $k = 7, a_7 = 1, a_3 = 1$ ($a_j = 0$ for all other j), and (i) $\nu = 1$ and $L = 7$ and (ii) $\nu = 3$ and $L = 7$.

3.7. Show how to initialize a GFSR of the form (3.6)–(3.7) so that it is equivalent to an LFSR of the form (3.4)–(3.5).

3.8. For $k = 7,\ldots,10$, determine how many different LFSR generators of the form (3.4)–(3.5) have a maximal period of $2^k - 1$.

3.9. Show that for a prime modulus m and $x \in \{1,\ldots,m-1\}$ we have that the inverse of x modulo m is given by $x^{-1} = x^{m-2} \bmod m$.

3.10. Describe an algorithm to compute $a^k \bmod m$ that requires $O(\log k)$ multiplications. (This problem is usually referred to as *modular exponentiation.*)

3.11. Generate the first 1000 points produced by the explicit inversive congruential generator of [274] with $m = 2^{31} - 1$, $a = 7$, and $b = 1$. Plot $\{(u_i, u_{i+1}), i = 0, \ldots, 999\}$.

3.12. Compute the value of d_2 from the spectral test for the MLCG with $m = 61$ and $a = 17$.

3.13. Show that if $\mathbf{h} \in L_s^*$, then the quantity $S(\mathbf{h})$ defined in (3.12) equals 1.

3.14. Compute the resolution ℓ_2 and the t-value (for $s = 2$) for the toy LFSR described in Prob. 3.6.

3.15. For each of the generators (i) to (iii) described below, compute the test statistics discussed in Section 3.5.3: $X^2, -H$, C, and B (from the birthday spacings test). Use $s = 2$ and try $r = 2^{15}$ and $r = 2^{16}$, and for d take $d = 8$ for X^2 and H, $d = r^2/16$ for C, and $d = r^{3/2}/2$ for B. The generators to test are (i) MRG32k3a, (ii) the explicit inversive congruential generator from Prob. 3.11, and (iii) the LCG defined by $m = 2^{31} - 1$ and $a = 65539$ (RANDU).

3.16. As a follow-up to Prob. 3.15, perform a second-level test for the generator MRG32k3a and the birthday spacings test. More precisely, generate a sample of 100 observations $B_1, \ldots, B_{100}$ of the test statistic B, and then perform a chi-square goodness-of-fit test based on the five bins corresponding to $B = i$ for $i = 0, \ldots, 3$ and $B \geq 4$. Compute the test statistic and p-value for this chi-square test.

Chapter 4
Variance Reduction Techniques

4.1 Introduction

In Chap. 1, we said that one way of improving the Monte Carlo integration error is to try reducing the variance σ^2 of the integrand f. More precisely, the goal is to find another function ϕ whose integral is equal to the integral of f but whose variance is smaller than that of f. Methods that achieve this are called *variance reduction techniques*, and we will be describing several of them in this chapter. This topic has been widely studied and is surveyed, for example, in [45, 165, 243, 247, 321, 391], which also give several other references.

In our presentation of these techniques, we go back and forth between the integration formulation and the more intuitive simulation setup. In preparation for this, we first recall the notation used when discussing these two different interpretations.

Following the terminology of Fig. 1.6, if the goal of the simulation study is to estimate the expectation μ of some output function $h(\mathbf{X})$, then we can write

$$\mu = \mathrm{E}(h(\mathbf{X})) = \int_{[0,1)^s} f(\mathbf{u})d\mathbf{u} = \mathrm{E}(f(\boldsymbol{U})). \tag{4.1}$$

The first equality in (4.1) states the problem using the simulation formulation, where $\mathbf{X}$ is the vector of random variables required to run the simulation. The second equality rewrites the problem as a multivariate integral over the unit hypercube. The third equality views μ as the expected value of f when evaluated at a randomly uniformly distributed point $\boldsymbol{U}$ in $[0,1)^s$. In addition, we also use the notation

$$Y = f(\boldsymbol{U}) = h(\mathbf{X}). \tag{4.2}$$

That is, Y is the random variable that represents the output measure of interest, written either as the valuation of f at a random input point $\boldsymbol{U}$ or the output of a simulation run driven by the random variables in $\mathbf{X}$.

C. Lemieux, *Monte Carlo and Quasi–Monte Carlo Sampling*,
Springer Series in Statistics 692, DOI: 10.1007/978-0-387-78165-5_4,

For instance, in Example 1.2, the random vector $\mathbf{X}$ can be defined as the vector $(A_1, S_1, A_2, S_2, \ldots)$ of interarrival and service times, and h in this case is

$$h(\mathbf{X}) = \sum_{j=1}^{N(\mathbf{X})} \mathbf{1}_{W_j(A_1,S_1,\ldots,A_j)>5},$$

where we wrote the waiting time W_j as a function of $(A_1, S_1, \ldots, A_j)$ and the total number $N = N(\mathbf{X})$ of clients that entered the bank during a day as a function of $\mathbf{X}$ to make the dependence on $\mathbf{X}$ as explicit as possible.

In practice, a realization $\mathbf{x}$ is generated from a uniform vector $\mathbf{u}$ using a random variate generation method such as those discussed in Chap. 2. Hence we can write $\mathbf{x} = g(\mathbf{u})$ for some function g. Therefore the relation between f and h is that $f(\mathbf{u}) = h(g(\mathbf{u}))$; that is, $f = h \circ g$. For instance, in Example 1.2, g was given by

$$g(\mathbf{u}) = (-\ln(1 - u_1), -0.75 \ln(1 - u_2), -\ln(1 - u_3), \ldots) = (a_1, s_1, a_2, \ldots).$$

As we mentioned at the end of Sect. 1.2 and as illustrated in Example 1.3, for a given μ there are several choices to make that will affect the definition of the function f in (4.1). With the notation we just introduced, we can be more precise about this and view g as representing our choice of random variate generation method and the pair $(h, \mathbf{X})$ as our description of the simulation model to be used for estimating μ. For instance, in Example 1.3, the three possibilities considered via the functions f_1 to f_3 respectively correspond to using (1) $\mathbf{X} = (X_1, X_2)$, where X_1 and X_2 are independent Exp(0.75) and $h(\mathbf{X}) = \mathbf{1}_{X_1+X_2>2.5}$; (2) $\mathbf{X} = N$, where $N \sim$ Poisson(10/3) and $h(\mathbf{X}) = \mathbf{1}_{N<2}$; and (3) $\mathbf{X} = X$, where $X \sim$ Gamma(2, 0.75) and $h(\mathbf{X}) = h(X) = \mathbf{1}_{X>2.5}$.

To summarize, we have the notation in Fig. 4.1:

$g : [0,1)^s \to \mathbb{R}^k$ is the function that transforms a vector of s i.i.d. uniform numbers into a vector $\mathbf{X} = (X_1, \ldots, X_k)$ of random variables used to describe the simulation model; $h : \mathbb{R}^k \to \mathbb{R}$ is the function that takes as input a vector $\mathbf{X}$ of random variables describing the simulation model and turns them into an observation of the quantity of interest; and $f : [0,1)^s \to \mathbb{R}$ is the composition of g and h (i.e., $f(\mathbf{u}) = h(g(\mathbf{u}))$) and represents the function that turns a vector of s i.i.d. uniform numbers into an observation of the quantity of interest. This is the integrand in the integration formulation of the problem.

Fig. 4.1 Different ways of describing a problem through the functions g, h, and f.

Before we begin our presentation of the most commonly used variance reduction techniques, we first briefly discuss the concept of efficiency.

4.2 Efficiency

Finding ways of constructing estimators with smaller variance can often lead to an improvement in the *efficiency* as well. The efficiency is a quality measure for estimators that takes into account both their variance and computation time [165]. Considering the efficiency rather than just the reduction in variance is certainly desirable, as we want to prevent the use of techniques that could only reduce the variance at the expense of a large increase in computation time. The concept of efficiency can be defined in different ways. The definition we chose to use comes from [247] and goes back to [165] in the case of unbiased estimators. It has the intuitive property that it is independent of n for a naive unbiased Monte Carlo estimator, as we will see shortly. A more general treatment of efficiency can be found in [153, 157].

Definition 4.1. The efficiency of an estimator $\hat{\mu}$ for a quantity μ is given by

$$\mathrm{Eff}(\hat{\mu}) = [\mathrm{MSE}(\hat{\mu}) \times C(\hat{\mu})]^{-1},$$

where $\mathrm{MSE}(\hat{\mu}) = \mathrm{Var}(\hat{\mu}) + B^2(\hat{\mu})$ is the *mean-square error* of $\hat{\mu}$, $B(\hat{\mu}) = \mathrm{E}(\hat{\mu}) - \mu$ is the bias of $\hat{\mu}$, and $C(\hat{\mu})$ is the expected computation time for $\hat{\mu}$.

The larger the efficiency, the better is the estimator. This definition also implies that if we have two unbiased estimators $\hat{\mu}_1$ and $\hat{\mu}_2$ that require the same computation time, then if $\mathrm{Var}(\hat{\mu}_1) < \mathrm{Var}(\hat{\mu}_2)$, we prefer $\hat{\mu}_1$ over $\hat{\mu}_2$.

If $\hat{\mu}$ is a naive unbiased Monte Carlo estimator for μ, then $\mathrm{Var}(\hat{\mu}) = \sigma^2/n$, where σ^2 is the variance of $f(\boldsymbol{U})$, and the expected computation time is cn for some constant $c > 0$. Since $\hat{\mu}$ is unbiased, the efficiency is thus $\mathrm{Eff}(\hat{\mu}) = 1/c\sigma^2$, which is independent of n. This means that for the naive Monte Carlo estimator, our definition of efficiency is such that the decrease in variance obtained by increasing the sample size is exactly offset by the increase in computation time. Therefore, in order to find more efficient estimators than the naive Monte Carlo estimator, we need to find ways of getting a q-fold reduction in variance while restricting the increase in computation time to a factor no larger than q.

For each of the variance reduction techniques presented in this chapter, we will mostly be discussing how and why they reduce the variance, but we will also use numerical examples to compare the efficiency of the corresponding estimators with the naive Monte Carlo method.

4.3 Antithetic variates

This method was introduced by Hammersley and Morton in 1956 [164]. It can be applied easily to most problems and often produces at least a modest variance reduction. In its simplest form, it is based on the idea that instead of estimating μ by the average of i.i.d. random variables having expectation μ, use *pairs* of negatively correlated random variables, again with expectation μ. Within each pair, the negative correlation should have the effect of "cancelling out" departures from μ. Therefore, if we approximate μ by the average of the pairs' average, we should get an estimator with smaller variance. More general ways of applying antithetic variates are discussed in [6, 7, 57, 122, 477].

In what follows, we make the assumption that n is even and apply antithetic variates in a way that preserves the total number of function evaluations. That is, we replace n independent observations by $n/2$ pairs of antithetic observations. In this way, comparisons based on the variance are more "fair". We could also double the number of function evaluations, replacing each observation by a pair of antithetic observations, but then the extra work would need to be taken into account when making variance comparisons.

Using the integration point of view, the method of antithetic variates consists in replacing the naive Monte Carlo estimator

$$Q_n = \frac{1}{n}\sum_{i=1}^{n} f(\mathbf{u}_i)$$

by the antithetic estimator

$$Q_{n,\text{ant}} = \frac{1}{n/2}\sum_{i=1}^{n/2} \frac{f(\mathbf{u}_i) + f(\tilde{\mathbf{u}}_i)}{2},$$

where $\tilde{\mathbf{u}}_i \sim U([0,1)^s)$ is negatively correlated with $\mathbf{u}_i$, for $i = 1, \ldots, n/2$. The most common way to induce this negative correlation is to define $\tilde{u}_{ij} = 1 - u_{ij}$, where u_{ij} and $\tilde{u}_{ij}$ are the jth coordinates of $\mathbf{u}_i$ and $\tilde{\mathbf{u}}_i$, respectively. From now on, we assume antithetic variates are applied with this particular choice of definition for $\tilde{\mathbf{u}}_i$.

From the simulation point of view and using the notation given in Fig. 4.1, the antithetic variates estimator can be written as

$$\hat{\mu}_{\text{ant}} = \frac{1}{n/2}\sum_{i=1}^{n/2} \frac{h(\mathbf{X}_i) + h(\tilde{\mathbf{X}}_i)}{2},$$

where $\tilde{\mathbf{X}}_i$ is generated using the random numbers in $\tilde{\mathbf{u}}_i = (1-u_{i1}, \ldots, 1-u_{is})$ as input. That is, for some function g, we have

$$\mathbf{X}_i = g(u_{i1}, \ldots, u_{is}),$$
$$\tilde{\mathbf{X}}_i = g(1 - u_{i1}, \ldots, 1 - u_{is}).$$

Let $\sigma^2 = \mathrm{Var}(f(\boldsymbol{U}))$. Then the variance of $Q_{n,\text{ant}}$ is given by

$$\begin{aligned}\mathrm{Var}(Q_{n,\mathrm{ant}}) &= \frac{0.25}{(n/2)}(\sigma^2 + \sigma^2 + 2\mathrm{Cov}(f(\mathbf{u}_i), f(\tilde{\mathbf{u}}_i))) \\ &= \frac{\sigma^2}{n} + \frac{1}{n}\mathrm{Cov}(f(\mathbf{u}_i), f(\tilde{\mathbf{u}}_i)),\end{aligned} \tag{4.3}$$

which is no larger than the naive Monte Carlo estimator's variance σ^2/n as long as $\mathrm{Cov}(f(\mathbf{u}_i), f(\tilde{\mathbf{u}}_i)) \leq 0$. Hence the performance of this technique depends on how much of the negative correlation between $\mathbf{u}_i$ and $\tilde{\mathbf{u}}_i$ is preserved after f is applied to these two points. Theorem 4.3 below addresses this question. Equivalently, using the simulation point of view, we can say that the method's ability to reduce the variance depends on how the negative correlation between $\mathbf{u}_i$ and $\tilde{\mathbf{u}}_i$ will be preserved (i) once these points are transformed into $\mathbf{X}_i$ and $\tilde{\mathbf{X}}_i$, respectively, and then (ii) after h is applied to $\mathbf{X}_i$ and $\tilde{\mathbf{X}}_i$. Theorem 4.4 below partly addresses this question.

Note that the variance of $Q_{n,\mathrm{ant}}$ can be estimated by the estimator

$$\hat{\sigma}^2_{n,\mathrm{ant}} = \frac{1}{(n/2)(n/2-1)} \sum_{i=1}^{n/2} (Z_i - Q_{n,\mathrm{ant}})^2,$$

where $Z_i = 0.5(f(\mathbf{u}_i) + f(\tilde{\mathbf{u}}_i))$, since these Z_i's are independent. It is important to observe that it would be incorrect to use the sample variance of $\{f(\mathbf{u}_1), \ldots, f(\mathbf{u}_{n/2}), f(\tilde{\mathbf{u}}_1), \ldots, f(\tilde{\mathbf{u}}_{n/2})\}$ to construct an estimator for the variance of $Q_{n,\mathrm{ant}}$, as this sample does not contain n independent observations but rather $n/2$ pairs of correlated observations.

We note that with the antithetic variates method, any linear function is integrated with zero error. Problem 4.4 at the end of the chapter asks you to prove this. The following example deals with a simple special case.

Example 4.2. Assume we want to estimate $I(f) = \int_0^1 f(u)du$, where $f(u) = au + b$ and a, b are some real constants. Of course, we know $I(f) = a/2 + b$ in this case. Consider the naive Monte Carlo estimator based on a sample of size n, where n is even,

$$Q_n = \frac{1}{n}\sum_{i=1}^{n}(au_i + b),$$

where $u_1, \ldots, u_n$ are i.i.d. $U(0,1)$. A simple calculation shows

$$\mathrm{Var}(Q_n) = \frac{a^2}{12n}.$$

Now, suppose we use the antithetic pairs $(au_i + b, a(1-u_i) + b)$ for $i = 1, \ldots, n/2$, where $u_1, \ldots, u_{n/2}$ are independent $U(0,1)$ and form the antithetic estimator

$$Q_{n,\text{ant}} = \frac{1}{n/2} \sum_{i=1}^{n/2} \left(\frac{au_i + b + a(1 - u_i) + b}{2} \right).$$

Then we can see that

$$Q_{n,\text{ant}} = \frac{1}{n/2} \sum_{i=1}^{n/2} \frac{a + 2b}{2} = a/2 + b = I(f),$$

and therefore $\text{Var}(Q_{n,\text{ant}}) = 0$. Hence, for this example, using antithetic variates gives us a perfect estimator.

Alternatively, using the simulation framework — which is very simplistic in this case — we can say the goal is to estimate $\text{E}(X)$, where $X \sim U(b, b+a)$. In that case, the Monte Carlo estimator is

$$\hat{\mu}_{\text{mc}} = \frac{1}{n} \sum_{i=1}^{n} X_i,$$

where the X_i are i.i.d. $U(b, b + a)$, while the antithetic estimator is

$$\hat{\mu}_{\text{ant}} = \frac{1}{n/2} \sum_{i=1}^{n/2} \frac{X_i + \tilde{X}_i}{2},$$

where $X_i = au_i + b$, $\tilde{X}_i = a(1 - u_i) + b = a + 2b - X_i$, and $u_1, \ldots, u_{n/2}$ are i.i.d. $U(0, 1)$. Hence

$$\hat{\mu}_{\text{ant}} = \frac{1}{n/2} \sum_{i=1}^{n/2} \frac{a + 2b}{2} = a/2 + b = I(f).$$

In this simple example, we exploited the fact that initially we have that $\tilde{u}_i$ is perfectly negatively correlated with u_i,

$$\rho(u_i, \tilde{u}_i) = \frac{\text{Cov}(u_i, \tilde{u}_i)}{\sigma_u^2} = -1,$$

since $\sigma_u^2 = \text{Var}(u_i) = 1/3 - 1/4 = 1/12$, and

$$\text{Cov}(u_i, \tilde{u}_i) = \text{E}(u_i - u_i^2) - \text{E}^2(u_i) = 1/2 - 1/3 - 1/4 = -1/12.$$

Then, since $f(u)$ is linear in u, this perfect negative correlation between u_i and $\tilde{u}_i$ is preserved when f is applied, which means $\text{Cov}(f(u_i), f(\tilde{u}_i)) = \text{Var}(f(u_i))$ (see Prob. 4.3). Thus, from (4.3), we see that $\text{Var}(Q_{n,\text{ant}}) = 0$ and $Q_{n,\text{ant}} = I(f)$. Figure 4.2 illustrates the application of antithetic variates for this example.

In general, antithetic variates do not work perfectly because the functions we deal with are usually not linear. In Fig. 4.3, we illustrate for two simple

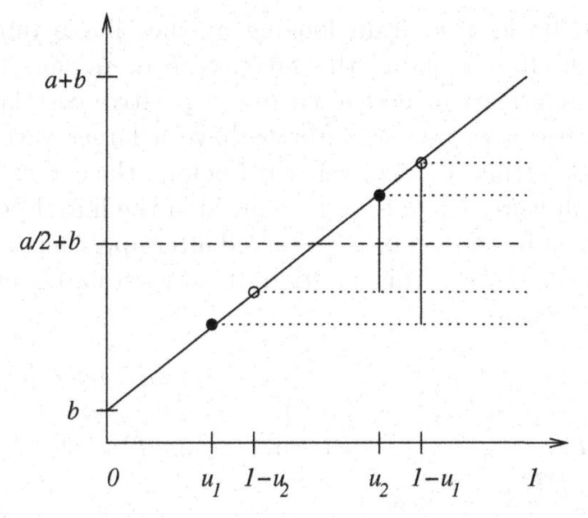

Fig. 4.2 Antithetic variates applied to $f(u) = au + b$. Each point u_i is paired with $1 - u_i$ so that the average $0.5(f(u_i) + f(1 - u_i))$ equals $I(f) = a/2 + b$.

functions the effect of nonlinearity. On the left-hand side of this figure, we consider $f(u) = u^2$. We see that, in this case, the average $0.5(f(u_i)+f(1-u_i))$ — shown by a tick on the line that joins the two evaluations of f for a pair of points — is not necessarily equal to the integral $I(f) = 1/3$ due to the convexity of the function. The right-hand side of Fig. 4.3 shows an even worse case, where for the function $f(u) = (1 - 2u)^2$, which is symmetric around $u = 0.5$, the two antithetic evaluations $f(u_i)$ and $f(1 - u_i)$ are equal. This results in a "waste" of half the function evaluations and an increase in the variance when applying antithetic variates compared with the naive Monte Carlo method.

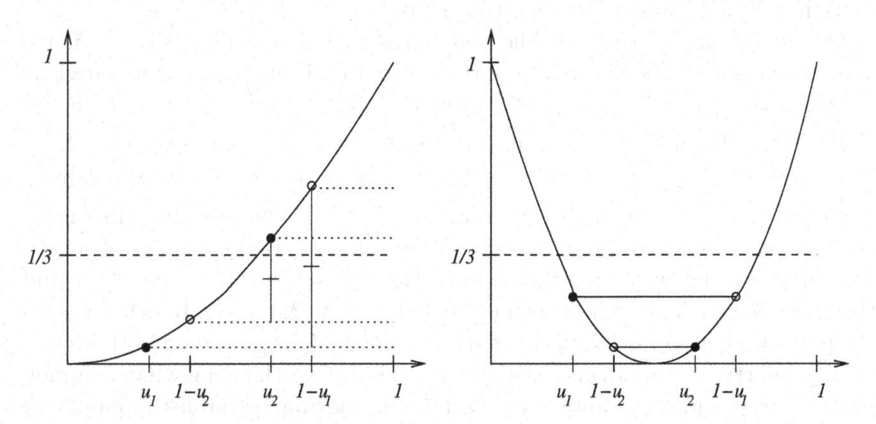

Fig. 4.3 Antithetic variates applied to $f(u) = u^2$ (left) and $f(u) = (1-2u)^2$ (right). Each point u_i is paired with $1 - u_i$.

The lesson to be learned from looking at the simple function $f(u) = (1-2u)^2$ is that after applying the function f to $\mathbf{u}_i$ and $1-\mathbf{u}_i$, a perfect negative correlation can be turned into a positive correlation and thus cause the antithetic variates estimator to have a larger variance than the naive Monte Carlo estimator. As mentioned before, the extent to which antithetic variates will work depends on how much of the initial perfect negative correlation between $\mathbf{u}_i$ and $1-\mathbf{u}_i$ is preserved after applying f. The following result offers some answers to this question. It comes from Lehmann [275] and is discussed, for instance, in [45].

Theorem 4.3. *[275] Let $f : [0,1)^s \to \mathbb{R}$ be a bounded and monotone function in each of its arguments. Suppose also that f is not constant in the interior of its domain. Let $\boldsymbol{U} = (U_1, \ldots, U_s) \sim U([0,1)^s)$ and $\tilde{\boldsymbol{U}} = (1-U_1, \ldots, 1-U_s)$. Then* $\mathrm{Cov}(f(\boldsymbol{U}), f(\tilde{\boldsymbol{U}})) < 0$.

This result says that if f is monotone in each of its arguments — and this does not mean that it has to be, say, increasing in all of its arguments; i.e., it can be increasing in the first argument, decreasing in the second, etc. — then in light of (4.3), using antithetic variates provides an estimator with a smaller variance than with the naive Monte Carlo estimator.

The following result is also relevant, especially within the simulation formulation.

Theorem 4.4. *[473] Let X be a random variable with CDF $F(\cdot)$. If $\tilde{U} = 1-U$, then $(F^{-1}(U), F^{-1}(\tilde{U}))$ has a minimum correlation among all pairs of random variables with marginal CDF given by $F(\cdot)$.*

What this result says is that to produce a pair (X, Y) of variables such that (i) F is the marginal CDF of each of X and Y and (ii) the correlation between X and Y is minimized, the optimal approach is to generate X by inverting F at U, and Y by inverting F at $1-U$.

One of the implications of Theorem 4.4 is that if $\mathbf{X} = (X_1, X_2, \ldots, X_s)$ is a vector of independent random variables and each of them is generated by inversion from $\mathbf{u} = (u_1, u_2, \ldots, u_s)$ by letting $X_j = F_j^{-1}(u_j)$, where F_j is the CDF of X_j, then using antithetic variates to generate a random vector $\tilde{\mathbf{X}}$ — that is, we let $\tilde{X}_j = F_j^{-1}(1-u_j)$ — with the same distribution as $\mathbf{X}$ produces pairs $(X_j, \tilde{X}_j)$ with minimum correlation. A good example that illustrates this property is to consider the case where $X \sim N(0,1)$. In that case, by symmetry of the normal pdf, we have that $\Phi^{-1}(1-U) = -\Phi^{-1}(U)$ and therefore $\tilde{X} = -X$, which in turn implies that $\rho(X, \tilde{X}) = -1$. In other words, the perfect negative correlation between U and $1-U$ is preserved in that case. This property of the normal distribution has led to the somewhat common practice that applying antithetic variates to normal variables amounts to pairing each X with $-X$, regardless of the nonuniform generation method used to generate X. However, taking $X = F^{-1}(U)$ and $Y = -F^{-1}(U) = -X$

is not generally correct. As a simple counterexample, consider the case where we want a marginal CDF that is exponentially distributed with mean β. Then

$$Y = -F^{-1}(U) = \beta \ln(1 - U) < 0$$

clearly does not have the correct distribution since we must have $Y > 0$.

We note that in light of Theorem 4.3 and using the fact that F^{-1} is monotone for any CDF F, the (minimum) correlation between X and $\tilde{X}$ mentioned in Theorem 4.4 can in fact be shown to be negative. However, since $\mathbf{X}$ and $\tilde{\mathbf{X}}$ are further transformed when h is applied, Theorem 4.4 does not guarantee that $Y = h(\mathbf{X})$ and $\tilde{Y} = h(\tilde{\mathbf{X}})$ will also have a minimum correlation, even if each pair $(X_j, \tilde{X}_j)$ does. Their correlation could actually be positive. Nevertheless, this theorem gives us at least some kind of "intermediate" optimality result.

Going back to the result stated in Theorem 4.3, it is important to point out that even if, for a given simulation study, we do not have an explicit definition of the corresponding function f such that (4.1) holds, it can still be feasible to check whether the monotonicity conditions given in this result hold or not. Here is an example illustrating how this can be done.

Example 4.5. In Example 1.2, let F_A be the CDF of the exponential distribution with mean 1, let F_S be the CDF of the exponential distribution with mean 0.75, and let $f_j(u) = f(u_1, \ldots, u_{j-1}, u, u_{j+1}, \ldots)$ as defined in (1.12). There are two cases to consider:

1. If u is used to generate an interarrival time (by inversion) — that is, $a_k = F_A^{-1}(u)$ for some $k \geq 1$ — then a_k increases with u, and therefore, for any $u_1, u_2, \ldots, u_{j-1}, u_{j+1}, \ldots$, $f_j(u)$ *decreases* with u because if an interarrival time increases and everything else remains the same, this can only decrease the waiting time of the clients from that point on and therefore decrease the number of clients that will wait more than 5 minutes.
2. If u is used to generate a service time (by inversion) — that is, $s_k = F_S^{-1}(u)$ for some $k \geq 1$ — then s_k increases with u, and therefore, for any $u_1, u_2, \ldots, u_{j-1}, u_{j+1}, \ldots$, $f_j(u)$ *increases* with u because if a service time increases, this can only increase the waiting time of the clients that come after and thus possibly increase the number of clients that will wait more than 5 minutes.

Hence, for this example, f satisfies the monotonicity conditions given in Theorem 4.3.

When the conditions of Theorem 4.3 hold, we can safely apply antithetic variates. That is, applying antithetic variates should reduce the variance compared to the naive Monte Carlo method. For some problems, it might be the case that f is monotone only in a certain subset of its arguments. If that is the case, then one can apply antithetic variates only to that subset. That is, if $\mathcal{J} \subseteq \{1, \ldots, s\}$ is such that f is monotone in u_j if and only if $j \in \mathcal{J}$, then antithetic variates can be applied as follows:

$$\frac{1}{n/2}\sum_{i=1}^{n/2}\frac{f(\mathbf{u}_i)+f(\tilde{\mathbf{u}}_{\mathcal{J},i})}{2},$$

where $\tilde{\mathbf{u}}_{\mathcal{J},i} = (\tilde{u}_{\mathcal{J},i,1},\ldots,\tilde{u}_{\mathcal{J},i,s})$, and

$$\tilde{u}_{\mathcal{J},i,j} = \begin{cases} 1-u_{ij} & \text{if } j \in \mathcal{J}, \\ w_{ij} & \text{if } j \notin \mathcal{J}, \end{cases}$$

where the variables $w_{ij} \sim U(0,1)$ are independent from the variables u_{ij}.

Finally, it is important to note that in order to apply Theorem 4.3, simulation with antithetic variates must be done so that we have *synchronization* [243, pp. 586ff.]. This means the jth uniform number u_j has to be used for the same purpose in the simulation based on $\mathbf{u}$ and the one based on $\tilde{\mathbf{u}}$. It is usually not too difficult to achieve this by carefully writing the simulation code. For example, the code given in Fig. 1.7 for Example 1.2 achieves synchronization. However, for this example, an implementation where service times would be generated only when the service starts would *not* achieve synchronization. The reason is that, for instance, in one simulation, customer 3 could start his service before customer 5 arrives and then in the antithetic simulation he could start *after* customer 5 arrives. If this happens, then the uniform number used for his service would be generated before the fifth interarrival time in one case and *after* it in the other case, which would break the synchronization.

Now that we have looked at the main theoretical aspects of antithetic variates, let us present two examples that will illustrate how to apply this method. These two examples will be used throughout the chapter to illustrate the use of the different variance reduction techniques discussed.

Example 4.6. This example is closely related to Example 1.2, but with the additional feature that the speed of the server is randomly determined at the beginning of the day [45]. More precisely, with probability 0.2, the mean service time is 35 seconds, with probability 0.7, it is 50 seconds, and with probability 0.1, it is 55 seconds. Figure 4.4 gives pseudocode for using antithetic variates in this example.

The second example has been used in [22] to illustrate the effectiveness of different variance reduction techniques and their combinations. We find it to be a useful example that is different from the more traditional queueing problems.

Example 4.7. A *stochastic activity network* (SAN) is a directed acyclic graph $(\mathcal{N},\mathcal{A})$, where the set of nodes $\mathcal{N}$ contains a *source* and a *sink* and the edges in $\mathcal{A}$ represent activities. Each activity $j \in \mathcal{A}$ is assumed to have a certain duration D_j, which is a random variable with a CDF $F_j(\cdot)$. Dummy activities with zero duration can be used to enforce precedence relations between other activities. Let $N(A)$ denote the number of activities with a nonzero duration,

```
BankAntit(n)
  mus ← [7/12,5/6,11/12]
  for i = 1 to n/2 do
     result(i) ← OneSimBankAntit()
  hw ← 1.96 ×√(var(result)/(n/2))
  print ("average is", ave(result))
  print ("95% CI half-width is", hw)

GenDisc(p,k,u)
  i ← 1
  done ← 0
  while(i ≤ k AND done=0)
     if u < p[k] then
        done ← 1
        return(i)
     else i ← i + 1

OneSimBankAntit()
  NbWait5 ← 0
  w ← 0
  u[1] ← Rand01()
  type ← GenDisc([0.2,0.9,1],3,u[1])
  v ← mus[type]
  u[2] ← Rand01()
  a ← GenExpon(1,u[2])
  time ← a
  // antithetic initialization
  aNbWait5 ← 0
  aw ← 0
  atype ← GenDisc([0.2,0.9,1],3,1 − u[1])
  av ← mus[atype]
  aa ← GenExpon(1,1 − u[2])
  atime ← aa
  j ← 3
  while (time < 300 or atime < 300) do
     u[j] ← Rand01()
     u[j + 1] ← Rand01()
     s ← GenExpon(v, u[j])
     a ← GenExpon(1,u[j + 1])
     time ← time + a
     w ← max(0, w + s − a)
     if ((time < 300) and (w > 5)) then
        NbWait5 ← NbWait5 + 1
     // antithetic simulation
     as ← GenExpon(av, 1 − u[j])
     aa ← GenExpon(1, 1 − u[j + 1])
     atime ← atime + aa
     aw ← max(0, aw + as − aa)
     if ((atime < 300) and (aw > 5)) then
        aNbWait5 ← aNbWait5 + 1
     j ← j + 2
  return 0.5(NbWait5+aNbWait5)
```

Fig. 4.4 Pseudocode for using antithetic variates in Example 4.6.

let $N(P)$ denote the number of directed paths from the source to the sink, and let $C_k \subseteq \mathcal{A}$ be the set of activities on path k for $1 \le k \le N(P)$. The *completion time T of the network* is the length of the longest path from the source to the sink. Figure 4.5 gives an example of a SAN.

Here we assume the goal is to estimate the probability that the completion time T will be smaller than some value $t_0 > 0$. Formally, we want to estimate

$$\mu = F_T(t_0) = P(T \le t_0),$$

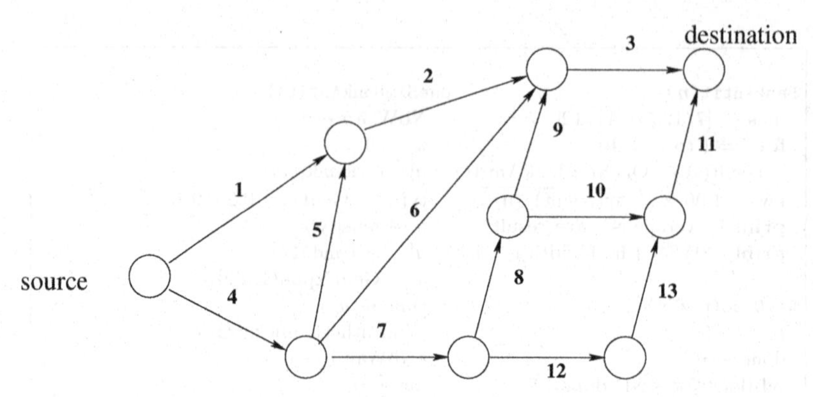

Fig. 4.5 SAN example from [22]. Adapted with permission from A. N. Wilson and J. R Wilson, Integrated Variance Reduction Strategies for Simulation, volume 44, number 2, March–April 1996. Copyright 1996, the Institute for Operations Research and the Management Sciences, 7240 Parkway Drive, Suite 300, Hanover, Maryland 21076.

where the completion time T is given by

$$T := T(D_1, \ldots, D_{N(A)}) = \max_{1 \le k \le N(P)} P_k,$$

and P_k is the length of the kth path. That is,

$$P_k = \sum_{j \in C_k} D_j.$$

The naive Monte Carlo estimator based on n i.i.d. simulations of a SAN is given by

$$\hat{\mu}_{\mathrm{mc}} = \frac{1}{n} \sum_{i=1}^{n} \mathbf{1}_{T_i \le t_0},$$

where

$$T_i = \max_{1 \le k \le N(P)} \sum_{j \in C_k} D_{i,j}$$

is the completion time for the ith simulation and $D_{i,j}$ is the simulated duration of the jth activity in the ith simulation.

The specific parameters used in our experiments are taken from [22] and used for the SAN shown in Fig. 4.5. Activities 1, 3, 4, 7 and 13 are normally distributed with respective means 5.5, 3.2, 13, 5.2, and 10.3 and standard deviation equal to 0.25 times the mean. Activities 2, 5, 6, 8, 9, 10, 11, and 12 are exponentially distributed with respective means 14.7, 7, 16.5, 6, 20, 4, 16.5, and 10.3. Also, we use $t_0 = 75$. Figure 4.6 gives pseudocode for using antithetic variates in this example.

```
SanAntit(n, t_0)
    N_A ← 13 //nb of activities
    N_P ← 6 // nb of paths
    for i = 1 to n/2 do
        max ← 0; amax ← 0
        for j = 1 to N_A do
            u[j] ← Rand01()
            D[j] ← GenF(j, u[j])
            aD[j] ← GenF(j, 1 − u[j])
        for k = 1 to N_P do
            L ← 0; aL ← 0
            for j = 1 to c_k do
                L ← L + D[C[k, j]]
                aL ← aL + aD[C[k, j]]
            if (L > max) then
                max ← L
            if (aL > amax) then
                amax ← aL
        indic ← 1; aindic ← 1
        if(max > t_0) then
            indic ← 0
        if(amax > t_0) then
            aindic ← 0
        result[i] ← 0.5(indic + aindic)
    hw = 1.96 × sqrt(var(result)/(n/2))
    print ("average is", ave(result))
    print ("95% CI half-width is", hw)
```

Fig. 4.6 Pseudocode for Example 4.7. We assume GenF(j, u) returns an observation from the distribution of the jth duration by inversion of the uniform number u, $C[k, j]$ returns the index of the jth arc on the kth path, and c_k is the number of arcs on path k.

Tables 4.1 and 4.2 give results comparing the efficiency of the naive Monte Carlo and antithetic estimators for Examples 4.6 and 4.7 with $n = 1024$. As is typically done in empirical studies on variance reduction techniques, the values reported in these tables are based on a certain number m of i.i.d. copies of each estimator ($m = 25$ in our case). That is, $\hat{\mu}$ is the average value of the estimator over the sample $\hat{\mu}_1, \ldots, \hat{\mu}_m$, and the half-width of, say, a 95% confidence interval is computed as

$$1.96\sqrt{\frac{1}{m(m-1)}\sum_{i=1}^{m}(\hat{\mu}_i - \hat{\mu})^2}.$$

The CPU time used to estimate the efficiency is based on the time required to run these m groups of n simulations and compute the estimator desired.

Table 4.1 Comparison of Monte Carlo and antithetic estimators for Example 4.6 (bank). HW is the half-width of a 95% confidence interval for μ.

Method	$\hat{\mu}$	HW	CPU(sec)	$\hat{\text{Eff}}(\hat{\mu})$
MC	73.04	0.788	11.9	0.521
Antithetic	73.35	0.530	7.76	1.766

Table 4.2 Comparison of Monte Carlo and antithetic estimators for Example 4.7 (SAN). HW is the half-width of a 95% confidence interval for μ.

Method	$\hat{\mu}$	HW	CPU(sec)	$\hat{\text{Eff}}(\hat{\mu})$
MC	0.7502	5.41e−3	0.197	667913
Antithetic	0.7521	5.16e−3	0.151	951511

As predicted, for both examples, the antithetic estimator has a smaller variance than the naive Monte Carlo estimator. Not surprisingly, the computation time of the antithetic estimator is smaller than the Monte Carlo one. This is due to the fact that we need to generate twice as many uniform random numbers for the Monte Carlo estimator. Hence the gain in efficiency is larger than the variance reduction, with an improvement factor of about 3.4 for the bank example and 1.4 for the SAN example. These efficiency gains are fairly typical for antithetic variates and are not as large as those that can be obtained by some other variance reduction techniques that can be applied in a more problem-specific way.

To conlude this section, we wish to show in a simplified version of Example 4.6 the effect of applying antithetic variates on the function f. That is, in light of the discussion at the beginning of this chapter, we can think of antithetic variates as transforming the function $f(\mathbf{u})$ into the function

$$\phi(\mathbf{u}) = \frac{f(\mathbf{u}) + f(1 - \mathbf{u})}{2},$$

where the notation $1 - \mathbf{u}$ refers to the vector whose jth coordinate is $1 - u_j$ for $j = 1, \ldots, s$.

Example 4.8. Consider Example 4.6, but where we are interested in estimating the mean waiting time ω_{30} for the first 30 clients. Note that the corresponding dimension here is 60 as we need to generate the service speed, 30 interarrival times, and 29 service times. To get a sense for what the corresponding 60-dimensional function f looks like (i.e., the function f such that $\mathrm{E}(f(\boldsymbol{U})) = \omega_{30}$, as in (4.1)), we can fix all but two of the coordinates and then plot f as a function of the two remaining (unfixed) coordinates. In Fig. 4.7 (top), we show the function f as u_{22} and u_{23} vary — interarrival and service times for the tenth client — and where all other coordinates u_j have been randomly chosen (and fixed, as u_{22} and u_{23} vary over $[0,1)^2$). Note that since all variables except u_{22} and u_{23} are fixed, the integral of the two-dimensional function shown in these graphs is not ω_{30} but instead is given by the conditional expectation of $\sum_{j=1}^{30} w_j/30$ given $u_1, \ldots, u_{21}, u_{24}, \ldots, u_{60}$.

On the bottom of Fig. 4.7, we show the corresponding graph for $\phi(\mathbf{u})$. Note that while f is monotonically decreasing in u_{22} and monotonically increasing in u_{23} (arguments similar to those used in Example 4.5 can be applied to verify why this holds), ϕ is not monotone.

4.4 Control variates

The method of control variates shares a common feature with the method of antithetic variates. They are both based on the idea of using correlation in order to reduce the variance of the naive Monte Carlo estimator. However, the way the correlation is induced is quite different here. With antithetic variates, we saw that (negative) correlation was induced directly on the sampling points $\mathbf{u}$. Using the notation $Y = h(\mathbf{X})$ introduced in (4.2), with control variables, we instead try to find a variable C — the *control variable* — that is related to our simulation model and correlated with Y but for which $\mu_c = \mathrm{E}(C)$ is known. By comparing the sample average of C obtained by simulation with the exact mean μ_c, one can then appropriately adjust the naive Monte Carlo estimator.

More precisely, suppose $Y_1, \ldots, Y_n$ and $C_1, \ldots, C_n$ are two i.i.d. samples, with Y_i and C_i obtained from the ith simulation run. First, suppose Y and C are positively correlated. In that case, we know that if

$$\hat{\mu}_c = \frac{1}{n} \sum_{i=1}^{n} C_i$$

is larger than μ_c, then the naive Monte Carlo estimator $\hat{\mu}_{\mathrm{mc}} = \sum_{i=1}^{N} Y_i/n$ is probably also larger than μ, and so we should adjust $\hat{\mu}_{\mathrm{mc}}$ by subtracting a certain (positive) value related to the difference observed, $\hat{\mu}_c - \mu_c$. If $\hat{\mu}_c$ is smaller than μ_c, then similarly we should add something positive to $\hat{\mu}_{\mathrm{mc}}$.

More precisely, a *control variate estimator* has the form

$$\hat{\mu}_{\mathrm{cv}} = \frac{1}{n} \sum_{i=1}^{n} (Y_i + \beta(\mu_c - C_i)), \tag{4.4}$$

where β is a constant to be determined. It is easy to see that, for a fixed β, the control variate estimator is unbiased since

$$\mathrm{E}(Y_i + \beta(\mu_c - C_i)) = \mathrm{E}(Y_i + \beta(\mu_c - \mathrm{E}(C_i))) = \mathrm{E}(Y) + \beta \times 0 = \mu.$$

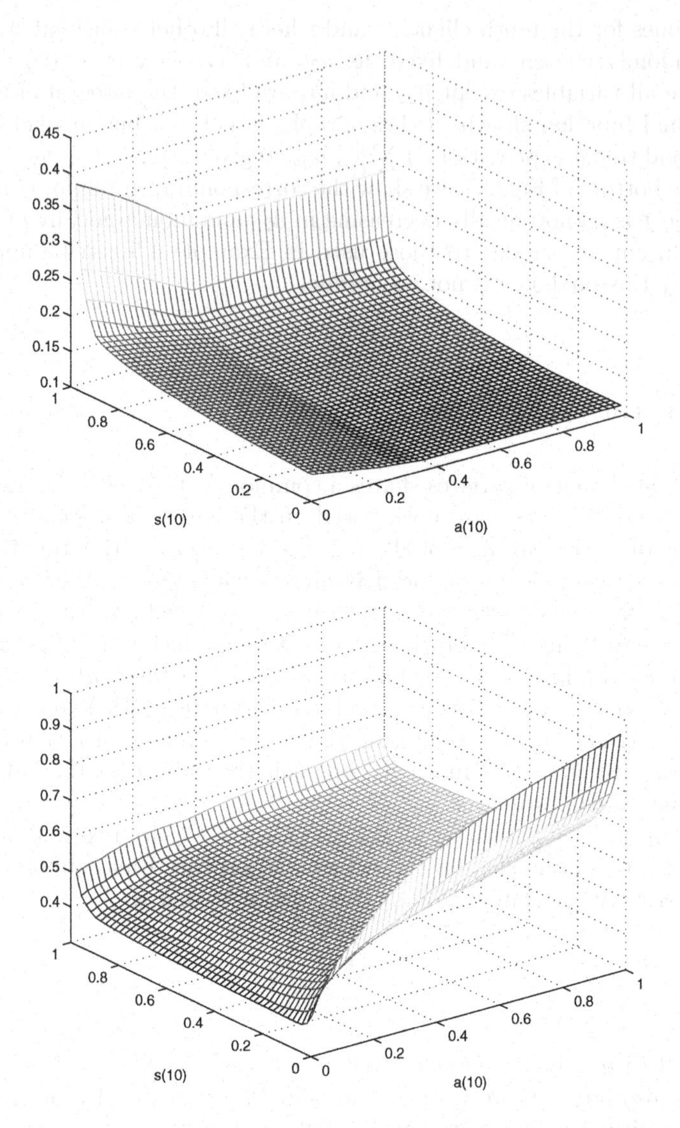

Fig. 4.7 Top: function f for simplified version of bank example; bottom: corresponding antithetic variates function ϕ. The axes are labeled with the variate generated by the corresponding uniform number.

To determine which value of β should be used, recall that our goal is to produce an estimator $\hat{\mu}_{\mathrm{cv}}$ whose variance is smaller than the naive Monte Carlo estimator $\hat{\mu}_{\mathrm{mc}}$. Hence we can find the value of β that minimizes $\mathrm{Var}(\hat{\mu}_{\mathrm{cv}})$. First, we write

$$\mathrm{Var}(\hat{\mu}_{\mathrm{cv}}) = \frac{1}{n}\left[\mathrm{Var}(Y_i) + \beta^2\mathrm{Var}(C_i) - 2\beta\mathrm{Cov}(Y_i, C_i)\right],$$

and so

$$\frac{\partial}{\partial\beta}\mathrm{Var}(\hat{\mu}_{\mathrm{cv}}) = \frac{1}{n}\left[2\beta\mathrm{Var}(C_i) - 2\mathrm{Cov}(Y_i, C_i)\right]. \tag{4.5}$$

By setting (4.5) to 0 (and verifying that the second derivative is positive), we see that $\mathrm{Var}(\hat{\mu}_{\mathrm{cv}})$ is minimized when

$$\beta = \beta^* := \frac{\mathrm{Cov}(Y_i, C_i)}{\mathrm{Var}(C_i)}.$$

Note that if we take $\beta = \beta^*$ in (4.4), then the corresponding estimator, denoted $\hat{\mu}_{\mathrm{cv},\beta^*}$, has variance

$$\begin{aligned}\mathrm{Var}(\hat{\mu}_{\mathrm{cv},\beta^*}) &= \frac{1}{n}\left[\mathrm{Var}(Y_i) + \frac{(\mathrm{Cov}(Y_i, C_i))^2}{\mathrm{Var}(C_i)} - 2\frac{(\mathrm{Cov}(Y_i, C_i))^2}{\mathrm{Var}(C_i)}\right] \\ &= (1-\rho^2)\mathrm{Var}(\hat{\mu}_{\mathrm{mc}}),\end{aligned} \tag{4.6}$$

where

$$\rho = \mathrm{Corr}(Y_i, C_i) = \frac{\mathrm{Cov}(Y_i, C_i)}{\sqrt{\mathrm{Var}(Y_i)\mathrm{Var}(C_i)}}$$

is the correlation coefficient between Y and C. Hence, if $\rho \pm 1$ — which happens when Y and C are linearly correlated — then the control variate estimator has zero variance. In general, the stronger the correlation (i.e., the closer $|\rho|$ is to 1), the better the improvement we get by using $\hat{\mu}_{\mathrm{cv}}$ instead of $\hat{\mu}_{\mathrm{mc}}$.

Of course, in practice we cannot compute β^* exactly because the covariance term $\mathrm{Cov}(Y, C)$ is usually unknown. (If it were known, then μ would also be known and we would not need to estimate it). We thus have to estimate it. We can do that by using the same sample $(Y_1, C_1), \ldots, (Y_n, C_n)$ as the one used to define $\hat{\mu}_{\mathrm{cv}}$. That is, we can take

$$\hat{\beta} = \frac{\sum_{i=1}^n Y_i C_i - n(\hat{\mu}_{\mathrm{mc}} \cdot \hat{\mu}_c)}{(n-1)\hat{\sigma}_c^2}, \tag{4.7}$$

where $\hat{\sigma}_c^2$ is the sample variance of $\{C_i, i = 1, \ldots, n\}$. If $\mathrm{Var}(C)$ is known exactly, then it can replace $\hat{\sigma}_c^2$ in (4.7).

One drawback of this approach is that the resulting estimator

$$\hat{\mu}_{\mathrm{cv},\hat{\beta}} = \hat{\mu}_{\mathrm{mc}} + \hat{\beta}(\mu_c - \hat{\mu}_c) \tag{4.8}$$

is not necessarily unbiased. This results from the fact that $\hat{\beta}$ now depends on $C_1, \ldots, C_n$, and is no longer independent of $\hat{\mu}_c$, so that $\mathrm{E}(\hat{\beta}(\mu_c - \hat{\mu}_c))$ is not necessarily equal to $\mathrm{E}(\hat{\beta})\mathrm{E}(\mu_c - \hat{\mu}_c) = 0$.

More generally, when replacing the optimal β^* by an estimate $\hat{\beta}$, the variance expression (4.6) no longer holds. Thus, it would be wrong to estimate $\mathrm{Var}(\hat{\mu}_{\mathrm{cv},\hat{\beta}})$ by, for instance, $(1-\hat{\rho}^2)\hat{\sigma}^2_{\mathrm{mc}}$, where $\hat{\rho}$ is the sample correlation between Y and C and $\hat{\sigma}^2_{\mathrm{mc}}$ is the estimated variance of the Monte Carlo estimator. Instead, $\mathrm{Var}(\hat{\mu}_{\mathrm{cv},\hat{\beta}})$ can be estimated in the standard way as

$$\frac{1}{n(n-1)}\sum_{i=1}^{n}(Y_{\mathrm{cv},i}-\hat{\mu}_{\mathrm{cv},\hat{\beta}})^2, \tag{4.9}$$

where $Y_{\mathrm{cv},i} = Y_i + \hat{\beta}(\mu_c - C_i)$. But even with this formula, we must note that if $\hat{\beta}$ is given by (4.7), then the variables $Y_{\mathrm{cv},i}$ are not independent, and therefore this sample variance estimator is not necessarily unbiased.

To get some insight on the impact of the bias introduced by replacing β^* with the estimate $\hat{\beta}$, it is useful to see the connection between control variates and regression [241, 242]. More precisely, let us write

$$Y = \mu + \beta(\mu_c - C) + \epsilon,$$

where $\mathrm{E}(\epsilon) = 0$. Then, if we make the assumption that (Y, C) has a bivariate normal distribution, standard results in regression imply that $\hat{\beta}$ and $\hat{\mu}_{\mathrm{cv},\hat{\beta}}$ as defined in (4.7) and (4.8), respectively, are the least-squares estimators of μ and β, which can in turn be used to construct an unbiased estimator $\hat{\sigma}^2_{\mathrm{cv},\beta}$ for $\mathrm{Var}(\hat{\mu}_{\mathrm{cv},\hat{\beta}})$ as in (4.9). Under this normality assumption, one can also show that $(\hat{\mu}_{\mathrm{cv},\hat{\beta}} - \mu)/\hat{\sigma}_{\mathrm{cv},\beta}$ has a Student t-distribution with $n-2$ degrees of freedom. Hence $\hat{\mu}_{\mathrm{cv},\hat{\beta}}$ is an unbiased estimator of μ in that case. Furthermore, it can be shown that the increase in variance that results from replacing β^* by its least-squares estimate $\hat{\beta}$ is such that

$$\frac{\mathrm{Var}(\hat{\mu}_{\mathrm{cv},\hat{\beta}})}{\mathrm{Var}(\hat{\mu}_{\mathrm{cv},\beta^*})} = \frac{n-1}{n-2}.$$

Hence the increase in variance becomes negligible as n tends to infinity.

Without the normality assumption, though, these results do not hold. In such cases, the bias can be eliminated by using a technique called *splitting*, which consists in using the estimator

$$\hat{\mu}_{\mathrm{cv},s} = \frac{1}{n}\sum_{i=1}^{n} Y_{i,cv,s},$$

where

$$Y_{i,cv,s} = Y_i - \hat{\beta}_{-i}(\mu_c - C_i),$$

and $\hat{\beta}_{-i}$ is the least-squares estimator for β, but where the results (Y_i, C_i) from the ith simulation are not included [45]. That is, $\hat{\beta}_{-i}$ is based on the sample $(Y_1, C_1), \ldots, (Y_{i-1}, C_{i-1}), (Y_{i+1}, C_{i+1}), \ldots, (Y_n, C_n)$. Since $\hat{\beta}_{-i}$ and C_i are independent, we have that $\mathrm{E}(Y_{i,cv,s}) = \mu$, and thus $\hat{\mu}_{\mathrm{cv},s}$ is unbiased.

Another possibility is to use a technique called *jackknifing* [45, 98, 99, 217]. In that case, the estimator

$$\hat{\mu}_{\mathrm{cv},j} = \frac{1}{n} \sum_{i=1}^{n} Y_{i,cv,j}$$

is used, where

$$Y_{i,cv,j} = n\hat{\mu}_{\mathrm{cv},\hat{\beta}} - (n-1)\hat{\mu}_{\mathrm{cv},\hat{\beta},-i},$$

and $\hat{\mu}_{\mathrm{cv},\hat{\beta},-i}$ represents the control variate estimator $\hat{\mu}_{\mathrm{cv},\hat{\beta}}$ in which the results (Y_i, C_i) from the ith simulation have been deleted.

The two preceding approaches manage to reuse the sample $(Y_1, C_1), \ldots, (Y_n, C_n)$ in a clever way in order to reduce (or eliminate) the bias. However, they both imply additional computational time in order to construct the values $Y_{i,cv,s}$ and $Y_{i,cv,j}$. As an alternative to these two approaches, we can instead use a small number r of *pilot simulations* and then compute $\hat{\beta}$ based on the resulting sample $(Y_1, C_1), \ldots, (Y_r, C_r)$. Since $\hat{\beta}$ is now independent of $\hat{\mu}_c$, the control variate estimator and the variance estimator (4.9) are unbiased. Here the additional computational effort is spent generating these pilot simulations.

However, it should be noted that thanks to a result of Nelson [331], regardless of the distribution of (Y, C), if we use the least-squares estimate (4.7) for $\hat{\beta}$, we have a central limit theorem for $\hat{\mu}_{\mathrm{cv},\hat{\beta}}$ of the form

$$\sqrt{n}(\hat{\mu}_{\mathrm{cv},\hat{\beta}} - \mu) \Rightarrow N(0, \sigma^2_{\mathrm{cv},\beta^*})$$

as n goes to infinity, where $\sigma^2_{\mathrm{cv},\beta^*} = \mathrm{Var}(\hat{\mu}_{\mathrm{cv},\beta^*})$ is given in (4.6). This result implies that, in practice, if n is large enough, then we can construct confidence intervals for μ based on the normal distribution, as will be done in Examples 4.9 and 4.10.

Before going further, let us go back to Examples 4.6 and 4.7 and see how control variates can be used in these two cases.

Example 4.9. For the bank example given in Example 4.6, a possible control variable is to use the average interarrival time $\sum_{i=1}^{N+1} a_i/(N+1)$, whose expectation is 60 seconds. (The reason why we take $N+1$ is because we include the interarrival time between the last client and the first one who arrives after 3 pm, which needs to be generated in order to determine N, as discussed on p. 19.) Another possibility is to use the average service time $\sum_{i=1}^{N} s_i/N$, whose expectation is $0.2 \times 35 + 0.7 \times 50 + 0.1 \times 55 = 47.5$ seconds. The former case is denoted CV-arrival in the results below, while the latter one is denoted CV-service. Figure 4.8 gives pseudocode for using the control variate based on the average service time.

```
SimCV                                          RunCVSim
  NbWait5 ← 0                                    for i = 1 to n
  w ← 0                                              y(i) ← SimCV[1]
  u ← Rand01()                                       c(i) ← SimCV[2]
  type ← GenDisc([0.2,0.9,1],3,u)                β ← cov(y,c)/var(c)
  v ← mus[type]                                  return (ave(y) + β(47.5/60 − ave(c)))
  a ← GenExpon(1,Rand01())
  time ← a
  sums ← 0
  nbcust ← 1
  while (time < 300) do
     s ← GenExpon(mu,Rand01())
     a ← GenExpon(1,Rand01())
     time ← time + a
     w ← max(0, w + s − a)
     if ((time < 300) and (w > 5)) then
        NbWait5 ← NbWait5 + 1
        sums ← sums + s
        nbcust ← nbcust + 1
  return [NbWait5, sums/nbcust]
```

Fig. 4.8 Pseudocode for using CV-service. `RunCVSim` returns the control variate estimator, the function `cov`(y, c) returns the sample covariance of the vectors y and c, `SimCV`$[j]$ contains the jth returned value of `SimCV`, and `GenDisc` is as described in Fig. 4.4.

Table 4.3 Comparison of Monte Carlo and control variate estimators for Example 4.6. HW is the half-width of a 95% confidence interval for μ.

Method	$\hat{\mu}$	HW	CPU(sec)	$\hat{\text{Eff}}(\hat{\mu})$
MC	73.04	0.788	11.9	0.5210
CV-arrival	74.90	0.732	11.98	0.5992
CV-service	73.14	0.500	11.95	1.2838

As is seen in Table 4.3, the control variate based on service time thus manages to reduce the variance by a factor $(0.788/0.5)^2 = 2.5$.

Note that since our results are based on 25 i.i.d. replications of the estimators, we circumvent the problem of using a biased estimator for the variance of $\hat{\mu}_{\text{cv},\hat{\beta}}$ such as the one presented in (4.9), which we would use if we had only performed one replication.

Example 4.10. For the SAN described in Example 4.7, a possible control variable is to use the length of the path with the largest expected length, which in our case is the path 4–7–12–13–11, for an expected length of 48.2. This is based on an idea used in [21, 22]. As can be seen in Table 4.4, the reduction in variance is marginal in this case.

Table 4.4 Comparison of Monte Carlo and control variate estimators for Example 4.7. HW is the half-width of a 95% confidence interval for μ.

Method	$\hat{\mu}$	HW	CPU(sec)	$\widehat{\text{Eff}}(\hat{\mu})$
MC	0.7502	5.41e−3	0.197	667913
CV	0.7500	5.34e−3	0.201	672314

Let us now say a few words on the kind of control variables that are typically used in practice. As we said at the beginning of this section, in theory, any variable C correlated with Y and whose expectation is known can be used as a control variable. What this typically translates to is that we use as control variables quantities that are closely related to the one for which we try to estimate the mean but that are in some sense simpler and thus for which the expectation is known.

A property that such functions often exhibit is that they are based on the same vector $\mathbf{X}$ of random variables as the quantity of interest Y, which means both Y and C can be computed at the same time. In other words, there exists a function h_c such that we can write $C = h_c(\mathbf{X})$, while $Y = h(\mathbf{X})$. The control variables used in Example 4.9 satisfy this. Control variables having this property are sometimes called *internal control variables* [243]. An example of a control variable that does not satisfy this property — an *external control variable* — is when we take C so that it represents the same quantity as Y but for a simpler model. For instance, if we are trying to estimate the mean waiting time in a complicated queueing model, we could use as a control variable the average waiting time for a simpler but related queueing model. For this to work, we need to make sure we have correlation between the two quantities Y and C. This can usually be achieved by using the same uniform numbers to generate the interarrival and service times in both models. That is, we need to use *common random numbers*, a technique discussed in Sect. 4.8. If we do so, we can assume there is a function c such that we can write $C = c(\mathbf{u})$, where c is defined in relation to the function f such that $Y = f(\mathbf{u})$ so that synchronization (see p. 96, Sect. 4.3) is achieved. The following example illustrates the use of an external control variable.

Example 4.11. Suppose that, in Example 4.8, rather than modeling the service times as exponential random variables with a varying mean, we instead use a Weibull distribution with a mean of 45 seconds. Then let f be the function

$$f(u_1,\ldots,u_{59}) = \frac{1}{30}\sum_{j=1}^{30} w_j(-\ln(1-u_1),\gamma(u_2),\ldots,-\ln(1-u_{2j-1})),$$

where γ represents the inverse CDF of the Weibull distribution used to model the service times. That is, in this definition of $f(\cdot)$, we wrote the jth waiting time w_j as a function of the previous interarrival times $a_1 = -\ln(1-u_1),\ldots,a_j = -\ln(1-u_{2j-1})$ and service times $s_1 = \gamma(u_2),\ldots,s_{j-1} = \gamma(u_{2(j-1)})$. In that case, $\omega_{30} = \int_{[0,1)^{59}} f(\mathbf{u})d\mathbf{u}$ cannot be computed exactly. However, if we use exponential service times with a mean of 45 seconds instead, then ω_{30} can be computed exactly [243, Example 11.11, p. 607] and is denoted as $\omega_{30,\mathrm{exp}}$ below. Hence we can use the average waiting time in the queue of the first 30 customers in the simpler model based on exponential service times as our (external) control variable. The corresponding function $c(\mathbf{u})$ representing this control variable is given by

$$c(u_1,\ldots,u_{59}) = \frac{1}{30}\sum_{j=1}^{30} w_j(-\ln(1-u_1),-0.75\ln(1-u_2),\ldots,-\ln(1-u_{2j-1})),$$

and the control variate estimator can then be written as

$$\frac{1}{n}\sum_{i=1}^{n}(f(\mathbf{u}_i)+\beta(\omega_{30,\mathrm{exp}}-c(\mathbf{u}_i))).$$

Our preceding remark about common random numbers and synchronization simply has to do with the fact that for both systems we use u_1 to generate the first interarrival time, u_2 for the first service time, and so on. A similar example is discussed in [243, Problem 11.14, p. 620].

In Example 4.9, we gave two possible control variables. It seems natural that, just like for regression, we should be able to use more than one control variable at the same time, with the hope that additional explanatory variables will contribute to further reducing the variance. More precisely, with the theory of *multiple control variables* [241, 242, 331], we are now looking at estimators of the form

$$\hat{\mu}_{\mathrm{cv}} = \frac{1}{n}\sum_{i=1}^{n} Y_i + \boldsymbol{\beta}^{\mathrm{T}}(\boldsymbol{\mu}_c - \mathbf{C}_i), \tag{4.10}$$

where $\mathbf{C}_i^{\mathrm{T}} = (C_{1i},\ldots,C_{qi})$ is a vector of q control variables, $\boldsymbol{\beta}^{\mathrm{T}} = (\beta_1,\ldots,\beta_q)$ is a vector of q coefficients, and $\boldsymbol{\mu}_c^{\mathrm{T}} = (\mathrm{E}(C_1),\ldots,\mathrm{E}(C_q))$ is the vector containing the expectation of the q control variables. Based on arguments similar to those used to derive the optimal β^* in the single control variate case, it can be shown that the vector of coefficients $\boldsymbol{\beta}$ that minimizes the variance of the multiple control variate estimator (4.10) is given by

$$\boldsymbol{\beta}^* = \Sigma_c^{-1}\Sigma_{y,c},$$

where Σ_c is the covariance matrix for the vector $\mathbf{C}$, and

$$\Sigma_{y,c}^{\mathrm{T}} = [\mathrm{Cov}(Y, C_1), \ldots, \mathrm{Cov}(Y, C_q)]$$

is the vector containing the covariances between Y and each of the control variables C_j for $j = 1, \ldots, q$. With this $\boldsymbol{\beta}^*$, the corresponding estimator $\hat{\mu}_{\mathrm{cv},\boldsymbol{\beta}^*}$ has variance

$$\mathrm{Var}(\hat{\mu}_{\mathrm{cv},\boldsymbol{\beta}^*}) = (1 - R_{y,c}^2)\mathrm{Var}(\hat{\mu}_{\mathrm{mc}}), \tag{4.11}$$

where

$$R_{y,c}^2 = \Sigma_{y,c}^{\mathrm{T}}\Sigma_c^{-1}\Sigma_{y,c}$$

is the *coefficient of determination* of Y and $\mathbf{C}$.

As in the case of a single control variate, here we are also faced with the fact that in practice $\boldsymbol{\beta}^*$ usually is not known exactly and that replacing it by its estimate

$$\hat{\boldsymbol{\beta}} = \hat{\Sigma}_c^{-1}\hat{\Sigma}_{y,c}$$

makes the corresponding estimator

$$\hat{\mu}_{\mathrm{cv},\hat{\boldsymbol{\beta}}} = \frac{1}{n}\sum_{i=1}^{n} Y_i + \hat{\boldsymbol{\beta}}^{\mathrm{T}}(\boldsymbol{\mu}_c - \mathbf{C}_i)$$

biased in general. Here again, though, under the assumption that $(Y, \mathbf{C})$ is multinormal, $\hat{\mu}_{\mathrm{cv},\hat{\boldsymbol{\beta}}}$ is unbiased and

$$\frac{\mathrm{Var}(\hat{\mu}_{\mathrm{cv},\hat{\boldsymbol{\beta}}})}{\mathrm{Var}(\hat{\mu}_{\mathrm{cv},\boldsymbol{\beta}^*})} = \frac{n-1}{n-q-1}, \tag{4.12}$$

where $\hat{\mu}_{\mathrm{cv},\boldsymbol{\beta}^*}$ is the control variate estimator based on the exact optimal $\boldsymbol{\beta}^*$, whose variance is given in (4.11) [241, 242]. What the ratio (4.12) suggests is that it may not always be beneficial to add control variables because the reduction in variance that is obtained through the factor $(1 - R_{y,c}^2)$ may be offset by the increase in the ratio $(n-1)/(n-q-1)$ when q increases. Intuitively speaking, this happens because each term of the form $\hat{\beta}_j(\mu_{c,j} - \hat{C}_j)$ adds noise to the control variate estimator, where $\hat{C}_j$ is the estimator $\sum_{i=1}^{n} C_{ji}/n$ for the expectation $\mu_{c,j}$ of the jth control variable C_j. Hence, if C_j does not help "explain" Y very much (that is, if Y and C_j are not highly correlated), then its "variance-reducing" effect may be outweighed by this noise. More generally, what we said about splitting, jackknifing, pilot simulations, and the central limit theorem of Nelson all apply to the multiple control variate case [45, 331].

The method of control variables can be used as a general framework to study other variance reduction techniques. For example, we can think of antithetic variates as using $f(\mathbf{u}_i) - f(\tilde{\mathbf{u}}_i)$ as a control variate, with $\beta = 1/2$

[155, 391]. Also, in [151], the theory of control variables is used to study an estimation method called *weighted Monte Carlo*, which has been proposed in the context of finance to calibrate models to market data [20]. The connection between control variates and weighted estimators is studied in a more general context in [177]. One of the tasks for which this connection can be helpful is quantile estimation. More connections between control variates and other variance reduction techniques are studied in [155].

Control variables can also be used in the following context ([165],[243, p. 610], and [45, Problem 2.3.9]). Suppose we have q unbiased estimators $\hat{\mu}_1, \ldots, \hat{\mu}_q$ for μ and want to use a linear combination $\sum_{j=1}^{q} w_j \hat{\mu}_j$ of them as our global estimator for μ. Then we can think of $\hat{\mu}_1$ as our naive Monte Carlo estimator and $(\hat{\mu}_1 - \hat{\mu}_j)$, for $j = 2, \ldots, q$, as our $q-1$ control variables. We can then use the theory of control variables to determine the coefficients w_j that will produce the estimator with the smallest variance in the linear combination.

Finally, if we look at the control variate method from the integration point of view, we can say that it amounts to replacing $f(\mathbf{u})$ by

$$\phi(\mathbf{u}) = f(\mathbf{u}) + \beta(\mu_c - c(\mathbf{u})),$$

where $c : [0,1)^s \to \mathbb{R}$ is the function such that $c(\mathbf{u}) = C$ and $\mathrm{E}(c(\boldsymbol{U})) = \mu_c$. Namely, just as we did for f, we can think of c as the function that turns the vector $\mathbf{u}$ of uniform numbers used for the simulation into an observation of the control variable. We described this formulation in Example 4.11. As a second illustration, going back to Example 4.9, in that case the control variable based on the average service time corresponds to the function

$$c(\mathbf{u}) = \frac{-v(u_1)}{29} \left(\sum_{j=1}^{29} \ln(1 - u_{2j+1}) \right),$$

where

$$v(u) = \begin{cases} 35/60 \text{ if } u < 0.2 \\ 50/60 \text{ if } 0.2 \leq u < 0.9 \\ 55/60 \text{ if } u \geq 0.9. \end{cases}$$

That is, the first random number u_1 is used to determine the mean service time, and then each of the 29 first service times are obtained by inverting the exponential CDF with the chosen mean. In Fig. 4.9, in a fashion similar to what was done in Fig. 4.7, we show the control variate integrand $\phi(\mathbf{u})$ as u_{22} and u_{23} vary over $[0,1)^2$, while other variables are fixed. We see that as u_{23} increases from 0 to 1, the corresponding service time s_{10} increases and apparently causes $\hat{\beta}(\mu_c - \hat{\mu}_c)$ to decrease faster than $\hat{\mu}_{\mathrm{mc}}$ increases, causing $\phi(\mathbf{u})$ to decrease.

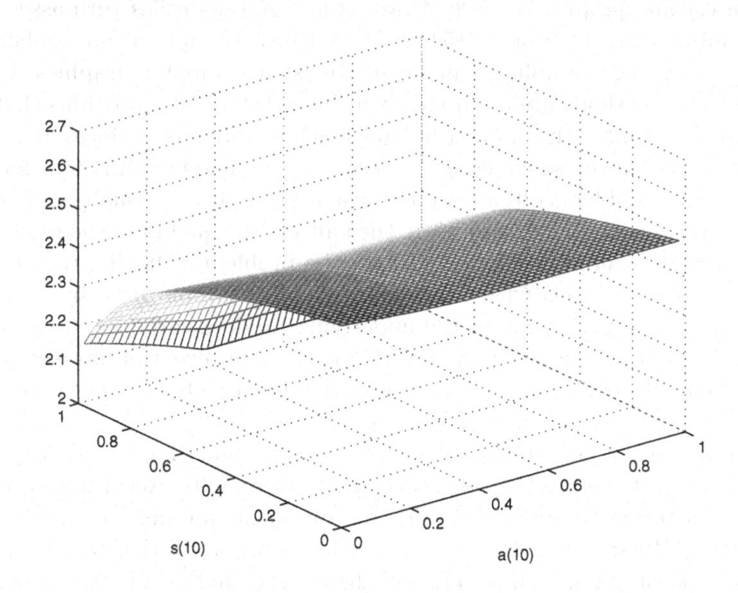

Fig. 4.9 Function $\phi(\mathbf{u})$ corresponding to the use of a control variable based on the average service time for the simplified version of the bank. The axes are labeled with the variate generated by the corresponding uniform number.

4.5 Importance sampling

Unlike the two variance reduction techniques previously discussed, importance sampling is a method that is not based on correlated sampling but instead tries to direct the sampling effort toward the most important regions of the integration domain. It is most useful for *rare event* simulation problems. That is, this method is typically used when we need to observe an unlikely event in order to estimate the quantity of interest. For such problems, we may observe the rare event of interest only a few times (or not at all) in our simulation runs, and therefore the estimate to be constructed might only be based on a small number of observations.

The first example of such problems is to consider the bank in Example 4.6 but where we want to estimate the expected waiting time for clients who wait more than 15 minutes. In a given simulation of a day at the bank, it is possible that there will not be any customer waiting more than 15 minutes, and thus the observation output for that run will be 0, although it is obvious that the quantity to estimate is not 0. A second example is when the goal is to estimate probabilities of losing information cells in communication networks [55, 258]. These probabilities are usually pretty small (e.g., less than 0.001), and naive simulation gives estimators with large relative errors that make them unreliable. More generally, importance sampling is an especially useful tool for estimating probabilities of rare events and quantiles with associated probabilities close to 0 or 1 [154, 177]. A third example of an application where importance sampling can be useful is in computer graphics, where Monte Carlo methods are often used within *path-tracing* algorithms that are designed to estimate the amount of light reaching different surfaces in a scene to be rendered in the so-called *global illumination problem* [213, 330, 460]. In that context, problems often arise because areas that are visually important do not receive enough light, which in turn affects the quality of the rendering.

Importance sampling tries to address the problem of having too small a number of observations where the event of interest took place by changing the probability distribution of the underlying random variables in the simulation, denoted by the vector $\mathbf{X}$ in (4.1), so that this event of interest occurs more often. The estimator is then appropriately corrected so that it remains unbiased.

Another technique often used for rare event simulation is *splitting* [149, 209, 260] and its companion approach, *Russian roulette.* According to Kahn [209], both terms are apparently due to von Neumann and Ulam. The idea of splitting/Russian roulette is to establish a certain criterion by which simulation runs can be valued in terms of their associated likelihood to enter an "interesting region". The interesting runs can then be "splitted", meaning that they are replicated in a certain number of copies. The counterpart is that uninteresting simulations can be eliminated. The decision to eliminate or not can itself be done using randomness, which is the "Russian roulette" part of this methodology. As we will see in Chap. 8, these are the very same ideas as those used in what is known as the *bootstrap filter.*

To describe the importance sampling estimator, we first write the quantity μ to be estimated as

$$\mu = \mathrm{E}(h(\mathbf{X})) = \int_{\Omega} h(\mathbf{x})\varphi(\mathbf{x})d\mathbf{x},$$

where $\varphi(\mathbf{x})$ is the pdf of $\mathbf{X}$. Now consider another pdf $\psi(\mathbf{x})$ for $\mathbf{X}$ and write

$$\mu = \int_{\Omega} h(\mathbf{x})L(\mathbf{x})\psi(\mathbf{x})d\mathbf{x}, \tag{4.13}$$

where

$$L(\mathbf{x}) = \frac{\varphi(\mathbf{x})}{\psi(\mathbf{x})}$$

is called the *likelihood ratio.* Based on (4.13), the idea of importance sampling is that rather than sampling $\mathbf{X}$ according to $\varphi(\mathbf{x})$ and using the naive Monte Carlo estimator

$$\hat{\mu}_{\mathrm{mc}} = \frac{1}{n}\sum_{i=1}^{n} h(\mathbf{x}_i),$$

we instead generate an i.i.d. sample $\tilde{\mathbf{x}}_1, \ldots, \tilde{\mathbf{x}}_n$ from the new pdf $\psi(\mathbf{x})$ and then use the importance sampling estimator

$$\hat{\mu}_{\text{is}} = \frac{1}{n} \sum_{i=1}^{n} h(\tilde{\mathbf{x}}_i) L(\tilde{\mathbf{x}}_i). \tag{4.14}$$

Before going further, we must verify that $L(\mathbf{x})$ is defined. A sufficient condition for that is to make sure that φ is *absolutely continuous* with respect to ψ (that is, $\varphi(E) = 0$ for every set E such that $\psi(E) = 0$) and to use the convention that $0/0 = 0$.

To verify that the importance sampling estimator $\hat{\mu}_{\text{is}}$ is unbiased, we simply write

$$\begin{aligned} \mathrm{E}(\hat{\mu}_{\text{is}}) = \mathrm{E}(h(\tilde{\mathbf{X}})L(\tilde{\mathbf{X}})) &= \int_{\Omega} h(\mathbf{x}) L(\mathbf{x}) \psi(\mathbf{x}) d\mathbf{x} \\ &= \int_{\Omega} h(\mathbf{x}) \frac{\varphi(\mathbf{x})}{\psi(\mathbf{x})} \psi(\mathbf{x}) d\mathbf{x} \\ &= \int_{\Omega} h(\mathbf{x}) \varphi(\mathbf{x}) d\mathbf{x} = \mu, \end{aligned}$$

as required.

Hence, whatever choice we make for the new pdf $\psi(\mathbf{x})$, as long as the absolute continuity condition is satisfied, we are guaranteed that $\hat{\mu}_{\text{is}}$ is an unbiased estimator. However, as we will see in the derivation of the variance, not all choices of $\psi(\mathbf{x})$ give us an estimator with reduced variance compared with $\hat{\mu}_{\text{mc}}$. This means that if the new pdf is not chosen carefully, we could actually *increase* the variance.

Taking a look at the variance, we first write

$$\mathrm{Var}(\hat{\mu}_{\text{is}}) = \frac{1}{n} \mathrm{Var}(h(\tilde{\mathbf{X}})L(\tilde{\mathbf{X}})).$$

Since $\mathrm{E}(h(\tilde{\mathbf{X}})L(\tilde{\mathbf{X}})) = \mu$, we can focus on $\mathrm{E}(h^2(\tilde{\mathbf{X}})L^2(\tilde{\mathbf{X}}))$. We have that

$$\begin{aligned} \mathrm{E}(h^2(\tilde{\mathbf{X}})L^2(\tilde{\mathbf{X}})) &= \int_{\Omega} h^2(\mathbf{x}) L^2(\mathbf{x}) \psi(\mathbf{x}) d\mathbf{x} \\ &= \int_{\Omega} h^2(\mathbf{x}) \frac{\varphi(\mathbf{x})}{\psi(\mathbf{x})} \varphi(\mathbf{x}) d\mathbf{x} \\ &= \mathrm{E}(h^2(\mathbf{X}) L(\mathbf{X})). \end{aligned}$$

Hence

$$\mathrm{Var}(\hat{\mu}_{\text{is}}) = \frac{1}{n} \left[\mathrm{E}(h^2(\mathbf{X})L(\mathbf{X})) - \mu^2 \right], \tag{4.15}$$

which means $\mathrm{Var}(\hat{\mu}_{\text{is}}) \leq \mathrm{Var}(\hat{\mu}_{\text{mc}})$ if and only if

$$\mathrm{E}(h^2(\mathbf{X})L(\mathbf{X})) \leq \mathrm{E}(h^2(\mathbf{X})). \tag{4.16}$$

Just as we did for antithetic variates and control variates, it is useful to determine if it is possible to obtain a zero-variance importance sampling estimator in some cases. From (4.15), this means we want to know if we can find $\psi(\mathbf{x})$ such that

$$\mathrm{E}(h^2(\mathbf{X})L(\mathbf{X})) = \mu^2.$$

In the case where $h(\mathbf{X}) \geq 0$ for all $\mathbf{X} \in \Omega$, we can take $\psi(\mathbf{x}) = h(\mathbf{x})\varphi(\mathbf{x})/\mu$ and then

$$\mathrm{E}(h^2(\mathbf{X})L(\mathbf{X})) = \mu \mathrm{E}(h(\mathbf{X})) = \mu^2,$$

as required. Obviously, since the optimal new density $\psi(\mathbf{x})$ requires the knowledge of μ, it cannot be determined in practice.

However, from this result and the inequality given in (4.16), we can get a good sense for the properties that $\psi(\mathbf{x})$ should have in order for the importance sampling estimator to have a smaller variance than the Monte Carlo estimator. When $h(\mathbf{x})$ is large, the new pdf should make $\mathbf{x}$ more likely, so that the likelihood ratio $L(\mathbf{x})$ is small. When $h(\mathbf{x})$ is small, then we can afford to have a likelihood ratio larger than one. Note that if $L(\mathbf{x})$ is never larger than one whenever $h(\mathbf{x})$ is nonzero, then we are guaranteed that the importance sampling estimator will have a smaller variance than the naive Monte Carlo estimator. But it is usually difficult to guarantee that this condition will hold.

The analysis above gives us some intuition to guide us in our choice of the new pdf $\psi(\mathbf{x})$, but there is generally no way of constructing a pdf $\psi(\mathbf{x})$ that will achieve the largest variance reduction, or even to construct one that will guarantee that the variance is reduced compared with the naive Monte Carlo estimator. In fact, the task of identifying a good new pdf $\psi(\mathbf{x})$ remains an important research problem.

One possibility is to use a technique called *exponential twisting/tilting*, which in the univariate case amounts to using a new pdf of the form

$$\psi_\theta(x) = e^{\theta x - G(\theta)}\varphi(x),$$

where $G(\theta) = \log \mathrm{E}(e^{\theta X})$ is the cumulant generating function of X. Furthermore, if the goal is to estimate $P(X > x)$ for a large value x, then the quantity $\mathrm{E}(h^2(\mathbf{X})L(\mathbf{X}))$ on the left-hand side of (4.16), which we should try to minimize in order to minimize the variance of the importance sampling estimator, is given by

$$\mathrm{E}(L(X)\mathbf{1}_{X>x}) = \mathrm{E}(e^{-\theta X + G(\theta)}\mathbf{1}_{X>x}) \leq e^{-\theta x + G(\theta)}. \tag{4.17}$$

This inequality suggests that to design an importance sampling estimator based on exponential twisting, we should use the value of θ that minimizes the upper bound above. Since $G(\theta)$ is the cumulant generating function of X, it is convex, and therefore the minimum of the upper bound in (4.17) is attained at $\theta = \theta_x$, where θ_x is the root of the equation $G'(\theta_x) = x$. An example where this idea is applied will be presented in Chap. 7.

Another approach to ease the process of identifying a good importance sampling distribution — which in some cases overlaps with exponential twisting — is to restrict our attention to pdfs $\psi(\mathbf{x})$ such that each random variable X_i in the problem follows the same type of distribution as in the original formulation with $\varphi(\mathbf{x})$ but with different parameters. The parameters of the new distribution are then chosen so that (hopefully) the variance of the resulting importance sampling estimator will be reduced. This can be done in a heuristic way using the reasoning discussed previously — trying to make the more "important" or "costly" events happen more often — or by using some kind of theoretical analysis where the parameters are derived by solving a certain optimization problem or by exploiting properties of the problem at hand. We give a few examples.

Using large deviations. In [146], the authors consider a certain type of parameter change for the application of importance sampling, and then, using large deviations asymptotics, they derive an approximately asymptotically optimal parameter change. This particular application of importance sampling will be discussed in more detail in Chap. 7.

Searching for the best parameter. An alternative to the approach above is to write out the problem of finding the parameters yielding the importance sampling estimator with the smallest variance as a parametric optimization problem that can then be solved using techniques such as *infinitesimal perturbation analysis* and *stochastic approximation* [133]. This typically requires more computational work than approaches like the one used in [146], but since this work is only done once, this is not an important disadvantage of this method. It is used in the context of option pricing in [430, 459]. This approach is discussed in more detail in Chap. 7 as well.

Exploiting properties of the problem. Asmussen uses importance sampling in the context of risk theory to estimate the *ruin probability* of an insurance company [13]. For a simple claim process model, he shows how to change the pdf of the claim sizes and interarrival times based on exponential twisting and the *Lundberg equation*, which is well-known in risk theory. He then proves that the importance sampling estimator thus obtained is optimal in the infinite-horizon case.

An alternative to the importance sampling estimator given in (4.14) is to use the *weighted importance sampling* estimator (also called *ratio estimate* in [176])

$$\hat{\mu}_{\mathrm{is},w} = \frac{\sum_{i=1}^{n} h(\tilde{\mathbf{x}}_i)L(\tilde{\mathbf{x}}_i)}{\sum_{i=1}^{n} L(\tilde{\mathbf{x}}_i)}.$$

A significant advantage of this estimator over the "usual" importance sampling estimator (4.14) is that since the weights that multiply the $h(\tilde{\mathbf{x}}_i)$ are given by the normalized likelihood ratios

$$\frac{L(\tilde{\mathbf{x}}_i)}{\sum_{i=1}^{n} L(\tilde{\mathbf{x}}_i)}, \qquad i = 1, \ldots, n, \tag{4.18}$$

they add up to 1 and are bounded between 0 and 1, which is not necessarily the case with the estimator (4.14). Also, this weighted version can be useful if φ and/or ψ are complicated pdfs with, for example, normalizing constants that cannot be evaluated exactly. By normalizing the likelihood ratios $L(\tilde{\mathbf{x}}_i)$ as in (4.18), these constants cancel out and thus do not need to be evaluated. The tradeoff is that $\hat{\mu}_{\mathrm{is},w}$ is no longer unbiased because although $\mathrm{E}(h(\tilde{\mathbf{x}}_i)L(\tilde{\mathbf{x}}_i)) = \mu$ and $\mathrm{E}(L(\tilde{\mathbf{x}}_i)) = 1$, in general $\mathrm{E}(X/Y)$ is not equal to $\mathrm{E}(X)/\mathrm{E}(Y)$. However, it can be shown that this estimator is *consistent*; i.e., its bias goes to 0 as n goes to infinity [176, 391, 423].

Finally, we should also point out that sometimes importance sampling is used simply because the alternative distribution $\psi(\mathbf{x})$ is easier to sample from and not necessarily because we are dealing with a rare event simulation.

We now illustrate the idea of importance sampling on our two examples. In both cases, we are not really dealing with rare event simulations. Nevertheless, we manage to reduce the variance by applying importance sampling in an ad hoc way, following the intuition explained above of making the "important" or "costly" events happen more often.

Example 4.12. In Example 1.2, one way to apply importance sampling is to change the parameter of the exponential distribution used to simulate the interarrival times. For example, we can use a mean of 58 seconds instead of 1 minute, which should increase the waiting times and therefore produce a larger number of clients that wait more than 5 minutes. In this case, let $\mathbf{X}_i$ be the vector $\mathbf{X}_i = (v_i, a_{i,1}, s_{i,1}, \ldots, a_{i,N_i}, s_{i,N_i}, a_{i,N_i+1})$, where v_i is the mean service time for the ith simulation, $a_{i,j}$ is the jth interarrival time in the ith simulation, and $s_{i,j}$ is the jth service time in the ith simulation. Then the likelihood ratio has the form

$$L(\tilde{\mathbf{x}}_i) = \prod_{j=1}^{\tilde{N}_i+1} \frac{e^{-\tilde{a}_{i,j}}}{(30/29)e^{-30\tilde{a}_{i,j}/29}} = \left(\frac{29}{30}\right)^{\tilde{N}_i+1} e^{\sum_{j=1}^{\tilde{N}_i+1} \tilde{a}_{i,j}/29},$$

where the interarrival times $\tilde{a}_{i,j}$ are generated according to an exponential distribution with mean 58 seconds, and the variable $\tilde{N}_i$ is the corresponding number of clients obtained under this new distribution for the ith simulation. Hence the importance sampling estimator in this case is given by

$$\hat{\mu}_{\mathrm{is}} = \frac{1}{n} \sum_{i=1}^{n} \left(\frac{29}{30}\right)^{\tilde{N}_i+1} \times e^{\sum_{j=1}^{\tilde{N}_i+1} \tilde{a}_{i,j}/29} \times \sum_{j=1}^{\tilde{N}_i+1} \mathbf{1}_{\tilde{w}_{i,j}>5}.$$

Figure 4.10 gives pseudocode for computing the importance sampling estimator above. Table 4.5 gives numerical results where we compare the performance of the importance sampling estimator against the naive Monte Carlo estimator. As we can see there, importance sampling reduces the size of the confidence interval half-width by about 30% and increases the efficiency by a factor of about two. The reason why the importance sampling estimator requires a bit more time is that the expected number of clients arriving in a day is larger than before as a result of the decrease in the expected interarrival times.

```
SimIS
  NbWait5 ← 0
  w ← 0
  u ← Rand01()
  type ← GenDisc([0.2,0.9,1],3,u)
  v ← mus[type]
  u ← Rand01()
  a ← GenExpon(58/60,u)
  time ← a
  sums ← 0
  nbcust ← 1
  L ← (29/30) × exp(a/29)
  while (time < 300) do
      s ← GenExpon(mu,Rand01())
      a ← GenExpon(58/60,Rand01())
      nbcust ← nbcust + 1
      time ← time + a
      L ← L × (29/30) × exp(a/29)
      w ← max(0, w + s − a)
      if ((time < 300) and (w > 5)) then
          NbWait5 ← NbWait5 + 1
  return (NbWait5 ×L)
```

Fig. 4.10 Pseudocode showing how to use importance sampling on the bank example.

Table 4.5 Comparison of Monte Carlo and importance sampling estimators for Example 4.6. HW is the half-width of a 95% confidence interval for μ.

Method	$\hat{\mu}$	HW	CPU(sec)	$\widehat{\mathrm{Eff}}(\hat{\mu})$
MC	73.04	0.788	11.9	0.521
IS	72.71	0.567	12.4	0.964

Example 4.13. For the SAN described in Example 4.7, one way of applying importance sampling is to decrease the expected duration of certain activities, so that the length of the longest path decreases and thus becomes smaller than T_0 more often. For instance, we chose activities 2, 6, 9, and 11 and changed their mean to 90% of their original value. (Note that each path contains at least one of these activities.) Since these four activities have an exponential distribution, the likelihood ratio for this change of measure is

$$L(\tilde{\mathbf{x}}_i) = 0.9^4 e^{(1/0.9-1)(\tilde{D}_{i,2}/d_2+\ldots+\tilde{D}_{i,11}/d_{11})},$$

where $\mathbf{x}_i = (\tilde{D}_{i,1}, \tilde{D}_{i,2}, \ldots, \tilde{D}_{i,13})$ is the vector containing the durations sampled under the new distribution and d_j is the original expected duration for activity j. The corresponding importance sampling estimator is then

$$\hat{\mu}_{\text{is}} = \frac{1}{n}\sum_{i=1}^{n} \mathbf{1}_{T(\tilde{\mathbf{x}}_i)\leq t_0} L(\tilde{\mathbf{x}}_i),$$

where we used the notation $T(\tilde{\mathbf{x}}_i)$ instead of T_i to emphasize the dependence on $\tilde{\mathbf{x}}_i$. Table 4.6 gives the results. As we see there, importance sampling reduces the half-width of the confidence interval by a modest factor of about 1.1.

Table 4.6 Comparison of Monte Carlo and importance sampling estimators for Example 4.7. HW is the half-width of a 95% confidence interval for μ.

Method	$\hat{\mu}$	HW	CPU(sec)	$\widehat{\text{Eff}}(\hat{\mu})$
MC	0.7502	5.41e−3	0.197	667,913
IS	0.7499	4.90e−3	0.214	748,862

As we did for the previously discussed variance reduction techniques, we now describe the function $\phi(\mathbf{u})$ corresponding to the integration formulation of the importance sampling estimator. We can write it as

$$\phi(\mathbf{u}) = h(\tilde{g}(\mathbf{u}))L(\tilde{g}(\mathbf{u})),$$

where $\tilde{g}$ corresponds to the function that transforms the vector of uniform numbers $\mathbf{u}$ into a vector $\mathbf{X}$ according to the new pdf $\psi(\mathbf{x})$. We use the $\tilde{g}$ notation to distinguish it from the function g used in Fig. 4.1, which has the same meaning but for the original pdf $\varphi(\mathbf{x})$. The function $h(\cdot)$ is as defined in Fig. 4.1, and $L(\cdot)$ is the likelihood ratio. Figure 4.11 shows the function $\phi(\mathbf{u})$ in our usual setting for the simplified bank example. Compared with the function $f(\mathbf{u})$ corresponding to the naive Monte Carlo estimator that is shown in Fig. 4.7 (top), the function $\phi(\mathbf{u})$ depicted in Fig. 4.11 seems to be larger in the "important" areas corresponding to larger waiting times.

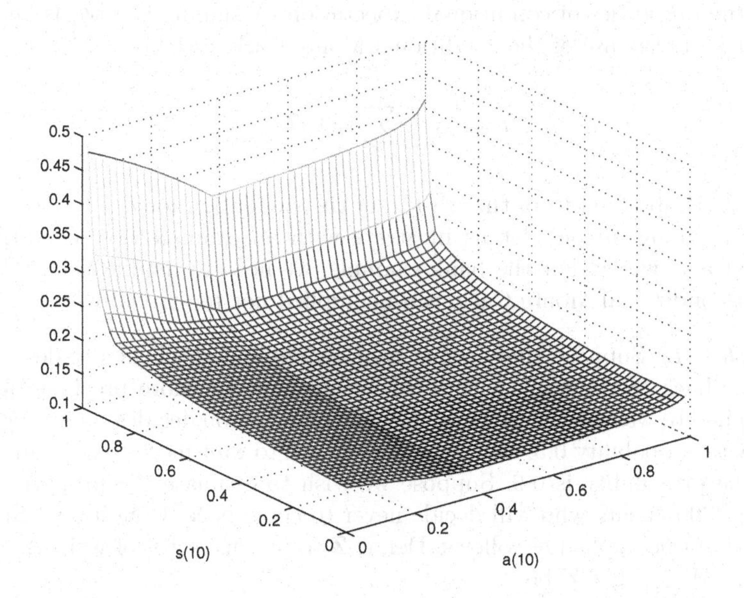

Fig. 4.11 Function $\phi(\mathbf{u})$ corresponding to the use of importance sampling for the simplified version of the bank. The axes are labeled with the variate generated by the corresponding uniform number.

4.6 Conditional Monte Carlo

This method was introduced by Trotter and Tukey in 1956 [451] and generalized shortly after [162, 165]. It shares a similarity with the control variates technique in that it tries to use auxiliary information in order to improve the quality of the naive Monte Carlo estimator. With conditional Monte Carlo, rather than using this information to adjust the Monte Carlo estimator with an appropriately chosen additive term as we do with control variates, we instead compute the expectation of the quantity of interest *conditioned* on the value taken by the auxiliary quantity.

More precisely, let $\mathbf{Z}$ be the *conditioning variable*, which is typically a vector that we can either view as being a function of $\mathbf{X}$ (that is, $\mathbf{Z} = z(\mathbf{X})$) or a function of $\boldsymbol{U}$ (that is, $\mathbf{Z} = \zeta(\boldsymbol{U})$). Typically, these functions (z or ζ) actually only depend on a subset of $\mathbf{X}$ or $\boldsymbol{U}$, and $\mathbf{Z}$ itself is usually a vector of random variables. Examples will be given shortly.

The idea is then to write

$$\mu = \mathrm{E}(Y) = \mathrm{E}(\mathrm{E}(Y|\mathbf{Z})) \tag{4.19}$$

using the properties of conditional expectation. Assuming $\mathrm{E}(Y|\mathbf{Z})$ is known, this suggests the use of the conditional Monte Carlo estimator

$$\hat{\mu}_{\mathrm{cmc}} = \frac{1}{n}\sum_{i=1}^{n}\mathrm{E}(Y|\mathbf{Z}_i),$$

where $\mathbf{Z}_i$ is obtained from the ith run of the simulation for $i = 1, \ldots, n$.

Before going further, let us illustrate with an example how conditional Monte Carlo works. For the bank example, we need to modify the problem so that conditional Monte Carlo can be applied not too trivially.

Example 4.14. Suppose that in our bank example each client decides never to come back to the bank with a certain probability. More precisely, if the person had to wait less than 5 minutes, then the person decides to never come back with probability 0.5, but if the person had to wait more than 5 minutes, then the probability is 0.9. Suppose we wish to estimate the proportion of the first 300 clients who will decide never to come back. Conditional Monte Carlo could be applied as follows. Define $\mathbf{Z}$ to be the vector of waiting times $(W_1, \ldots, W_{300})$. Let Y be

$$Y = \frac{1}{300}\sum_{j=1}^{300} B_j,$$

where B_j is an indicator function whose value is 1 if the jth person decides never to come back and 0 otherwise. Note that $1 + 3 \times 300 - 1 = 900$ uniform numbers are required in order to evaluate Y, while only 600 are required in order to compute $\mathbf{Z}$, the difference being due to the fact that to evaluate Y we need to generate a decision of never coming back or not for each of the 300 clients.

Based on the fact that

$$\mathrm{E}(B_j|W_j) = \begin{cases} 0.5 & \text{if } W_j \le 5 \\ 0.9 & \text{if } W_j > 5, \end{cases}$$

we have that

$$\mathrm{E}(Y|\mathbf{Z} = (w_1, \ldots, w_{300})) = \frac{1}{300}\sum_{j=1}^{300}\left(0.5 \times \mathbf{1}_{w_j<5} + 0.9 \times \mathbf{1}_{w_j \ge 5}\right).$$

Hence the conditional Monte Carlo estimator is

$$\hat{\mu}_{\text{cmc}} = \frac{1}{n}\sum_{i=1}^{n}\frac{1}{300}\sum_{j=1}^{300}\left(0.5 \times \mathbf{1}_{w_{i,j}<5} + 0.9 \times \mathbf{1}_{w_{i,j}\geq 5}\right),$$

where $w_{i,j}$ is the waiting time of the jth client on the ith simulation run. In other words, with the conditional Monte Carlo estimator, rather than randomly generating a decision (0 or 1) of coming back or not for each client, we instead output the *probability* of deciding never to come back: 0.5 if the waiting time was less than 5 minutes and 0.9 otherwise.

Table 4.7 contains numerical results comparing the efficiency of the naive Monte Carlo and conditional Monte Carlo estimators for this example.

Table 4.7 Comparison of Monte Carlo and conditional Monte Carlo estimators. HW is the half-width of a 95% confidence interval for μ.

Method	$\hat{\mu}$	HW	CPU(sec)	$\widehat{\text{Eff}}(\hat{\mu})$
MC	0.5943	1.19e−3	16.3	166,910
CMC	0.5938	7.78e−4	11.8	535,333

Hence, conditional Monte Carlo manages to reduce the half-width of the 95% confidence interval by a factor of about 2/3, and with a significantly smaller computation time due to a reduction from 900 to 600 variables to generate, the gain in efficiency is by a factor larger than three.

Let us now study the bias and variance of the conditional Monte Carlo estimator. We have that

$$\text{E}(\hat{\mu}_{\text{cmc}}) = \frac{1}{n}\sum_{i=1}^{n}\text{E}(\text{E}(Y|\mathbf{Z}_i)) = \text{E}(\text{E}(Y|\mathbf{Z}_i)) = \text{E}(Y) = \mu,$$

where we used (4.19) for the second-to-last equality. Thus the conditional Monte Carlo estimator is unbiased. Now, for the variance, we have

$$\begin{aligned}\text{Var}(\hat{\mu}_{\text{cmc}}) &= \frac{1}{n^2}\sum_{i=1}^{n}\text{Var}(\text{E}(Y|\mathbf{Z}_i)) = \frac{1}{n}\text{Var}(\text{E}(Y|\mathbf{Z}_i)) \\ &= \frac{1}{n}[\text{Var}(Y) - \text{E}(\text{Var}(Y|\mathbf{Z}_i))] \leq \frac{1}{n}\text{Var}(Y),\end{aligned}$$

where the inequality follows from the fact that $\text{Var}(Y|\mathbf{Z}_i) \geq 0$. Hence the conditional Monte Carlo estimator has a variance no larger than the Monte Carlo estimator's variance. In fact, we have a strict inequality as long as Y is not completely determined by $\mathbf{Z}$, which happens, for example, if we take $\mathbf{Z} = Y$. This also tells us that we could in theory get a zero variance with our conditional Monte Carlo estimator if $\text{E}(\text{Var}(Y|\mathbf{Z}_i)) = \text{Var}(Y)$, which can happen if

Y and $\mathbf{Z}$ are independent. However, if this is true, then $\mathrm{E}(Y|\mathbf{Z}) = \mathrm{E}(Y) = \mu$, which cannot be computed, and so conditional Monte Carlo cannot be applied in this case. These two extreme cases suggest that, to get a good variance reduction, we should try to choose a conditioning variable $\mathbf{Z}$ that "explains" Y as little as possible but is still such that $\mathrm{E}(Y|\mathbf{Z})$ can be computed.

Note that since $\mathbf{Z} = \zeta(\mathbf{u})$ usually depends on fewer variables u_j than f does, conditional Monte Carlo usually contributes to reducing the dimension of the problem. Often, this translates into savings for the computation time as well, as we saw in Table 4.7. This feature will also be useful when this technique is combined with quasi–Monte Carlo sampling.

We now describe how conditional Monte Carlo can be used effectively on the SAN example [22, 50].

Example 4.15. For this problem, conditional Monte Carlo can be applied as follows. Choose a *uniformly directed cutset* $\mathcal{L}$ that is a subset of $\mathcal{A}$ such that each path j from the source to the sink contains exactly one activity l_j in $\mathcal{L}$. Then, take $\mathbf{Z} = (D_j : j \in \mathcal{A} \backslash \mathcal{L})$, that is, $\mathbf{Z}$ is the vector of durations for activities that are not in the directed cutset $\mathcal{L}$. In what follows, $Y = \mathbf{1}_{T \le t_0}$ is the indicator function that is equal to 1 when the completion time T is no larger than t_0. We have that

$$\begin{aligned}\mathrm{E}(Y|\mathbf{Z} &= (D_j, j \in \mathcal{A}\backslash\mathcal{L})) \\ &= \mathrm{E}\left(\prod_{j=1}^{N(P)} \mathbf{1}_{P_j \le t_0}|\mathbf{Z}\right) = \mathrm{E}\prod_{j=1}^{N(P)} \mathbf{1}_{D_{l_j} \le \tilde{t}_0^j} = \prod_{j=1}^{N(P)} F_{l_j}(\tilde{t}_0^j), \quad (4.20)\end{aligned}$$

where

$$\tilde{t}_0^j = \max\left(0, t_0 - \sum_{k \in C_j, k \ne l_j} D_k\right) \quad (4.21)$$

is the maximum duration that l_j can have in order for the length of the jth path to be smaller than or equal to t_0, given the durations of the activities that are not in $\mathcal{L}$. Hence, the conditional Monte Carlo estimator is given by

$$\hat{\mu}_{\mathrm{cmc}} = \frac{1}{n}\sum_{i=1}^{n}\prod_{j=1}^{N(P)} F_{l_j}(\tilde{t}_0^j).$$

The pseudocode describing how to use conditional Monte Carlo for this example is given in Fig. 4.12. The results obtained using the uniformly directed cutset given by $\mathcal{L} = \{2, 6, 9, 10, 11\}$ as in [22] are given in Table 4.8.

As is seen in Table 4.8, conditional Monte Carlo increases the efficiency by a factor larger than three for this example. Note that the conditioning variable $\mathbf{Z}$ only requires thet eight uniform numbers be generated; i.e., if we write $\mathbf{Z} = \zeta(\mathbf{u})$, then $\mathbf{u} \in [0, 1)^8$, while for the original formulation $Y = f(\mathbf{u})$ is defined over $[0, 1)^{13}$.

```
SanCMC
 N_A ← 13 //nb of activities
 N_P ← 6 // nb of paths
 N_L ← 5 // size of directed cutset
 for j = 1 to N_A − N_L do
     D[L̃[j]] ← GenF(L̃[j], Rand01())
 esp ← 1
 for k = 1 to N_P do
     L ← 0
     for j = 1 to c_k do
         if (C[k, j] is not in 𝓛) then
             L ← L + D[C[k, j]]
     esp ← esp ×CDF(l_k, max(0, t_0 − L))
 return (esp)
```

Fig. 4.12 Pseudocode to use conditional Monte Carlo for the SAN example. We assume that $\tilde{L}[j]$ returns the index of the jth activity that is not in $\mathcal{L}$, $CDF(l_k, t)$ evaluates at t the CDF of the activity l_k in the kth path that is in the directed cutset, $C[k, j]$ returns the index of the jth arc on the kth path, and c_k is the number of arcs on path k.

Table 4.8 Comparison of Monte Carlo and conditional Monte Carlo estimators for Example 4.7. HW is the half-width of a 95% confidence interval for μ.

Method	$\hat{\mu}$	HW	CPU(sec)	$\widehat{\text{Eff}}(\hat{\mu})$
MC	0.7502	5.41e−3	0.197	667,913
CMC	0.7509	3.13e−3	0.195	1,995,301

More generally, if we look at the function $\phi(\mathbf{u})$ corresponding to the application of conditional Monte Carlo, we have that

$$\phi(\mathbf{u}) = \mathrm{E}(Y|\zeta(\mathbf{u})),$$

where $\mathbf{u} = (u_1, \ldots, u_t)$ is the vector of uniform numbers required to evaluate ζ. As we just saw for the SAN example, the size t of this vector is typically smaller than the original dimension s. Also, in the SAN example, we have

$$\zeta(\mathbf{u}) = (F_{i_1}^{-1}(u_1), \ldots, F_{i_t}^{-1}(u_t)),$$

where $\{i_1, \ldots, i_t\} = \mathcal{A} \backslash \mathcal{L}$ is the set of indices of durations that are not in the uniformly directed cutset $\mathcal{L}$. Furthermore, following (4.20), we have

$$\phi(\mathbf{u}) = \prod_{j=1}^{N(P)} F_{l_j}(\tilde{t}_0^j(u_1, \ldots, u_t)),$$

where we wrote $\tilde{t}_0^j = \tilde{t}_0^j(u_1, \ldots, u_t)$ as a function of $u_1, \ldots, u_t$ to emphasize the dependence on $\mathbf{u}$ that occurs through the durations $D_j = F_j^{-1}(u_r)$ in (4.21) for whichever u_r has been assigned to D_j, where $1 \leq r \leq t$.

Figure 4.13 (top) shows the function $f(\mathbf{u})$ corresponding to the SAN example as u_4 and u_{11} vary and all other coordinates are fixed (at some random point in $[0,1)^{11}$). The function f is defined so that u_i is used to generate the duration of activity i for $i = 1, \ldots, 13$. Hence the two coordinates u_4 and u_{11} correspond to the durations of activities 4 and 11, respectively.

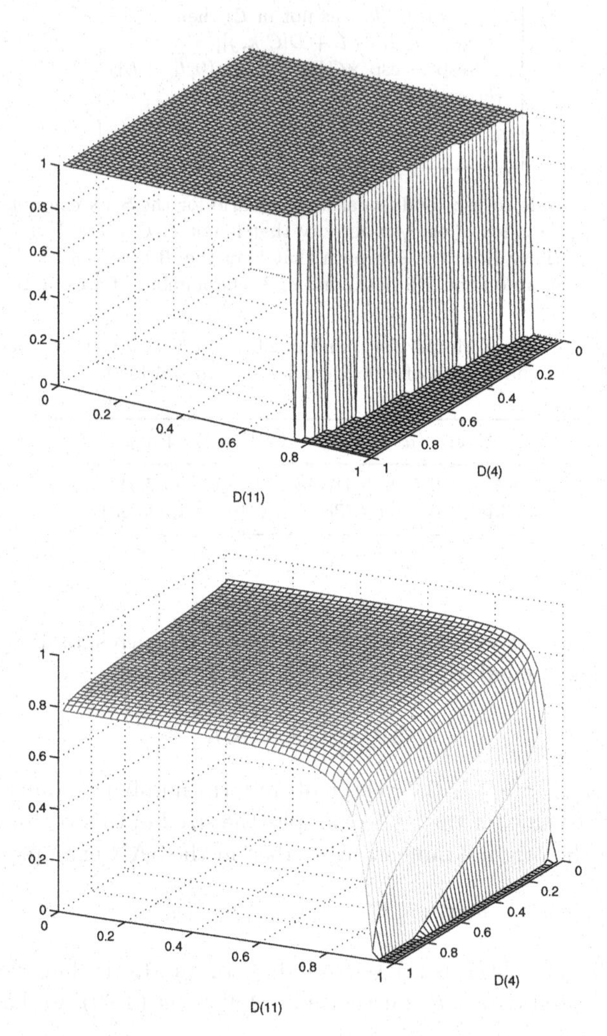

Fig. 4.13 Top: function f for SAN example; bottom: corresponding function ϕ for CMC. The axes are labeled with the variate generated by the corresponding uniform number.

The function $\phi(\mathbf{u})$ corresponding to the application of conditional Monte Carlo is shown on the bottom of Fig. 4.13 as u_3 and u_7 vary and all six remaining coordinates are fixed (randomly, but to the same value as they were for our depiction of $f(\mathbf{u})$). The function $\phi(\mathbf{u})$ is defined so that u_3 and u_7 generate the durations of activities 4 and 11, respectively, so that the comparison with $f(\mathbf{u})$ makes more sense. What we see there is that with naive Monte Carlo the corresponding function f is a step function that is 0 on part of its domain (when large numbers u_j generate long durations) and then 1 elsewhere. By contrast, the function ϕ corresponding to the conditional Monte Carlo estimator is 0 only on a small portion of its domain (corresponding to cases where one of the values $\tilde{t}_0^j$ is zero) and then increases smoothly to 1.

4.7 Stratification

Stratification is a variance reduction technique that builds on ideas that are commonly used in statistical sampling [60]. Namely, the main idea here is to partition the sample space Ω (or the unit cube $[0,1)^s$) into M strata and estimate μ separately for each stratum. For instance, suppose we write $\Omega = S_1 \cup \ldots \cup S_M$, where the strata S_l are disjoint (i.e., $S_l \cap S_m = \emptyset$ for any $l \neq m$). Let $p_j = P(\mathbf{X} \in S_j)$ for $j = 1, \ldots, m$. Then we can write

$$\mu = \mathrm{E}(h(\mathbf{X})) = \sum_{j=1}^{m} p_j \mathrm{E}(h(\mathbf{X})|\mathbf{X} \in S_j).$$

This suggests the use of the *stratified* estimator

$$\hat{\mu}_{\text{str}} = \sum_{j=1}^{m} p_j \hat{\mu}_j, \tag{4.22}$$

where

$$\hat{\mu}_j = \frac{1}{N_j} \sum_{i=1}^{N_j} h(\mathbf{X}_{i,j}), \tag{4.23}$$

$\mathbf{X}_{i,j}$ is distributed according to the conditional density of $(\mathbf{X}|\mathbf{X} \in S_j)$, and N_j is the number of draws that we take with $\mathbf{X}$ in the jth stratum. Thus we have that $N_1 + \ldots + N_m = n$. Since $\mathrm{E}(\hat{\mu}_j) = \mathrm{E}(h(\mathbf{X}_{i,j})) = \mathrm{E}(h(\mathbf{X})|\mathbf{X} \in S_j)$, we have that $\hat{\mu}_{\text{str}}$ is unbiased.

The variance of $\hat{\mu}_{\text{str}}$ is given by

$$\mathrm{Var}(\hat{\mu}_{\text{str}}) = \sum_{j=1}^{m} \frac{p_j^2}{N_j} \sigma_j^2,$$

where $\sigma_j^2 = \text{Var}(h(\mathbf{X})|\mathbf{X} \in S_j)$. To analyze this variance further, we need to say more about the variables N_j and how they are chosen. Here are two possibilities.

1. Proportional allocation: Choose $N_j = np_j$. This gives us

$$\text{Var}(\hat{\mu}_{\text{str}}) = \frac{1}{n}\sum_{j=1}^{m} p_j\sigma_j^2,$$

which can be shown to be smaller than the variance $\text{Var}(h(\mathbf{X}))/n$ of the naive Monte Carlo estimator (see Prob. 4.13).

2. Optimal allocation (also called *Neyman allocation*): Find values for the N_j such that the variance of $\hat{\mu}_{\text{str}}$ is minimized. Hence we need to solve the optimization problem

$$\begin{aligned} &\text{minimize } \text{Var}(\hat{\mu}_{\text{str}}) = \sum_{j=1}^{m} \frac{p_j^2}{N_j}\sigma_j^2, \\ &\text{s.t. } N_1 + \ldots + N_m - n = 0. \end{aligned}$$

We can use a Lagrange multiplier λ and rewrite the problem as being the minimization of

$$\sum_{j=1}^{m} p_j^2 \frac{\sigma_j^2}{N_j} + \lambda(N_1 + \ldots + N_m - n).$$

Taking the derivative with respect to each N_j and putting them equal to 0, we get

$$-\frac{p_j^2}{N_j^2}\sigma_j^2 + \lambda = 0 \text{ for } j = 1, \ldots, m,$$

which means we must have

$$N_j = \frac{p_j\sigma_j}{\sqrt{\lambda}} \text{ for } j = 1, \ldots, m,$$

and λ is determined so that $N_1 + \ldots + N_m = n$, yielding

$$\lambda = \left(\frac{p_1\sigma_1 + \ldots + p_m\sigma_m}{n}\right)^2.$$

Thus the optimal allocation is to choose

$$N_j = \frac{np_j\sigma_j}{\sum_{l=1}^{m} p_l\sigma_l}, \qquad j = 1, \ldots, m. \tag{4.24}$$

Here the value of N_j is determined both by the probability p_j associated with the jth stratum and the variability of $h(\mathbf{X}|\mathbf{X} \in S_j)$. Based on this, we can think of stratification as a way of doing importance sampling where we make sure that the "important" strata are sampled more often than the "unimportant" strata.

Note that, with optimal allocation we get the variance

$$\mathrm{Var}(\hat{\mu}_{\mathrm{str}}) = \frac{1}{n}\left(\sum_{j=1}^{m} p_j \sigma_j\right)^2,$$

which can be shown to be smaller than the variance of $\hat{\mu}_{\mathrm{str}}$ under proportional allocation (see Prob. 4.14). However, since the values σ_j are not known, they must be estimated in order to determine the optimal values N_j given by (4.24). This can be done by using a different (smaller) random sample $\mathbf{X}_1, \ldots, \mathbf{X}_N$ and either proportional allocation or poststratification, which we now describe.

An alternative to the approach described above is to use *poststratification*, where instead of choosing the sample sizes N_j a priori for each stratum and then generating $\mathbf{X}_i$ conditionally on the stratum to which it belongs, we can generate $\mathbf{X}_i$ as usual and determine *afterward* how many of them belong to each stratum. More precisely, the poststratified estimator is given by

$$\hat{\mu}_{\mathrm{pstr}} = \sum_{j=1}^{m} p_j \left(\frac{1}{N_j}\sum_{i=1}^{n} h(\mathbf{X}_i) B_{i,j}\right), \tag{4.25}$$

where

$$B_{i,j} = \begin{cases} 1 & \text{if } \mathbf{X}_i \in S_j \\ 0 & \text{else} \end{cases}$$

and $N_j = \sum_{i=1}^{n} B_{i,j}$ is the number of vectors $\mathbf{X}_i$ that belong to S_j for $j = 1, \ldots, m$. Hence N_j is a random variable with $(N_1, \ldots, N_m)$ having a multinomial distribution with parameters $(n, p_1, \ldots, p_m)$. In the definition (4.25), we assume that if $N_j = 0$, then the estimator $\sum_{i=1}^{n} h(\mathbf{X}_i) B_{i,j}/N_j$ is set to 0.

The advantage of using poststratification is that the vectors $\mathbf{X}_i$ are generated as usual rather than by using the conditional distribution of $\mathbf{X}$ given $\mathbf{X} \in S_j$, which might be difficult to sample from in some cases. The drawback is that using a random N_j introduces more variability in the estimator $\hat{\mu}_{\mathrm{pstr}}$ compared with the estimator $\hat{\mu}_{\mathrm{str}}$ given in (4.22). Also, the poststratified estimator is only *conditionally* unbiased on the event that all $N_j \geq 1$ [45, 394]. To see why, let A denote this event. Then we have

$$\mathrm{E}(\hat{\mu}_{\mathrm{pstr}}|A) = \sum_{j=1}^{m} p_j \mathrm{E}\left(\mathrm{E}\left(\frac{1}{N_j}\sum_{i=1}^{n} h(\mathbf{X}_i) B_{i,j} | B_{i,j}, i = 1, \ldots, n\right) |A\right)$$

$$= \sum_{j=1}^{m} p_j \mathrm{E}\left(\frac{1}{N_j}(N_j \mathrm{E}(h(\mathbf{X}_i)|\mathbf{X}_i \in S_j)|A)\right)$$
$$= \sum_{j=1}^{m} p_j \mathrm{E}(h(\mathbf{X}_i)|\mathbf{X}_i \in S_j) = \mu.$$

But this implies that $\mathrm{E}(\hat{\mu}_{\text{pstr}}) = \mu P(A) \leq \mu$. For the variance, we first compute the conditional variance

$$\mathrm{Var}\left(\sum_{j=1}^{m} \frac{p_j}{N_j} \sum_{i=1}^{n} h(\mathbf{X}_i) B_{i,j} | B_{i,j}, i = 1, \ldots, n\right) = \sum_{j=1}^{m} \frac{p_j^2}{N_j} \sigma_j^2. \quad (4.26)$$

We then use the formula

$$\mathrm{Var}(\hat{\mu}_{\text{pstr}}|A) = \mathrm{Var}(\mathrm{E}(\hat{\mu}_{\text{pstr}}|B_{i,j}, i = 1, \ldots, n)|A) + \mathrm{E}(\mathrm{Var}(\hat{\mu}_{\text{pstr}}|B_{i,j}, i = 1, \ldots, n)|A)$$

to compute the (conditional) variance of $\hat{\mu}_{\text{pstr}}$. First, note that the inner conditional expectation in the first term is independent of N_j (conditioned on A) and thus the first term vanishes. We can use (4.26) to compute the second term and get

$$\mathrm{Var}(\hat{\mu}_{\text{pstr}}|A) = \sum_{j=1}^{m} p_j^2 \sigma_j^2 \mathrm{E}\left(N_j^{-1}|A\right). \quad (4.27)$$

Hence we just need to compute $\mathrm{E}(N_j^{-1}|A)$ where $(N_1, \ldots, N_m)$ has a multinomial distribution, as noted before. As was done for example in [60, pp. 134–135], we can estimate this expectation as follows:

1. For a multinomial distribution, we have that $\mathrm{E}(N_j) = np_j$ and $\mathrm{Var}(N_j) = np_j(1 - p_j)$.
2. If we condition on A, then these expectation and variance formulas are only slightly different (since $P(A)$ is almost 1 for n large), and so the ones given in the previous item can be used as approximations even when we condition on A (see Prob. 4.16).
3. For a random variable X such that $P(X = 0) = 0$, we can write

$$\mathrm{E}\left(\frac{1}{X}\right) = \frac{1}{\mathrm{E}(X)} \mathrm{E}\left(1 + \frac{X - \mathrm{E}(X)}{\mathrm{E}(X)}\right)^{-1}.$$

4. Using Taylor series, we can write

$$\frac{1}{1+y} \approx 1 - y + y^2.$$

Combining these four steps, we write

$$\mathrm{E}\left(N_j^{-1}|A\right) \approx \frac{1}{np_j}\left(1 + \frac{np_j(1-p_j)}{n^2p_j^2}\right) = \frac{1}{np_j}\left(1 + \frac{1-p_j}{np_j}\right).$$

Substituting this approximation into (4.27) gives

$$\mathrm{Var}(\hat{\mu}_{\mathrm{pstr}}|A) \approx \sum_{j=1}^{m} \frac{p_j}{n}\sigma_j^2 \left(1 + \frac{1-p_j}{np_j}\right).$$

Comparing this with proportional allocation, we see that the price to pay for using poststratification is that we have an (approximate) extra term

$$\frac{1}{n^2}\sum_{j=1}^{m}(1-p_j)\sigma_j^2$$

added to the variance. If n is large, this $O(n^{-2})$ term is pretty small.

Finally, to understand the differences between poststratification and naive Monte Carlo, we can rewrite the latter in a form similar to (4.25), so that we get

$$\hat{\mu}_{\mathrm{mc}} = \sum_{j=1}^{m} \frac{N_j}{n}\left(\frac{1}{N_j}\sum_{i=1}^{n} f(\mathbf{u}_i)B_{i,j}\right).$$

Hence the poststratified estimator replaces the weights N_j/n used in the Monte Carlo estimator by their expected value p_j, which intuitively should result in a reduction of the variance [217].

Let us now illustrate how stratification works on our bank example.

Example 4.16. For the bank example, we can use stratification on the variable v that determines the mean service time [45]. That is, we let $S_j = \{\mathbf{X} = (v, a_1, s_1, \ldots) \in \Omega : v = v_j\}$ for $j = 1, 2, 3$. Hence the stratified estimator is

$$\hat{\mu}_{\mathrm{str}} = 0.2\hat{\mu}_1 + 0.7\hat{\mu}_2 + 0.1\hat{\mu}_3,$$

where $\hat{\mu}_j$ is the estimator obtained when the mean service time is v_j minutes, with $v_1 = 35/60$, $v_2 = 50/60$, and $v_3 = 55/60$. If we take $n = 1024$, then with proportional allocation we get $N_1 = 205$, $N_2 = 717$, and $N_3 = 102$. To use optimal allocation, we generate a sample of size $N = 100$ and estimate σ_1, σ_2, and σ_3 using poststratification. We get $\hat{\sigma}_1 = 6.87$, $\hat{\sigma}_2 = 62.1$, and $\hat{\sigma}_3 = 62.1$. Hence we can perform optimal allocation with $N_1 = 28, N_2 = 872$, and $N_3 = 124$. Results comparing the three forms of stratification with the naive Monte Carlo estimator are given in Table 4.9. We see that the efficiency is increased by a factor of about two for all three stratification methods.

Table 4.9 Results for the modified bank example using stratification. HW is the half-width of a 95% confidence interval for μ.

Method	$\hat{\mu}$	HW	CPU(sec)	$\hat{\text{Eff}}(\hat{\mu})$
MC	73.04	0.788	11.9	0.521
post	73.07	0.493	11.9	1.328
prop	73.01	0.596	11.7	0.924
opt	72.95	0.530	11.9	1.147

Pseudocode describing how to use stratification with fixed N_j or post-stratification is given in Fig. 4.14, where the function `OneSimBank`$(\cdot,\cdot)$ is as defined on p. 15 of Chap. 1.

```
SimFixedStr(N1, N2, N3)
  str ← 0
  for l = 1 to 3
     v ← mus[l]
     for i = 1 to N_i
        x[i] ← OneSimBank(1,v)
     str ← str + p_l × ave(x)
  return(str)
```

```
SimPostStr
  N ← [0, 0, 0]
  x ← [0, 0, 0]
  for i = 1 to n
     U ← Rand01()
     l ← GenDisc([0.2,0.9,1],3,u)
     v ← mus[l]
     result ← OneSimBank(1,v)
     x[l] ← x[l] +result
     N[l] ← N[l] + 1
  return p1 x[1]/N[1] + p2 x[2]/N[2] + p3 x[3]/N[3]
```

Fig. 4.14 Pseudocode for stratification with the N_j fixed ahead of time (left) or a posteriori (right).

As was done in the previous sections, we conclude our discussion on stratification with a description of the function $\phi(\mathbf{u})$ corresponding to the integration formulation of this technique. When N_j is fixed ahead of time, we can break down the problem into m integration problems where the goal is to estimate the integral of

$$\phi_j(\mathbf{u}) = h(\tilde{g}_j(\mathbf{u})), \tag{4.28}$$

where $\tilde{g}_j(\cdot)$ is the function that transforms $\mathbf{u} = (u_1, \ldots, u_t)$ into an observation $\mathbf{X}$ from the conditional pdf of $\mathbf{X}$ given $\mathbf{X} \in S_j$ for $j = 1, \ldots, m$. Then each of these m integration problems are tackled using a sample of N_j i.i.d. points $\mathbf{u}_1, \ldots, \mathbf{u}_{N_j}$. Note that the dimension t of these points might be smaller than the original dimension s due to the fact that we generate observations

$\mathbf{X}$ under the conditional pdf. For instance, in the bank example, we need one less uniform number since we stratify on the mean service time and therefore do not need to generate an observation for the random mean service time under the conditional distribution.

For the poststratified estimator, we can first say that we are still integrating the function $f(\mathbf{u})$ but now replace the equal $1/n$ weights of the Monte Carlo estimator by the weights $w_i = p_j/N_j$, where j is the index of the stratum to which $\mathbf{X}_i = g(\mathbf{u}_i)$ belongs for $i = 1, \ldots, n$, and assuming all N_j are at least 1. A second interpretation is to say that we wish to integrate the function

$$\sum_{j=1}^{m} p_j \phi_j(\mathbf{u}), \tag{4.29}$$

where $\phi_j(\cdot)$ is as defined in (4.28), but we want to use a sample $\mathbf{X}_1, \ldots, \mathbf{X}_n$ generated under the unconditional distribution rather than work with conditional distributions. That is, we wish to use the sample $g(\mathbf{u}_1), \ldots, g(\mathbf{u}_n)$ rather than having m separate samples, each created according to $\tilde{g}_j(\mathbf{u})$, for $j = 1, \ldots, m$. Hence we can think of poststratification as applying importance sampling to (4.29), where for each j the "original" distribution is the conditional pdf $\varphi(\mathbf{x}|\mathbf{x} \in S_j)$, and the new one used for importance sampling is the unconditional pdf $\varphi(\mathbf{x})$. Using the fact that

$$\varphi(\mathbf{x}|\mathbf{x} \in S_j) = \frac{\varphi(\mathbf{x})\mathbf{1}_{\mathbf{x} \in S_j}}{p_j},$$

we have that the likelihood ratio for each j has the form

$$L(\mathbf{x}_i) = \frac{\varphi(\mathbf{x}_i|\mathbf{x}_i \in S_j)}{\varphi(\mathbf{x}_i)} = \frac{B_{i,j}}{p_j} = \frac{\mathbf{1}_{g(\mathbf{u}_i) \in S_j}}{p_j}.$$

If we use the usual importance sampling estimator, we get

$$\sum_{j=1}^{m} p_j \left(\frac{1}{n} \sum_{i=1}^{n} \frac{f(\mathbf{u}_i)\mathbf{1}_{g(\mathbf{u}_i) \in S_j}}{p_j} \right),$$

which is just the naive Monte Carlo estimator $\hat{\mu}_{\text{mc}}$. Instead, the poststratified estimator is obtained using the weighted importance sampling estimator $\hat{\mu}_{\text{is},w}$ described on p. 115. That is,

$$\hat{\mu}_{\text{pstr}} = \sum_{j=1}^{m} p_j \sum_{i=1}^{n} \frac{f(\mathbf{u}_i)\mathbf{1}_{g(\mathbf{u}_i) \in S_j}/p_j}{\sum_{i=1}^{n} \mathbf{1}_{g(\mathbf{u}_i) \in S_j}/p_j} = \sum_{j=1}^{m} \frac{p_j}{N_j} \sum_{i=1}^{n} f(\mathbf{u}_i)\mathbf{1}_{g(\mathbf{u}_i) \in S_j}.$$

4.8 Common random numbers

This last technique is typically used when the goal is to estimate the difference between two related quantities. More generally, it can be used when several systems having common features need to be simulated in order to estimate a function that depends on the response obtained from each system. For instance, in the context of sensitivity analysis, we may want to find out whether a parameter change will significantly affect the value of a certain quantity of interest.

In what follows, we assume that the goal is to estimate

$$\mu = \mu_1 - \mu_2,$$

where $\mu_j = \mathrm{E}(f_j(\boldsymbol{U}))$, $j = 1, 2$.

As the name suggests, the idea of common random numbers is to use the same uniform random numbers to estimate μ_1 and μ_2. Intuitively speaking, using the same source of randomness should have the effect that the differences observed are due to intrinsic differences between the two functions f_1 and f_2 rather than variations in the random numbers used.

The common random numbers estimator is thus given by

$$\hat{\mu}_{\mathrm{crn}} = \frac{1}{n}\sum_{i=1}^{n}(f_1(\mathbf{u}_i) - f_2(\mathbf{u}_i)),$$

where $\mathbf{u}_1, \ldots, \mathbf{u}_n$ are i.i.d. uniform over $[0, 1)^s$. It is obvious that this estimator for μ is unbiased. Its variance is given by

$$\begin{aligned}\mathrm{Var}(\hat{\mu}_{\mathrm{crn}}) &= \frac{1}{n}\left(\mathrm{Var}(f_1(\boldsymbol{U})) + \mathrm{Var}(f_2(\boldsymbol{U})) - 2\mathrm{Cov}(f_1(\boldsymbol{U}), f_2(\boldsymbol{U}))\right)\\ &= \frac{1}{n}(\sigma_1^2 + \sigma_2^2 - 2\sigma_{1,2}),\end{aligned}$$

where $\sigma_{1,2} = \mathrm{Cov}(f_1(\boldsymbol{U}), f_2(\boldsymbol{U}))$. By contrast, if instead we use independent samples $\{\mathbf{u}_{1,i} : 1 \le i \le n\}$ and $\{\mathbf{u}_{2,i} : 1 \le i \le n\}$ to estimate μ_1 and μ_2, respectively, then the corresponding estimator

$$\hat{\mu}_{\mathrm{ind}} = \frac{1}{n}\sum_{i=1}^{n} f_1(\mathbf{u}_{1,i}) - \frac{1}{n}\sum_{i=1}^{n} f(\mathbf{u}_{2,i})$$

has variance

$$\mathrm{Var}(\hat{\mu}_{\mathrm{ind}}) = \frac{1}{n}(\sigma_1^2 + \sigma_2^2).$$

Hence the common random numbers estimator has a smaller variance than the independent estimator if and only if $\sigma_{1,2} > 0$. Results closely related to those seen for antithetic variates in Sect. 4.3 can be used to determine if this condition holds. More precisely, we have the following theorem [473].

Theorem 4.17. *Let X be a random variable with CDF $F(\cdot)$, and let Y be a random variable with CDF $G(\cdot)$. Then among all pairs with marginal CDFs given by $F(\cdot)$ and $G(\cdot)$, $(F^{-1}(U), G^{-1}(U))$ is the one with maximal correlation.*

What this tells us is that if X represents a certain random variable in our system and Y represents the same random variable but in the alternative system — for which μ_2 is estimated — then the best way to induce positive correlation between X and Y is to use inversion with the same uniform random number U to generate both of them. Similar to Theorem 4.4, this result gives us a way to guarantee "intermediate" positive correlation between the two measures to be estimated. To make sure this correlation actually holds for f_1 and f_2, we can use the following result [45, 275], which is similar to Theorem 4.3 in Sect. 4.3.

Theorem 4.18. *Assume that f_1 and f_2 are bounded functions not constant everywhere and that, for all $j \geq 1$, they are either both increasing or both decreasing as a function of their jth argument u_j. Then* $\mathrm{Cov}(f_1(\boldsymbol{U}), f_2(\boldsymbol{U})) \geq 0$.

The conditions of that theorem are the same as those given in Theorem 4.3, but this time they are used to verify that common random numbers can reduce the variance compared with using independent simulations.

We now illustrate with our two examples how common random numbers can be applied.

Example 4.19. For the bank example, let μ_1 be the expectation of the number of clients that will wait more than 5 minutes when the mean interarrival time is 60 seconds (as in Example 1.2), and let μ_2 be the same quantity but when the mean interarrival time is 55 seconds. Here, f_1 is the same function f as in Example 4.5, and f_2 is almost identical to f_1 except that it transforms the uniform numbers designated for interarrival times into exponential random variates with mean 55 seconds instead of 60 seconds. Using the same arguments as in Example 4.5, we can easily verify that both f_1 and f_2 satisfy the conditions of Theorem 4.18. Figure 4.15 gives the pseudocode for using common random numbers in this case.

Numerical results comparing the efficiency of common random numbers and the Monte Carlo estimator based on independent runs for Example 4.19 are given in Table 4.10. The common random numbers estimator not only decreases the half-width of the 95% confidence interval by a factor larger than four but also significantly reduces the computation time since the same random numbers are used to simulate both systems. Overall, common random numbers improves the efficiency by a factor of about 26.

Example 4.20. Suppose that in our SAN example we want to know if reducing the mean duration of activities 2, 5, 6, 10, and 11 will significantly increase

Table 4.10 Comparison of independent (ind.) and common random numbers (CRN) estimators for Example 4.19. HW is the half-width of a 95% confidence interval for μ.

Method	$\hat{\mu}$	HW	CPU(sec)	$\hat{\text{Eff}}(\hat{\mu})$
ind.	−50.70	1.86	24.7	4.48e−2
CRN	−49.67	0.47	14.8	1.154

the probability of completing the network within t_0. More precisely, we want to estimate the difference $\mu = p_0 - p_1$, where p_0 is $P(T \leq t_0)$ under the original model assumptions, while $p_1 = P(\tilde{T} \leq t_0)$, where $\tilde{T}$ is the completion time when the duration means for activities 2, 5, 6, 9, and 11 are 90% of their original values. Table 4.11 compares common random numbers with independent simulations for this problem.

```
OneSimBankCRN
 NbWait5_1 ← 0; NbWait5_2 ← 0
 w1 ← 0
 w2 ← 0
 type ← GenDisc([0.2,0.9,1],3,Rand01())
 v ← mus[type]
 u ← Rand01()
 a1 ← GenExpon(1,u)
 a2 ← GenExpon(55/60,u)
 time1 ← a1
 time2 ← a2
 while (min(time1,time2) < 300) do
     s ← GenExpon(v,Rand01())
     u ← Rand01()
     a1 ← GenExpon(1,u)
     a2 ← GenExpon(55/60,u)
     time1 ← time1 + a1
     time2 ← time2 + a2
     w1 ← max(0, w1 + s − a1)
     w2 ← max(0, w2 + s − a2)
     if ((time1 < 300) and (w1 > 5)) then
         NbWait5_1 ← NbWait5_1 + 1
     if ((time2 < 300) and (w2 > 5)) then
         NbWait5_2 ← NbWait5_2 + 1
 return (NbWait5_1 − NbWait5_2)
```

Fig. 4.15 Pseudocode showing how to apply common random numbers for the bank example.

Table 4.11 Comparison of independent (ind.) and common random numbers (CRN) estimators for Example 4.20. HW is the half-width of a 95% confidence interval for μ.

Method	$\hat{\mu}$	HW	CPU(sec)	$\hat{\text{Eff}}(\hat{\mu})$
ind.	−0.0696	6.64e−3	0.305	1,473,409
CRN	−0.0660	2.92e−3	0.406	214,710

Hence common random numbers reduces the size of the 95% confidence interval half-width by a factor larger than two and increases the efficiency by a factor of about seven compared with independent simulations.

We should mention that, for the same reasons we gave in Sect. 4.3, synchronization is extremely important for common random numbers to work. That is, we must make sure that, as much as possible, each uniform number u_j must be used for the same purpose in both simulations. Also, just like antithetic variates, common random numbers do not need to be applied on *all* uniform numbers. For instance, if, while verifying the conditions of Theorem 4.18, we realize that f_1 and f_2 are not both increasing or both decreasing for a subset $\mathcal{J}$ of their arguments, then this suggests that common random numbers should not be used for this subset.

It is fairly easy to see that, from the integration point of view, the use of common random numbers as described above amounts to building an approximation for the integrand $\phi(\mathbf{u}) = f_1(\mathbf{u}) - f_2(\mathbf{u})$.

As mentioned at the beginning of this section, common random numbers can be used in a much more general context, where we need to study J measures of performance corresponding to functions $f_1, \ldots, f_J$, and we need to estimate M functions involving $f_1, \ldots, f_J$. For instance, f_1 might correspond to the reference system and $f_2, \ldots, f_J$ to alternate configurations and the goal is to simultaneously estimate $\mathrm{E}(f_1(\boldsymbol{U}) - f_j(\boldsymbol{U})$ for $j = 2, \ldots, J$. Another example is in the context of control variates, where $f_2(\mathbf{u})$ might represent an external control variate and the goal is to estimate $f_1(\mathbf{u}) + \beta(\mu_2 - f_2(\mathbf{u}))$, as discussed on p. 108. A third example is in the context of *regenerative simulation*, where we are typically interested in estimating ratios of the form $\mathrm{E}(f_1(\mathbf{u}))/\mathrm{E}(f_2(\mathbf{u}))$. An example of regenerative simulation will be given in Sect. 7.3.

4.9 Combinations of techniques

Now that we have seen all these variance reduction techniques, it is natural to wonder if we have to use them separately or if we can combine them. The answer is that we can combine them, but care must be taken when doing so. For instance, combining antithetic variates and common variates may not necessarily reduce the variance even if each method does so separately [218]. The reason is that when we write out the variance of the combined estimator, some cross-covariance terms have a sign that cannot be predicted by Theorems 4.3 and 4.18. This is because even if f is originally monotone in each of its arguments, when we apply antithetic variates, as we saw in Fig. 4.7, this has the effect of transforming f into a function that no longer has these monotonicity properties. That transformed function thus fails to satisfy the conditions of Theorem 4.18.

Some combinations have been studied in more detail than others. For example, Avramidis and Wilson look at the three pairwise combinations arising from control variates, conditional Monte Carlo, and *correlation induction techniques* — this includes antithetic variates and another technique called *Latin hypercube sampling*, which will be discussed in Chaps. 6 and 8 — and show how to adapt results that hold for *one* of the techniques to the case where it is combined with another one. In [146], importance sampling, stratification, and conditional Monte Carlo are combined successfully. Hesterberg [177] studies the combination of control variates and importance sampling in the context of bootstrap simulations. In general, questions related to combining variance reduction techniques and relating them to each other remain an active research area.

Problems

4.1. Suppose $\hat{\mu}_1$ and $\hat{\mu}_2$ are estimators for μ. Assume $\hat{\mu}_1$ has bias c/n, variance σ^2/n, and expected computation time $d \times n$ for some constants $c, d, \sigma^2 > 0$. If $\hat{\mu}_2$ has a variance twice as small as $\hat{\mu}_1$ and a bias twice as big as $\hat{\mu}_1$, then asymptotically what is the largest factor by which its computation time can exceed that of $\hat{\mu}_1$ for its efficiency to remain larger than $\hat{\mu}_1$?

4.2. Let $f(u) = au^2 + bu + c$, where a, b, and c are some real constants. (a) Give expressions for the variance of the naive Monte Carlo estimator and the antithetic estimator for μ based on a total of n function evaluations. (b) Give conditions in terms of a and b under which the antithetic variates estimator has a smaller variance than the naive Monte Carlo estimator.

4.3. Let $f(u) = au + b$, where a and b are some real constants. Show that $\rho(f(U), f(1-U)) = -1$, where $U \sim U(0,1)$.

4.4. Consider the function $f(\mathbf{u}) = a_1 u_1 + \ldots + a_s u_s + b$. Show that the antithetic estimator $Q_{n,\text{ant}}$ has a zero variance for this function for any real constants $a_1, \ldots, a_s, b$.

4.5. Formulate the joint distribution function described in Theorem 4.4 using a copula.

4.6. Apply the method of antithetic variates to estimate the quantity p_K described in Prob. 1.12 of Chap. 1. Use a total of $n = 1000$ function evaluations, and compute the ratio of the 95% confidence interval half-width you get with antithetic variates over what was obtained with naive Monte Carlo in Prob. 1.12 of Chap. 1.

4.7. Show that for the function $f(\mathbf{u}) = \sum_{j=1}^{s}(1 - 2u_j)^2$, the antithetic estimator increases the variance by a factor of two compared with the Monte Carlo estimator.

4.8. Repeat the experiment outlined at the end of Sect. 4.4 for the bank example, but using the two control variates simultaneously. Compare the variance of the estimator obtained with that of each of the two single-control variate estimators.

4.9. For the bank example and the control variable given by the average service time, compare the estimate for c_5 and the estimated variance of the control variate estimator obtained in Example 4.9 with the ones based on (i) splitting and (ii) jackknifing. Use $m = 25$ groups of $n = 1024$ simulations to establish these comparisons.

4.10. Find a new distribution $\psi(\mathbf{x})$ for importance sampling such that, when $h(\mathbf{x}) < 0$ for all $\mathbf{x} \in \Omega$, the resulting importance sampling estimator has zero variance.

4.11. For the experiment outlined at the end of Sect. 4.5, repeat the experiment, but with interarrivals having a mean of 50 seconds instead of 58 seconds (under the new probability distribution). Estimate the variance of the importance sampling estimator in this case and compare it with the variance of the naive Monte Carlo estimator.

4.12. In Prob. 1.12 of Chap. 1, assume that if $S(T)/S(0) > 1.15$, you sell the stock at $T = 1$ with probability 0.75 and otherwise you keep it with probability 0.8. (a) Design a conditional Monte Carlo estimator for the probability of selling the stock at $T = 1$. (b) Estimate the variance of your conditional Monte Carlo estimator using $n = 1000$ runs and compare it with the variance of the naive Monte Carlo estimator.

4.13. Show that the stratified estimator with proportional allocation has a variance no larger than the naive Monte Carlo estimator's variance.

4.14. Show that optimal allocation gives a stratified estimator with smaller variance than proportional allocation by directly comparing the two variances.

4.15. Consider the SAN problem from Example 4.7. One way of applying stratification is to choose a subvector of r duration $D_{\mathcal{S}} = (D_{j_1}, \ldots, D_{j_r})$ and partition the range of $D_{\mathcal{S}}$ into $M = k^r$ equiprobable strata. That is, we have strata of the form $S_{l_1,\ldots,l_r} = [q_{1,l_1}, q_{1,l_1+1}) \times \ldots \times [q_{r,l_r}, q_{r,l_r+1})$, where q_{j,l_j} corresponds to the $100(l_j/k)\%$ percentile of D_j's distribution, and $1 \leq l_j \leq k$. (a) Prove that if inversion is used to generate the D_j, then $D_{\mathcal{S}} \in S_{l_1,\ldots,l_r}$ if and only if the uniform u_{j_v} used to generate D_{j_v} satisfies $u_{j_v} \in [l_v/k, (l_v+1)/k)$ for $v = 1,\ldots,r$. (b) Using the result proved in (a), use stratification (poststratification, then proportional, then optimal) based on $j_1 = 2, j_2 = 6$, $j_3 = 9$, and $k = 2$ to compute $\mu = P(T \leq t_0)$. Compare the variance obtained for each method with $n = 1024$ and $m = 25$ repetitions with the naive Monte Carlo estimator's variance for which results were presented in Table 4.2. (You can find in [146] a similar idea used in the context of finance.)

4.16. Show that if $(N_1,\ldots,N_m)$ has a multinomial distribution with parameters $(n, p_1,\ldots,p_m)$, where $p_j = 1/m$ for all j, and A represents the event where all $N_j \geq 1$, then $\mathrm{E}(N_j|A) = \mathrm{E}(N_j) = np_j$ and $\mathrm{Var}(N_j|A) = \mathrm{Var}(N_j) = np_j(1-p_j)$. Show that this does not necessarily hold if the probabilities p_j are not all equal.

4.17. Let $f_1(u) = au + b$ and $f_2(u) = (a+\delta)u + b$, where a, b, and δ are some constants. (a) Give an expression for the variance of the common random numbers estimator for $\mu_1 - \mu_2$ and compare it with the variance of the estimator based on independent simulations. (b) For a given a and b, find the smallest value for $|\delta|$ such that the (theoretical) 95% confidence interval for $\mu_1 - \mu_2$ based on common random numbers will not contain 0. (c) Repeat (b) but for the estimator based on independent simulations.

4.18. Apply common random numbers with $n = 1000$ to estimate the difference $p_{K,\sigma=0.2} - p_{K,\sigma=0.3}$ in the probability p_K for $\sigma = 0.2$ and $\sigma = 0.3$ in Prob. 1.12 of Chap. 1. Compute a 95% confidence interval for $p_{K,\sigma=0.2} - p_{K,\sigma=0.3}$. Compare it with the 95% confidence interval obtained with independent simulations.

4.19. Consider the functions $f_1(u) = au + b$ and $f_2(u) = cu + d$, where a, b, c, and d are some real constants. Give an expression in terms of a, b, c, and d for (i) the variance of the estimator $\hat{\mu}_{\text{crn+ant}}$ that combines common random numbers and antithetic variates based on the i.i.d. sample points $\mathbf{u}_1,\ldots,\mathbf{u}_{n/2}$ and (ii) the variance of the naive Monte Carlo estimator based on two i.i.d. samples $\{\mathbf{u}_{1,1},\ldots,\mathbf{u}_{1,n}\}$ and $\{\mathbf{u}_{2,1},\ldots,\mathbf{u}_{2,n}\}$ (for f_1 and f_2, respectively).

4.20. Consider the combined antithetic variates and control variate estimator $\sum_{i=1}^{n} \phi(\mathbf{u}_i)/n$, where $\phi(\mathbf{u}) = 0.5(f(\mathbf{u})+f(\tilde{\mathbf{u}}))+\beta(\mu_c - 0.5(c(\mathbf{u})+c(\tilde{\mathbf{u}})))$. What is the optimal β for that estimator?

Chapter 5
Quasi–Monte Carlo Constructions

5.1 Introduction

In this chapter and the following one, we discuss the use of *low-discrepancy sampling* to replace the pure random sampling that forms the backbone of the Monte Carlo method. Using this alternative sampling method in the context of multivariate integration is usually referred to as *quasi–Monte Carlo.* A low-discrepancy sample is one whose points are distributed in a way that approximates the uniform distribution as closely as possible. Unlike for random sampling, points are not required to be independent. In fact, the sample might be completely deterministic.

Any attempt to construct such samples requires a precise way of measuring their "uniformity", so that we can compare different constructions and also make sure that we are indeed improving on random sampling. In fact, we are already familiar with the idea of measuring the uniformity of a point set from our discussion in Sect. 3.5 on theoretical tests for random number generators. Recall that there we were looking at the s-dimensional set Ψ_s representing all possible sequences of s successive numbers that can be produced by the generator, and our goal was to make sure this set was "as uniform as possible". We saw that sets Ψ_s arising from MRGs had a lattice structure that could be assessed via the spectral test, whereas $\mathbb{F}_2$-linear generators were producing sets Ψ_s whose uniformity could be measured via the concept of equidistribution through the resolution and t-value. As we will see later in this chapter, these uniformity measures can also be used for assessing the quality of low-discrepancy samples designed for quasi–Monte Carlo. But we will also see that many other measures can be used for that purpose.

As a first step, let us introduce a way of measuring the uniformity of a point set that is not specific to a particular type of construction. More precisely, the idea is to measure the distance between the empirical distribution induced by the point set and the uniform distribution via the Kolmogorov-Smirnov statistic. The concept of *discrepancy*, which is heavily used in the

C. Lemieux, *Monte Carlo and Quasi–Monte Carlo Sampling*,
Springer Series in Statistics 692, DOI: 10.1007/978-0-387-78165-5_5,

quasi–Monte Carlo community — among other things in the terminology *low-discrepancy point set/sequence* — looks precisely at such distance measures. To present these ideas, let us first consider the one-dimensional case.

Consider samples P_n of size n over the unit interval $[0,1)$. An obvious choice for a low-discrepancy sample P_n is $\{0, 1/n, 2/n, \ldots, (n-1)/n\}$, or maybe $\{1/2n, 3/2n, \ldots, (2n-1)/2n\}$. Alternatively to these two deterministic choices, one could also use a *randomized* version,

$$P_n(v) := \{v \bmod 1, (1/n + v) \bmod 1, \ldots, ((n-1)/n + v) \bmod 1\},$$

where $v \sim U(0,1)$. The higher uniformity of these one-dimensional samples can be stated in various ways that more or less all relate to the fact that the distance between adjacent pairs of points in those samples is equal to $1/n$. As a consequence, if we look at the empirical CDF induced by these samples, it is always within $1/n$ of the CDF of the uniform distribution over $[0,1)$. That is, consider the quantity

$$D^*(P_n) = \sup_{x\in[0,1)} |F(x) - \hat{F}_n(x)|, \tag{5.1}$$

where for $0 \le x < 1$, $F(x) = x$ is the CDF of a $U(0,1)$ random variable and $\hat{F}_n(x)$ is the empirical CDF induced by P_n. That is,

$$\hat{F}_n(x) = \frac{1}{n}\sum_{i=1}^{n} \mathbf{1}_{u_i \le x},$$

which is the proportion of the numbers u_i that are smaller than or equal to x. Then we have that

$$D^*(\{0, 1/n, 2/n, \ldots, (n-1)/n\}) = 1/n,$$
$$D^*(\{1/2n, 3/2n, \ldots, (2n-1)/2n\}) = 1/2n,$$

and

$$D^*(P_n(v)) = \max\left(v - \frac{\lfloor nv \rfloor}{n}, \frac{(\lfloor nv \rfloor + 1)}{n} - v\right) \le \frac{1}{n}.$$

We illustrate in Fig. 5.1 how, for the point set $\{1/2n, 3/2n, \ldots, (2n-1)/2n\}$ with $n = 5$, the distance between $\hat{F}_n(x)$ and $F(x)$ is never more than $1/2n$.

Comparing this with a truly random sample P_n, we see that if we are unlucky, $D^*(P_n)$ could be much larger than $1/n$ in that case. For instance, for given integers $k \in \{1, \ldots, n-1\}$ and $j \in \{0, \ldots, n-k\}$, with probability $((n-k)/n)^n$, a given interval of the form $[j/n, (j+k)/n)$ will contain no point, hence creating a difference of at least $k/2n$ with the uniform distribution.

Looking at the one-dimensional case helps give an idea of what low-discrepancy sampling is and how it differs from random sampling. However, the real challenge arises in the multidimensional case, where we need to find a way of improving on random sampling without resorting to grids of the

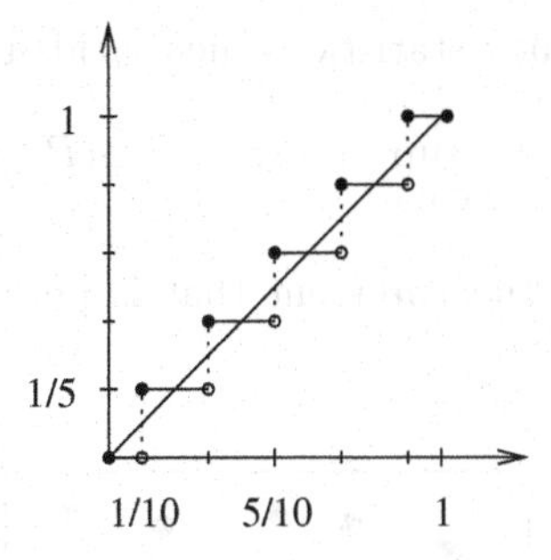

Fig. 5.1 Empirical distribution induced by $\{1/10, 3/10, 5/10, 7/10, 9/10\}$ compared with the uniform CDF of a $U(0,1)$. The dotted line shows the distance to $F(x) = x$.

form

$$\underbrace{P_N \times \ldots \times P_N}_{s \text{ times}},$$

where P_N is a one-dimensional low-discrepancy point set. As a special case of this type of construction, consider the point set given by the rectangular grid

$$P_n = \left\{ \left(\frac{l_1}{N}, \ldots, \frac{l_s}{N} \right), l_j = 0, \ldots, N-1, j = 1, \ldots, s \right\}, \qquad (5.2)$$

where $n = N^s$. (Note that this is slightly different from the point set used with the trapezoidal rule in Chap. 1 simply because here we exclude the coordinate 1 from the one-dimensional version, which is why we obtain N^s points rather than $(N+1)^s$.) As we discussed in Chap. 1, such constructions do not work well for multivariate integration unless s is very small. An alternative way to understand the problem with (5.2) is to look at how it departs from the uniform distribution via the concept of discrepancy.

More precisely, we consider the multivariate version of (5.1), also called the *star discrepancy* in the quasi–Monte Carlo literature [339]. To define this quantity, we first consider all sets of the form

$$B(\mathbf{v}) = \{\mathbf{u} \in [0,1)^s : 0 \le u_j \le v_j, 1 \le j \le s\},$$

where $\mathbf{v} = (v_1, \ldots, v_s) \in [0,1)^s$. We can think of such sets as hyper-rectangles with a corner at the origin. For a point set P_n, we then count how many of its points $\mathbf{u}_i$ fall in that box. That is, we determine the cardinality of the set

$$\{\mathbf{u}_i : 0 \le u_{i,j} \le v_j, i = 1, \ldots, n\}$$

and denote it by $\alpha(P_n, \mathbf{v})$. The empirical distribution induced by P_n assigns a probability of $\alpha(P_n, \mathbf{v})/n$ to this box instead of the value $\prod_{j=1}^s v_j$ assigned by the uniform distribution over $[0,1)^s$. We can thus measure the departure (or discrepancy) of P_n from uniformity by comparing $\alpha(P_n, \mathbf{v})/n$ and $\prod_{j=1}^s v_j$

via the Kolmogorov-Smirnov statistic, which yields the star discrepancy

$$D^*(P_n) = \sup_{\mathbf{v}\in[0,1)^s} |v_1 \dots v_s - \alpha(P_n, \mathbf{v})/n|.$$

Figure 5.2 illustrates the measurement that is performed when computing the star discrepancy.

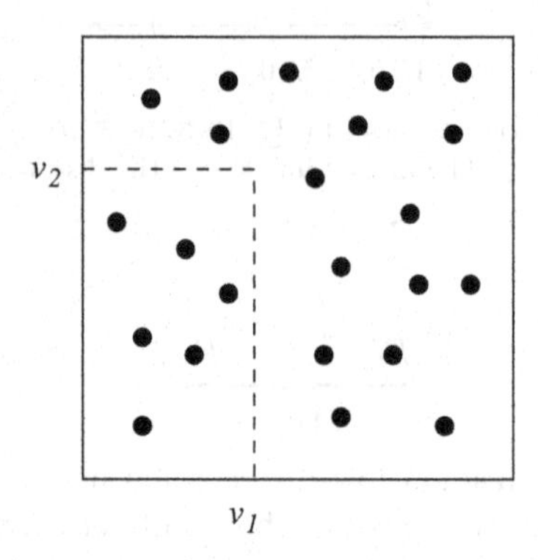

Fig. 5.2 The dotted lines show a box $B(\mathbf{v})$ with $v_1 = 0.4$ and $v_2 = 0.7$. We see that $\alpha(P_n, \mathbf{v}) = 6$ out of $n = 23$ points fall in the box, thus producing a difference $|v_1 v_2 - 6/23| = 0.019$.

For the rectangular grid (5.2), it can be shown that [339, pp. 41–42]

$$D^*(P_n) = 1 - (1 - 1/N)^s, \tag{5.3}$$

and therefore $D^*(P_n) \in O(n^{-1/s})$. Hence, although the star discrepancy of the rectangular grid goes to 0 with n, the convergence is quite slow. By contrast, using the law of iterated logarithms, it can be shown that for a random point set P_n, we have $D^*(P_n) \in O(\sqrt{\log\log n}/\sqrt{n})$ with probability 1, which for $s > 2$ converges to 0 faster than the rectangular grid's star discrepancy (5.3). Note that, when talking about asymptotic rates for the discrepancy, we are implicitly assuming that we are working with a *sequence* of points whose first n points form the set P_n.

Although $D^*(P_n)$ goes to 0 with probability 1 for random point sets, our goal here is to find constructions that avoid the gaps and clusters that typically arise with random point sets, as can be seen in Fig. 5.3 (top left). Now, the question is: By how much can we improve on the random sampling's discrepancy if we use a deterministic construction instead? A widely believed result is that the best possible bound attainable for a deterministic sequence is

$$D^*(P_n) \geq B_s n^{-1} (\log n)^{s-1},$$

where B_s is a constant independent of n [339, p. 32]. This result was proved in the case $s = 2$ by Schmidt [399] but is still a conjecture for $s \geq 3$. Several examples of point sets and sequences of points achieving this bound are known and are typically referred to as *low-discrepancy point sets/sequences.* That is, a sequence of points $\mathbf{u}_1, \mathbf{u}_2, \ldots$ is called a *low-discrepancy sequence* if $D^*(P_n) \in O(n^{-1}(\log n)^s)$, and finite point sets P_n obtained from such constructions are called *low-discrepancy point sets.*

Informally speaking, in this text we think of low-discrepancy point sets as sets of points P_n designed so that for a certain measure of uniformity — not necessarily given by the star discrepancy — they are more uniform than a random point set. The reason why we do not restrict ourselves to the star discrepancy is that this measure is used mostly to look at the asymptotic behavior of sequences of points and is very difficult to compute as soon as the dimension becomes moderately large. We also use the term *quasi–Monte Carlo sampling* (or *low-discrepancy sampling* or *quasi-random sampling*) to refer to the process by which a low-discrepancy point set is used to sample a function, typically for the purpose of integration but possibly for other reasons.

In this chapter and the following one, our goal is to present the main tools required to use low-discrepancy sampling, with an emphasis on topics that are essential to correctly and successfully apply this approach in practice. The current chapter is entirely devoted to presenting the main constructions that are used to perform quasi–Monte Carlo sampling.

The basic principles for constructing low-discrepancy point sets/sequences are presented in Sect. 5.2, and then two main families of constructions — *lattices* and *digital nets and sequences* — are covered in Sects. 5.3 and 5.4, respectively. In addition, the subclass of *recurrence-based point sets* is described in Sect. 5.5. Then we discuss in Sect. 5.6 different uniformity/discrepancy measures that can be used to assess the quality of these point sets. This allows us to make several connections between the two main families of constructions for low-discrepancy point sets/sequences. These measures are also used to present results on the integration error that arises in the context of quasi–Monte Carlo integration.

5.2 Main constructions: basic principles

There are two main families of constructions for low-discrepancy point sets and sequences: *lattices* and *digital nets/sequences.* Before explaining each of them in detail, let us first give the intuition behind these two approaches and describe the basic principles used to define them.

First, the rectangular grid described by (5.2) — and for which an example is shown in Fig. 5.3 (top right) — suffers from the same problem as the point sets used by the trapezoidal rule, which is that when we look at the *projections* of these point sets on each axis, several points map onto each other. That is, in (5.2), if we fix one of the values l_j, then we can find N^{s-1}

points in P_n whose jth coordinate is l_j/N. The impact of this defect on the integration error as s increases was discussed in Chap. 1.

From these observations, it seems clear that one of the properties that a low-discrepancy point set should have is that its projections should also have a low discrepancy. In particular, for a set P_n, it is best if each projection $P_n(I)$ contains n different points. Point sets with this property are said to be *fully projection-regular* [264, 407]. Here, for a given subset $I = \{j_1, \ldots, j_d\} \subseteq \{1, \ldots, s\}$ of indices, the notation $P_n(I)$ refers to the d-dimensional point set

$$P_n(I) = \{(u_{i,j_1}, \ldots, u_{i,j_d}), i = 1, \ldots, n\}.$$

For instance, suppose we have the point set

$$P_n = \{(0,0,0), (1/5, 2/5, 4/5), (2/5, 4/5, 3/5), (3/5, 1/5, 2/5), (4/5, 3/5, 1/5)\}.$$

Then, for $I = \{1, 3\}$, we have

$$P_n(I) = P_n(\{1,3\}) = \{(0,0), (1/5, 4/5), (2/5, 3/5), (3/5, 2/5), (4/5, 1/5)\},$$

and for $I = \{2\}$ we have

$$P_n(I) = P_n(\{2\}) = \{0, 2/5, 4/5, 1/5, 3/5\}.$$

This small point set is fully projection-regular since all its projections contain five points. It is easy to check the cases $I = \{1\}, \{3\}, \{1,2\}, \{2,3\}$ in addition to the two cases shown above. Summarizing, we have the following definition.

Definition 5.1. A point set P_n is *fully projection-regular* if all its projections $P_n(I)$ contain n distinct points.

Note that if P_n is such that each one-dimensional projection $P_n(\{j\})$ contains n points for $j = 1, \ldots, s$, then it is fully projection-regular since by definition $P_n(I)$ has at least as many points as $P_n(\{j\})$ if $j \in I$.

Looking again at the rectangular grid shown in Fig. 5.3 (top right), one way of modifying it so that it can become fully projection-regular would be to work with vectors that are not parallel to the axes when generating the points. That is, one way of building the rectangular grid with 64 points shown in Fig. 5.3 is to look at the two vectors $\mathbf{v}_1 = (1/8, 0)$ and $\mathbf{v}_2 = (0, 1/8)$ and then take all the combinations

$$z_1\mathbf{v}_1 + z_2\mathbf{v}_2, 0 \le z_1, z_2 < 8.$$

Instead, consider for instance the vector $\mathbf{v} = (1/64, 11/64)$. If we take all the multiples $z\mathbf{v}$ mod 1 for $z = 0, \ldots, 63$, where the modulo 1 operation is applied componentwise, then we obtain the point set shown on the lower left of Fig. 5.3. On this graph, we provide the value of z for the first few points just to show the "wraparound" that occurs as a result of the modulo 1 operation. As opposed to the rectangular grid shown in the top-right corner of that figure, we now have 64 points that all map to a different coordinate of the form $i/64$ for $i = 0, \ldots, 63$ on each axis. This particular construction is an example of a *Korobov point set*, introduced by Korobov [224] and Hlawka [191] around 1960, which in turn is a special case of a *lattice point set*. These constructions are discussed in Sect. 5.3. A related construction proposed in 1951 (even before Korobov point sets) for quasi–Monte Carlo integration is the Richtmyer sequence [383], which we will also briefly discuss in Sect. 5.3.

The foundation of digital nets and sequences is based on a completely different idea, which is to define $\mathbf{u}_i$ by looking at the expansion of the index i in a given base $b \geq 2$. More precisely, for a nonnegative integer i, we first write

$$i = \sum_{l=0}^{\infty} a_l(i) b^l,$$

where we assume infinitely many coefficients $a_l(i)$ are zero. We then use the *radical-inverse function in base* b, denoted ϕ_b and defined as

$$\phi_b(i) = \sum_{l=0}^{\infty} a_l(i) b^{-l-1}.$$

Hence $\phi_b(i) \in [0, 1)$. This function is used to define the *van der Corput sequence in base* b, which dates back to 1935 and is the building block for digital nets and sequences [456]. More precisely, the ith term of this sequence is simply given by $\phi_b(i-1)$ for $i \geq 1$. For example, to compute the first few terms of the van der Corput sequence in base 5, we write

$$0 = 0 \times 5^0 + 0 \times 5;\ 1 = 1 \times 5^0 + 0 \times 5;\ \ldots; 5 = 0 \times 5^0 + 1 \times 5; 6 = 1 \times 5^0 + 1 \times 5, \ldots,$$

and then get $u_1 = 0$, $u_2 = 1/5$, $u_3 = 2/5$, $u_4 = 3/5$, $u_5 = 4/5$, $u_6 = 1/25$, $u_7 = 6/25$, $u_8 = 11/25$.

It is useful to notice at this early stage of our discussion how the points in this sequence fill in the interval [0,1). We first place five equidistant points at $j/5$ for $j = 0, \ldots, 4$, then we go back to the origin and place one additional point in each subinterval formed at the previous stage, again spaced at a distance of 1/5, and then repeat this process over and over, with a different position for the first point of the sequence of five.

Also, if we compare the van der Corput sequence with a regular grid, we see at least two big differences. The first is that with the van der Corput sequence we do not need to decide ahead of time how many points n we need. With a grid, P_n is not a subset of P_{n+1} for $n \geq 2$, so if we need more points, we may have to completely reconstruct the point set. The second difference is that the points in the van der Corput sequence are placed in an order that in some sense attempts to never leave wide intervals in $[0, 1)$ containing no points. Such considerations usually do not appear when point sets are constructed with a prefixed cardinality. We will come back to this "space-filling" property in Sect. 5.4.

We give in Fig. 5.3 (bottom right) an example of a digital net based on the Sobol' sequence [415]. Just like for the lattice shown on the left of this point set, here we have 64 points that all map to a different coordinate of the form $i/64$ for $i = 0, \ldots, 63$. The uniformity of this point set does not show up as a lattice structure, but one definitely observes a deterministic pattern when looking at this point set. As we will see in Sect. 5.4, the uniformity is instead measured using the concept of equidistribution.

5.3 Lattices

In Fig. 5.3 (bottom left), we depicted a two-dimensional example of a Korobov point set and briefly described that construction. The more general class to

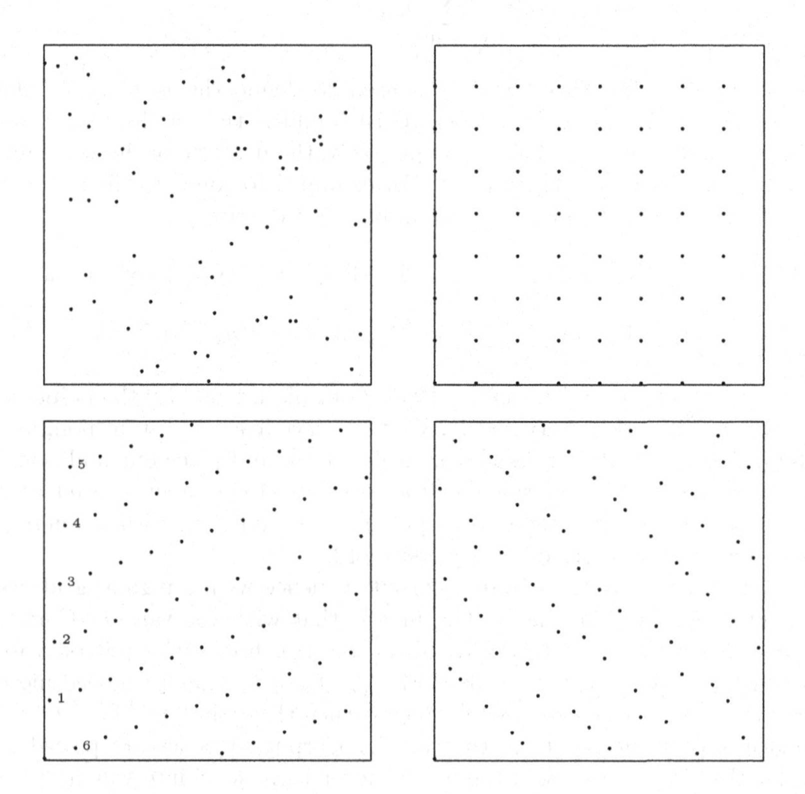

Fig. 5.3 Four different point sets with $n = 64$: random (top left), rectangular grid (top right), Korobov lattice (bottom left), and Sobol' (bottom right).

which this construction belongs is the one that yields *lattice rules*, described in detail in [197, 339, 407, 467]. Because the word "rule" usually refers to an approximation of the form $\sum_{i=1}^{n} f(\mathbf{u}_i)/n$, to describe the actual point set on which these rules are based, we use instead the term "lattice point set", which we now define.

Definition 5.2. For a given dimension s, a *lattice point set* P_n is defined by an *integration lattice* L_s of the form

$$L_s = \{v_1\mathbf{w}_1 + \ldots + v_s\mathbf{w}_s, \mathbf{v} \in \mathbb{Z}^s\},$$

where the s vectors $\mathbf{w}_1, \ldots, \mathbf{w}_s$ in $\mathbb{R}^s$ — which form a *basis* — are linearly independent over the rational numbers and are such that $\mathbb{Z}^s \subseteq L_s$. The corresponding point set is obtained as

$$P_n = L_s \cap [0,1)^s.$$

In other words, the points in P_n are obtained by taking all integer linear combinations of the vectors that fall in $[0,1)^s$. Note that different bases $\mathbf{w}_1, \ldots, \mathbf{w}_s$ can lead to the same point set.

The resulting number of points n in P_n can be shown to be equal to $1/|\det(\mathbf{W})|$, where $\mathbf{W}$ is the $s \times s$ matrix whose ith row is $\mathbf{w}_i$ [407]. The quantity $|\det(\mathbf{W})|$ is called the *determinant of* L and is independent of the basis $\mathbf{W}$ chosen. For instance, for a Korobov point set based on the generator a, we can use

$$\begin{aligned}
\mathbf{w}_1 &= \frac{1}{n}(1, a, a^2 \bmod n, \ldots, a^{s-1} \bmod n)\\
\mathbf{w}_2 &= (0,1,0,\ldots,0)\\
&\vdots\\
\mathbf{w}_s &= (0,\ldots,0,1).
\end{aligned}$$

In this case, it is fairly easy to see that $|\det(\mathbf{W})| = 1/n$, as required.

It can be shown that requiring L_s to be an integration lattice implies that the components of the basis vectors must be rational numbers. In fact, the basis vectors can all be written as fractions of the form l/n, where n is the cardinality of the corresponding lattice point set P_n.

To reduce the number of possible bases, we can use the notion of *rank* r and *invariants* $n_1, \ldots, n_r$, where r is the smallest integer such that we can find invariants satisfying (1) $n_l | n_{l+1}$ for all $l < r$; (2) $n_1 \ldots n_r = n$; and (3) we can write P_n as

$$P_n = \left\{ \left(\frac{i_1}{n_1}\mathbf{z}_1 + \ldots + \frac{i_r}{n_r}\mathbf{z}_r \right) \bmod 1, 0 \le i_l < n_l, l = 1, \ldots, r \right\} \tag{5.4}$$

for some vectors $\mathbf{z}_1, \ldots, \mathbf{z}_r$ in $\mathbb{Z}^s$. Here again, there is not a unique choice for the vectors $\mathbf{z}_1, \ldots, \mathbf{z}_r$, but the rank and invariants are uniquely determined. Hence, in the context of parameter searches for lattice point sets, it is typical to first fix n, s, and the rank r and then search for "good" vectors $\mathbf{z}_1, \ldots, \mathbf{z}_r$. Examples with $r = 2$ and $r = s$ can be found in [412] and [86, 407], respectively. More recent work with $r = 2, 3$ has been done in [231].

Although lattice point sets of higher rank can work well in some settings, in practice rank-1 lattices are more often used. Examples of applications include [40, 132, 214, 354, 402]. Based on the representation (5.4), a rank-1 lattice is determined by a *generating vector* $\mathbf{z} = (z_1, \ldots, z_s)$ of s integers and is then defined as

$$P_n = \left\{ \frac{i}{n}(z_1, \ldots, z_s) \bmod 1, i = 0, \ldots, n-1 \right\}.$$

One advantage that rank-1 lattices have over higher-rank lattices is that they can be made fully projection-regular simply by choosing the integers z_j to be relatively prime with n [264, 407]. That is, we should have $\gcd(z_j, n) = 1$ for each $j = 1, \ldots, s$. By contrast, higher-rank lattices cannot be made fully projection-regular (see Prob. 5.2). For functions that are sums of univariate functions, this difference turns out to be in favor of rank-1 lattices [277]. Figure 5.4 shows a comparison of rank-1 and rank-s lattices for $s = 2$. As can be seen in this figure, one way of constructing a rank-s lattice is to take a small rank-1 lattice, scale it appropriately, and then copy it in each of the 2^s subcubes obtained by partitioning each of the s axes of $[0, 1)^s$ in two. Using point sets defined in this way is usually referred to as a *copy rule* in the context of multivariate integration [86, 407].

As mentioned in Sect. 5.2, a special case of rank-1 lattice is a construction due to Korobov [224] and Hlawka [191] that we call a *Korobov point set.* It has also been called the "good lattice points" method by several authors.

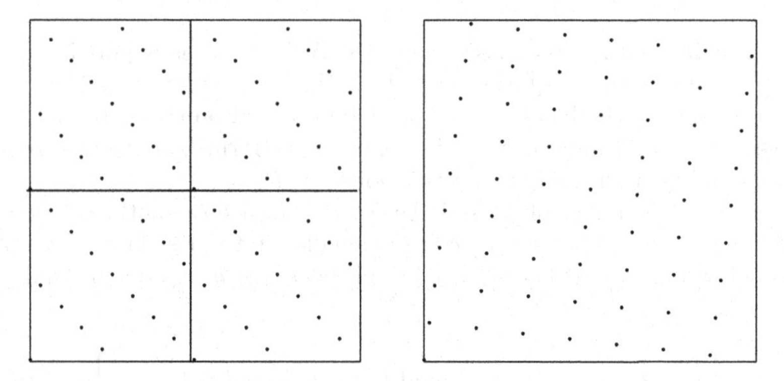

Fig. 5.4 Left: a rank-2 lattice; right: a rank-1 lattice. Both have 64 points. The rank-2 lattice contains the same scaled point set in each of the four squares shown and is therefore not fully projection-regular.

For a given n and dimension s, it is defined by a *generator* a, which is chosen to be an integer between 1 and $n-1$. The point set is then defined as

$$P_n = \left\{ \frac{i}{n}(1, a, a^2 \bmod n, \ldots, a^{s-1} \bmod n) \bmod 1, i = 0, \ldots, n-1. \right\}.$$

Hence, once a is chosen, we simply need to compute the s-dimensional integer vector formed by the successive powers of a (reduced modulo n), and then the n points in P_n are obtained by multiplying this vector by the n numbers $0, 1/n, \ldots, (n-1)/n$. Figure 5.5 gives an example of a small Korobov point set with $n = 10$. Figure 5.6 gives pseudocode to generate the n points of a Korobov point set.

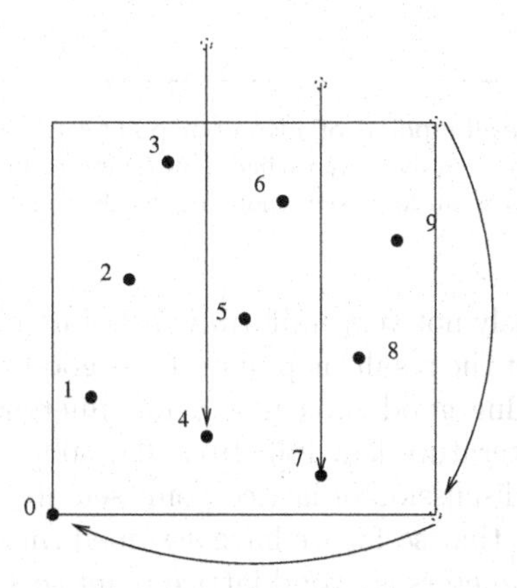

Fig. 5.5 A small Korobov point set with $n = 10$, $a = 3$, and $s = 2$. Arrows show the effect of the modulo 1 operation, including on the point corresponding to $i = 10$.

As can be seen from the code in Fig. 5.6, generating a Korobov point set is simple. And once n and s are chosen, only one parameter — the generator a — must be chosen. The main consideration when choosing a is that first it should be relatively prime with n because otherwise the point set is not fully projection-regular [264, 407]. For example, if $a = 2$ and $n = 8$, we have that

$$P_n(\{2\}) = \{0, 2/8, 4/8, 6/8, 0, 2/8, 4/8, 6/8\},$$

and so $P_n(\{2\})$ only contains four distinct points rather than eight. Because of this requirement, n is often chosen to be a prime number or a power of two. In the first case, a can be any integer between 1 and $n-1$, and in the second case a can be any odd integer between 1 and $n-1$.

Obviously, satisfying the requirement of being relatively prime to n should not be the only criterion used for choosing a. For instance, taking $a = 1$ guarantees that a is relatively prime with n, but the resulting point set ends up having all its points on the diagonal line joining $(0,\ldots,0)$ and $(1,\ldots,1)$ in

```
InitKorobov(a, n, s, z)
    z(1) ← 1
    for j = 2 to s
        z(j) ← a × z(j − 1) mod n
//
NextKorobov(n, z, u) // u is the previous point
    return ((u + z/n) mod 1)
//
GenKorobov(a, n, s)
    u ← 0
    InitKorobov(a, n, s, z)
    for i = 1 to n − 1
        u ← NextKorobov(n, z, u)
```

Fig. 5.6 Code to generate all n points of a Korobov point set. Notice how we compute the successive powers of a in a recursive way rather than trying to first raise it to the power j and then reduce it modulo n, an approach that may easily cause numerical overflows.

$[0,1)^s$, which is obviously not very uniformly distributed. Instead, a should be carefully chosen so that the resulting point set has good uniformity properties. Several tables containing good choices of a for different values of n (and s) can be found in the literature [35, 160, 197, 264, 300].

Going back to our discussion of lattice point sets in general, an important observation to make is that so far we have assumed that the size n was fixed. The tables giving parameters for good lattice point sets that were mentioned in the previous paragraph are usually built by searching, for a fixed n, the best generators according to some quality measure. Using point sets from such tables has the obvious drawback of forcing a user who wants more precision — and thus needs more evaluation points — to start over with a bigger point set. To overcome this problem, Hickernell and his collaborators have proposed a way of constructing *extensible lattice sequences* where, just like for digital sequences, it is possible to increase the number of evaluation points without discarding points previously used [182, 184]. Such sequences are based on rank-1 lattices and are defined so that, for a given base b (usually a prime number), the first b^k points of the sequence form a lattice point set. The key idea in order to define an extensible sequence is to make use of the radical-inverse function in base b, which was used in the definition of the van der Corput sequence on p. 145. More precisely, we have the following definition.

Definition 5.3. An *extensible rank-1 lattice sequence* based on a generating vector $\mathbf{z} = (z_1, \ldots, z_s) \in \mathbb{Z}^s$ has its ith point given by

$$\mathbf{u}_i = \phi_b(i-1)\mathbf{z} \bmod 1, \qquad i \geq 1,$$

where $\phi_b(i)$ is the radical-inverse function in base b applied to i.

Since the radical-inverse function is used to specify the order in which the points occur in the extensible sequence, this order will be different from the standard ordering used in the corresponding finite lattice point sets. This has the advantage that if we use a number of points in the sequence that is not a power of b, then the order of the points in the sequence is such that the corresponding point set is typically more uniform than the first n points of the lattice point set with cardinality equal to the smallest power of b larger than n. We illustrate this with the following example.

Example 5.4. Consider the vector $\mathbf{z} = (1, 7, 49)$ and base $b = 2$. Its corresponding extensible lattice sequence in dimension 3 starts off as

$$\begin{aligned}
\phi_2(0)(1,7,49) \bmod 1 &= (0,0,0),\\
\phi_2(1)(1,7,49) \bmod 1 &= (1/2,1/2,1/2),\\
\phi_2(2)(1,7,49) \bmod 1 &= (1/4,3/4,1/4),\\
\phi_2(3)(1,7,49) \bmod 1 &= (3/4,1/4,3/4),\\
\phi_2(4)(1,7,49) \bmod 1 &= (1/8,7/8,1/8),\\
\phi_2(5)(1,7,49) \bmod 1 &= (5/8,3/8,5/8),\\
\phi_2(6)(1,7,49) \bmod 1 &= (3/8,5/8,3/8),\\
\phi_2(7)(1,7,49) \bmod 1 &= (7/8,1/8,7/8).
\end{aligned}$$

By contrast, if we use the "standard" ordering given by $(i/n)\mathbf{z} \bmod 1$, for $n = 8$, we instead have

$$\begin{aligned}
&(0,0,0),(1/8,7/8,1/8),(1/4,3/4,1/4),(3/8,5/8,3/8),\\
&(1/2,1/2,1/2),(5/8,3/8,5/8),(3/4,1/4,3/4),(7/8,1/8,7/8).
\end{aligned}$$

We get the same eight points, but in a different order. In the second case, note that the first four points have their first and third coordinates smaller than 1/2. This means that these eight points start by filling the dyadic box $[0,1/2] \times [0,1] \times [0,1/2]$. This is not the case with the extensible lattice, which instead alternates nicely between the two half-intervals $[0,1/2)$ and $[1/2,1)$ for each coordinate. Hence, if we were to use only the first, say, five points in each case, we would get a point set with better properties by using the first five points of the sequence rather than the first five points of the finite lattice of size 8 based on a standard ordering.

Let us now turn to the choice of the generating vector $\mathbf{z}$ for extensible lattice sequences. A practical way to choose it is to first restrict the search to *extensible Korobov lattices* — which are extensible lattices based on a generating vector of the form $(1, a, \ldots, a^{s-1})$ — and then to fix the dimension s and a range $[l_1, \ldots, l_2]$ of powers of b to examine [186]. Then, by defining a global measure that assesses the quality of P_{b^k} for $l_1 \le k \le l_2$ in dimension s, computer searches aimed at finding an optimal generator a can be performed. More recent work in this area that also provides a few examples of good generators can be found in [141].

An extensible lattice sequence has some similarities with the *Richtmyer sequence*, which was one of the early constructions proposed for quasi–Monte Carlo integration [383]. This sequence can be described as follows. Choose a vector $\boldsymbol{\alpha} = (\alpha_1, \ldots, \alpha_s)$ of irrational real numbers such that $1, \alpha_1, \ldots, \alpha_s$ are linearly independent over the rationals. Then use the sequence

$$\mathbf{u}_i = (i-1)\boldsymbol{\alpha} \bmod 1, \qquad i \geq 1,$$

where the modulo 1 operation is applied componentwise. For instance, if $s = 2$ and we take

$$(\alpha_1, \alpha_2) = (2\cos 2\pi/7, 2\cos 4\pi/7) = (1.247, 2.494),$$

then we get the points $\mathbf{u}_1 = (0,0), \mathbf{u}_2 = (0.247, 0.494), \mathbf{u}_3 = (0.494, 0.988)$, and so on [333, p. 994]. One of the differences with the extensible Korobov sequence is that here the "generating vector" $\boldsymbol{\alpha}$ is based on irrational real numbers. Note that if $\boldsymbol{\alpha}$ were made up of rational numbers — that is, if we had $\alpha_j = p_j/q_j$ for some integers p_j, q_j — then the jth coordinate in the sequence $\mathbf{u}_1, \mathbf{u}_2, \ldots$ would only map to the q_j different values $0, 1/q_j, \ldots, (q_j - 1)/q_j$.

A question related to the ability of increasing the number of points in a lattice is the notion of lattices that are extensible in the dimension. That is, one might be interested in rank-1 lattices with generating vectors $\mathbf{z} \in \mathbb{Z}^s$ to which additional coordinates $z_{s+1}, z_{s+2}, \ldots$ can be added if needed while preserving the good quality of the lattice. Such *component-by-component* constructions have been devised by Sloan and his collaborators in several papers over the last few years [229, 230, 231, 409, 410, 411]. Specific parameters for dimensions up to $d = 100$ can be found in [409, 410, 411]. Typically, these constructions are such that a certain quality measure — usually related to an error bound for a certain class of functions — remains bounded as the dimension of the lattice increases. The development of these component-by-component constructions makes heavy use of existence results for lattice rules that were derived in the context of *tractability*, as we will discuss at the end of Chap. 6. That is, once it is known that it is possible to find a lattice rule, say based on a rank-1 lattice, such that the corresponding integration error "behaves well" for a certain class of functions, then the idea is to devise an algorithm that can actually find that lattice.

The component-by-component approach can also be used for extensible lattice sequences [62, 84] by applying an important existence result shown by Hickernell and Niederreiter [188]. In particular, parameters for extensible rank-1 lattices (not restricted to Korobov) are given in [62].

5.4 Digital nets and sequences

As we mentioned in Sect. 5.2, digital sequences are constructed so that their ith point makes use of the expansion of $i-1$ in a certain base. Recall that for $b \geq 2$ we first defined the *radical-inverse function in base* b, denoted ϕ_b, as

$$\phi_b(i) = \sum_{l=0}^{\infty} a_l(i) b^{-l-1},$$

where the coefficients $a_l(i)$ come from the expansion

$$i = \sum_{l=0}^{\infty} a_l(i) b^l$$

and where we assume infinitely many coefficients $a_l(i)$ are zero. Recall also that the ith term of the *van der Corput sequence in base* b is given by $\phi_b(i-1)$, so that, for example, the first few terms of the van der Corput sequence in base 5 are $u_1 = 0$, $u_2 = 1/5$, $u_3 = 2/5$, $u_4 = 3/5$, $u_5 = 4/5$, $u_6 = 1/25$, $u_7 = 6/25$, $u_8 = 11/25$.

A few remarks are in order before going further.

(1) With sequences like that, the order in which the points are defined matters because it affects the *space-filling* performance of the sequence. For instance, if the sequence instead started as

$$0, 2/5, 4/5, 1/5, 3/5, 11/25, 21/25,$$

gaps where there are no points would shrink faster.

(2) For the van der Corput sequence, when the base b gets larger, the space-filling performance of the sequence gets worse because we move more slowly from 0 to 1 when placing a cycle of b points.

(3) Following the discussion in item (1), a possibility for improving the space-filling performance is to try to change the order of the points in the sequence, and a natural way to do this is to permute the base b digits of i used to construct the points. More details on this approach will be given in Sects. 5.4.4 and 6.2.3.

Now, to extend the van der Corput sequence to a multidimensional sequence in $[0, 1)^s$, two approaches come to mind. The first idea is to use a different base b for each of the s coordinates. This is precisely what the *Halton sequence* does [161], where typically the jth prime number is used as the base b_j for the jth coordinate. Hence the ith term in this sequence is given by

$$\mathbf{u}_i = (\phi_{b_1}(i-1), \ldots, \phi_{b_s}(i-1)),\ i \geq 1.$$

The star discrepancy of this sequence can be shown to be in $O(n^{-1}(\log n)^s)$ [161], which implies that the Halton sequence qualifies as a low-discrepancy sequence. Related to the Halton sequence is the *Hammersley point set* [163], which for a given n is defined as

$$P_n = \left\{ \mathbf{u}_i = \left(\frac{i-1}{n}, \phi_{b_1}(i-1), \ldots, \phi_{b_{s-1}}(i-1) \right), i = 1, \ldots, n \right\}$$

and for which the star discrepancy is in $O(n^{-1}(\log n)^{s-1})$.

From the remarks we made earlier, one can already suspect that the large bases used by the Halton sequence in high dimensions might cause this sequence not to have such good space-filling properties. This is illustrated in Fig. 5.7, where we show the 49th and 50th coordinates of the first 1000 points of the Halton sequence. As we can see there, for this projection, the first 1000 points are concentrated on the main diagonal of $[0,1)^2$, with very large areas in $[0,1)^2$ containing no points.

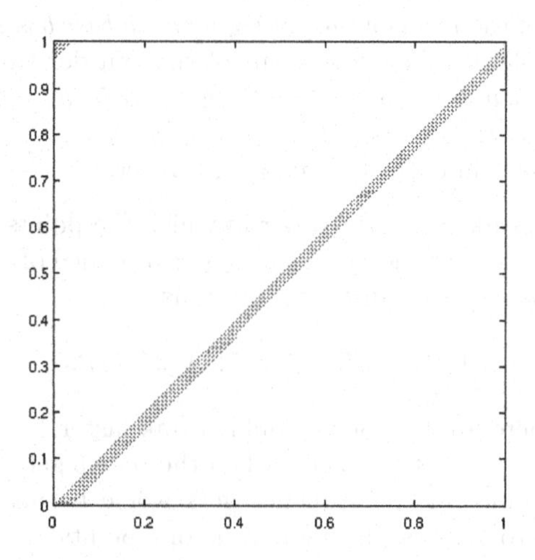

Fig. 5.7 First 1000 points of the Halton sequence, for the 49th and 50th coordinates, based on $b_{49} = 227$ and $b_{50} = 229$.

To overcome this problem, a second idea is to try to use the same — possibly small — base for each coordinate. To do that, we need to use something other than just the radical-inverse function to determine each coordinate because otherwise all the coordinates of a given point would be the same. One possibility is to apply a linear transformation to the digits $a_l(i)$ coming from the expansion of i in base b before they are input into the radical-inverse function. This is the idea that was used by Sobol' in 1967 to define his *LP_τ-sequence* [415], where the base b is 2 and different carefully chosen linear transformations are used for each coordinate. The Faure sequence [112] is also based on this idea, but for any prime base b.

The generalization of these constructions is what is now known as *digital sequences* and was introduced by Niederreiter in 1987 [335]. A recent survey that also contains new results can be found in [341]. To keep things simple, here we define a special case of the general definition of these sequences found in [339] and [441]. By doing so, we wish to emphasize how this is just the idea mentioned above of applying linear transformations to the digits $a_l(i)$ of i before using the radical-inverse function. The parameters required to define a digital sequence are a base b and s *generating matrices* of infinite size.

Definition 5.5. Let b be a prime number and $s \geq 1$ and $k \geq 1$ be integers. Assume we have s *generating matrices* $C_1, \ldots, C_s$ of dimension $\infty \times \infty$ with entries in $\mathbb{Z}_b$. Let

$$i = \sum_{l=0}^{\infty} a_l(i) b^l$$

with $a_l(i) \in \mathbb{Z}_b$ be the digit expansion of i in base b, and define the vector

$$(\tilde{a}_{j,0}(i), \tilde{a}_{j,1}(i), \ldots)^{\mathrm{T}} = C_j \cdot (a_0(i), a_1(i), \ldots)^{\mathrm{T}}$$

for each $j = 1, \ldots, s$. The jth coordinate of the ith point of the *digital sequence* based on $C_1, \ldots, C_s$ is given by

$$u_{ij} = \sum_{l=0}^{\infty} \tilde{a}_{j,l}(i-1) b^{-l-1}$$

for $i \geq 1$ and $j = 1, \ldots, s$.

The more general definition does not restrict b to be a prime and assumes the generating matrices are defined over a commutative ring R with cardinality b. It also applies bijections T_l from $\mathbb{Z}_b$ to R to the digits $a_l(i)$ before multiplying them by the matrices C_j and then other bijections $T_{j,l}$ from R to $\mathbb{Z}_b$ to the digits $\tilde{a}_{j,l}(i)$ before defining u_{ij}. That is, we need a ring R to perform additions and multiplications, and this can be viewed as our "working space". The set $\mathbb{Z}_b$ is then used just for handling input (the index i) and output (the digits defining $u_{i,j}$).

An important result that probably at least partly motivated the definition of these sequences is that they can be shown to have a star discrepancy $D^*(P_n)$ in $O(n^{-1}(\log n)^s)$, meaning that the sequences thus produced can be considered low-discrepancy sequences [335].

A *digital net* is a point set P_n based on the same principles as digital sequences, the only difference being that the generating matrices now only need a finite number of columns. That is, if the number of points is $n = b^k$ for some $k \geq 1$, then the generating matrices only need k columns since the expansion of i in base b requires at most k digits $a_0(i), \ldots, a_{k-1}(i)$ in this case. Most digital nets used in practice come from digital sequence constructions, although there are some specific net constructions that have been proposed. They will be discussed in Sect. 5.4.5.

When referring to digital sequences, in addition to their base b and dimension s, they are often labeled according to the t-value discussed in Chap. 3 for the case $b = 2$. To introduce this quality parameter in a general base b, we need the following definition, which is simply the generalization of Def. 3.9 introduced for $b = 2$ in Chap. 3.

Definition 5.6. Let $q_1, \ldots, q_s$ be nonnegative integers, and let $q=q_1+\ldots+q_s$. A point set P_n with $n = b^k$ points is $(q_1, \ldots, q_s)$-*equidistributed in base* b if every cell (or elementary interval) of the form

$$J(\mathbf{r}) := \prod_{j=1}^{s} \left[\frac{r_j}{b^{q_j}}, \frac{r_j + 1}{b^{q_j}} \right), \tag{5.5}$$

for $0 \leq r_j < b^{q_j}, j = 1, \ldots, s$, contains b^{k-q} points from P_n.

Also, for a given vector $(q_1, \ldots, q_s)$, the set of all cells of the form $J(\mathbf{r})$ is called a $(q_1, \ldots, q_s)$-*partition*.

We can now define the concepts of (t, k, s)-nets and (t, s)-sequences.

Definition 5.7. A set P_n containing $n = b^k$ points is called a (t, k, s)-*net in base* b if it is $(q_1, \ldots, q_s)$-equidistributed in base b whenever $q \leq k - t$ [335]. A (t, s)-sequence is a sequence of points for which each b-ary segment of the form $\mathbf{u}_{lb^k}, \ldots, \mathbf{u}_{(l+1)b^k-1}$ with $k \geq t$ and $l \geq 0$ is a (t, k, s)-net in base b. We refer to the smallest value of t for which P_n is a (t, k, s)-net as the t-*value* of P_n and similarly for sequences.

Sobol' was the first to introduce the concept of t-value (in base 2) as a way of characterizing the uniformity of his sequence. The constructions that were proposed later were often motivated by the desire to improve this quality measure. For example, Faure proposed sequences with $t = 0$ [112], and Niederreiter first proposed sequences with better bounds on the t-value than those for the Sobol' sequence [336] and later proposed with Xing (t, s)-sequences with $t \in O(s)$, which is the optimal convergence order in an arbitrary base for t as a function of s [345, 346, 347]. Because of the important historic role that this quality parameter has played in the development of digital nets and sequences, we chose to discuss it now rather than waiting until Sect. 5.6, where we describe quality measures for low-discrepancy point sets.

In particular, the t-value (or at least upper bounds on it) appears in upper bounds for the implied constant c_s of the star discrepancy $D^*(P_n)$ of the first n points of a (t, s)-sequence, as can be seen for instance in [336, p. 53]. There, the following upper bound is given on the discrepancy of the first n points of a (t, s)-sequence:

$$D^*(P_n) \leq c_s (\log n)^s + O((\log n)^{s-1}), \tag{5.6}$$

where for $s \geq 4$ we have the general formula

$$c_s = \frac{1}{s!} b^t \frac{b-1}{2\lfloor b/2 \rfloor} \left(\frac{\lfloor b/2 \rfloor}{\log b} \right)^s .$$

(These bounds have recently been improved by a factor of 1/2 or 1/3 — depending on b and s — in [225]). Thus it is important to be able to assess how the t-value behaves as s increases in order to be able to say something about the behavior of c_s as s increases. A useful tool for finding constructions with a small t-value is the MinT database created by Wolfgang Schmid and Rudolf Schürer [400, 489]. (This database also contains very detailed and valuable information on a large number of constructions and is updated regularly.)

In what follows, we start by discussing sequences and then talk about net constructions. We already described the Halton sequence, which can be thought of as a precursor to digital sequences, although it does not exactly fit their framework since a different base b is used in each dimension. In chronological order, the digital sequences that were first proposed were the Sobol' sequence, the Faure sequence, and the Niederreiter sequences. They are discussed next.

5.4.1 Sobol' sequence

As mentioned before, this sequence is defined in base $b = 2$. For each coordinate j, it first requires a primitive polynomial in $\mathbb{F}_2$ that we denote $p_j(z)$ and write out as

$$p_j(z) = z^{d_j} + a_{j,1} z^{d_j - 1} + \ldots + a_{j,d_j},$$

where each $a_{j,l} \in \mathbb{F}_2$ and d_j is the degree of $p_j(z)$. We then need d_j *direction numbers* of the form

$$v_{j,r} = \frac{m_{j,r}}{2^r},$$

where $m_{j,r}$ is an odd integer between 1 and $2^r - 1$ for $r = 1, \ldots, d_j$. The binary expansion of the numbers $v_{j,r}$ is used to determine the generating matrices of this sequence and is written as

$$v_{j,r} = v_{j,r,1} 2^{-1} + v_{j,r,2} 2^{-2} + \ldots + v_{j,r,d_j} 2^{-d_j}.$$

Once these d_j direction numbers are chosen, the following ones are obtained through the recurrence

$$v_{j,r} = a_{j,1} v_{j,r-1} \oplus \ldots \oplus a_{j,d_j-1} v_{j,r-d_j+1} \oplus v_{j,r-d_j} \oplus (v_{j,r-d_j}/2^{d_j}), \quad (5.7)$$

where $\oplus$ represents the addition of vectors with components in $\mathbb{F}_2$ or, computationally speaking, the exclusive-or operation on binary vectors.

The rth column of C_j is then formed by the base 2 expansion of $v_{j,r}$. That is, each direction number is assigned to a column of C_j and fills it with its binary representation. By the definition of the initial vectors $v_{j,1}, \ldots, v_{j,d_j}$ and the recurrence (5.7) used to obtain the next ones, one can see that each C_j is a nonsingular upper-triangular matrix. In turn, this implies that each one-dimensional projection of the Sobol' sequence is a $(0,1)$-sequence [415, Remark 3.5], and thus the corresponding (t,k,s)-nets are fully projection-regular for all $k \geq 0$. Figure 5.8 shows the first 256 points of the two-dimensional Sobol' sequence.

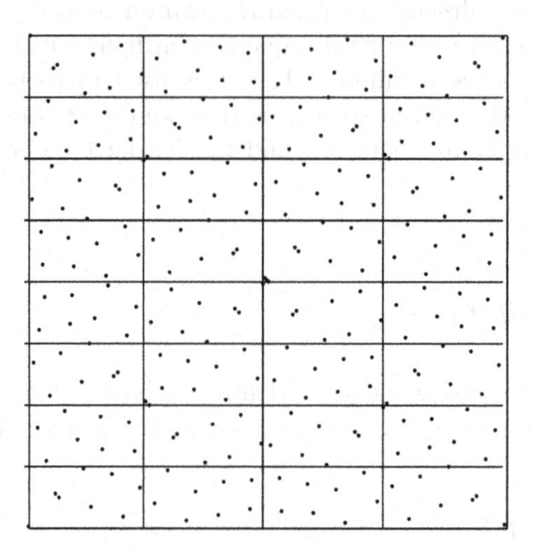

Fig. 5.8 Dyadic partition induced by $q_1 = 2$ and $q_2 = 3$. For the first 256 points of the Sobol' sequence, we have eight points in each box.

Let us look at a simple case to illustrate how this construction works. Suppose that, for $j = 3$, we take $p_3(z) = z^2 + z + 1$. Since the degree of $p_3(z)$ is two, we need to choose two direction numbers. Take $v_{3,1} = 1/2$ and $v_{3,2} = 3/4$. In vector representation, it means $v_{3,1} = (1,0)$ and $v_{3,2} = (1,1)$. Then, from the definition of $p_3(z)$, we have that $a_{3,1} = a_{3,2} = 1$, and so

$$v_{3,3} = \begin{pmatrix} 1 \\ 1 \end{pmatrix} \oplus \begin{pmatrix} 1 \\ 0 \end{pmatrix} \oplus \begin{pmatrix} 0 \\ 0 \\ 1 \end{pmatrix} = \begin{pmatrix} 0 \\ 1 \\ 1 \end{pmatrix},$$

$$v_{3,4} = \begin{pmatrix} 0 \\ 1 \\ 1 \end{pmatrix} \oplus \begin{pmatrix} 1 \\ 1 \end{pmatrix} \oplus \begin{pmatrix} 0 \\ 0 \\ 1 \\ 1 \end{pmatrix} = \begin{pmatrix} 1 \\ 0 \\ 0 \\ 1 \end{pmatrix}.$$

Therefore, the first four rows and four columns of C_3 are

$$\begin{pmatrix} 1\,1\,0\,1 \\ 0\,1\,1\,0 \\ 0\,0\,1\,0 \\ 0\,0\,0\,1 \end{pmatrix},$$

and the corresponding sequence thus starts as 0, 1/2, 3/4, 1/4, 3/8, 7/8, 5/8, 1/8, 9/16, 1/16, 5/16, 13/16, 15/16, 7/16, 3/16, 11/16.

As shown in [415, Thm. 3.3], the Sobol' sequence in dimension s based on primitive polynomials $p_1(z), \ldots, p_s(z)$ of respective degrees $d_1, \ldots, d_s$ is a (t, s)-sequence with

$$t = \sum_{j=1}^{s} (d_j - 1). \tag{5.8}$$

For this reason, the primitive polynomials $p_j(z)$ are sorted by increasing degree. With that assumption, we have that $t \in O(s \log s)$. Sobol' gives conditions under which the bound (5.8) on t is tight [415, Sect. 4]. That is, the right-hand side of (5.8) is equal to the t-value in those cases. Note that it is not tight for the jth one-dimensional projection of the sequence because the t-value is given by 0 in that case rather than by $d_j - 1$.

Although the direction numbers do not affect the value of this upper bound on t, their choice affects the quality of portions of the sequence of finite size (see Prob. 5.6). The direction numbers given by Sobol' in [420] satisfy an equidistribution criterion that he calls "Property A", which means the first 2^s points of the sequence are $(1, 1, \ldots, 1)$-equidistributed. Similarly, his Property A′ means the first 2^{2s} points are $(2, 2, \ldots, 2)$-equidistributed. Alternatively, using the terminology introduced in Chap. 3, we can say that Property A means the resolution of the first 2^s points is 1 and thus maximal. Similarly, Property A′ means the resolution of the first 2^{2s} points is 2. Direction numbers for up to $s \leq 50$ are given by Sobol' and his collaborators in [420, 422]. Direction numbers for $s > 50$ that also satisfy certain equidistribution properties can be found in [203, 207, 208, 279]. A detailed implementation of the Sobol' sequence is provided in [43], but direction numbers only up to $s = 40$ are given there.

A last word about the Sobol' sequence: To make the implementation more efficient, Antonov and Saleev [8] have shown that permuting the order of the points according to a Gray code is very helpful. More precisely, rather than using the binary expansion $(a_0(i), a_1(i), \ldots)$ of i to determine the $(i + 1)$th point in the sequence, the binary expansion $(g_0(i), g_1(i), \ldots)$ of $g(i) \in \mathbb{N}_0$ is used, where $g(\cdot)$ is the Gray code function. This function satisfies $g(0) = 0$, and $g(i + 1)$ is such that its binary expansion differs from that of $g(i)$ in only one position: If c is the smallest index such that $a_c(i) \neq b - 1$, then $g_c(i)$ is the digit whose value changes, and it becomes $g_c(i) + 1$ in the expansion for

$g(i+1)$ [441, Theorem 6.6]. That is, $g_c(i+1) = g_c(i) + 1$. We illustrate the use of the Gray code with the following example.

Example 5.8. Consider the first eight points of the Sobol' sequence. We have that $g(0) = 0$. Since $a_0(0) = 0$, then $c = 0$ is the smallest index such that $a_c(0) < 1$, and thus $g(1)$ has the expansion $(g_0(1) = 1, g_1(1) = 0, g_2(1) = 0, \ldots)$. Similarly, we find that $c = 1$ is the smallest index such that $a_c(1) < 1$, so that $g(2)$ has the expansion $(1, 1, 0, \ldots)$. For $i = 3$, $c = 0$ is the smallest index such that $a_c(2) < 1$, and so $g(3)$ has the expansion $(0, 1, 0, 0, \ldots)$. In a similar manner, we get the expansions

$$\begin{aligned} g(4) &: (0,1,1,0,\ldots), \\ g(5) &: (1,1,1,0,\ldots), \\ g(6) &: (1,0,1,0,\ldots), \\ g(7) &: (0,0,1,0,\ldots). \end{aligned}$$

Thus, with the Gray code, we enumerate the points from the original ordering of the Sobol' sequence in the order 1, 2, 4, 3, 7, 8, 6, 5.

Given the fact that using a Gray code only modifies the order of the points over the first 2^i points for each $i \geq 0$, it can be shown that the sequence thus obtained is still a (t, s)-sequence with the same value of t as the Sobol' sequence [8, 441]. In fact, as shown by Tezuka in [441], using a Gray code in base b is equivalent to premultiplying from the left the vector of coefficients $(a_0(i), a_1(i), \ldots,)^{\mathrm{T}}$ by a matrix G given by

$$G = \begin{pmatrix} 1 & b-1 & 0 & \ldots\ldots \\ 0 & 1 & b-1 & 0 \ \ldots \\ & & \ldots & \end{pmatrix}. \tag{5.9}$$

This is the same as multiplying the generating matrices C_j from the right by G for each $j = 1, \ldots, s$. Verifying again with the first eight numbers in the Sobol' sequence, we see that

$$G \begin{pmatrix} a_0(i) \\ a_1(i) \\ \vdots \end{pmatrix},$$

is successively given by

$$\begin{aligned} G(0,0,\ldots,0)^{\mathrm{T}} &= (0,0,\ldots)^{\mathrm{T}}, \\ G(1,0,\ldots)^{\mathrm{T}} &= (1,0,0,\ldots)^{\mathrm{T}}, \\ G(0,1,0,\ldots)^{\mathrm{T}} &= (1,1,0,\ldots)^{\mathrm{T}}, \end{aligned}$$

$$\begin{aligned}
G(1,1,0,\ldots)^{\mathrm{T}} &= (0,1,0,\ldots,)^{\mathrm{T}},\\
G(0,0,1,0,\ldots)^{\mathrm{T}} &= (0,1,1,0,\ldots)^{\mathrm{T}},\\
G(1,0,1,0,\ldots)^{\mathrm{T}} &= (1,1,1,0,\ldots)^{\mathrm{T}},\\
G(0,1,1,0,\ldots)^{\mathrm{T}} &= (1,0,1,0,\ldots)^{\mathrm{T}},\\
G(1,1,1,0,\ldots)^{\mathrm{T}} &= (0,0,1,0,\ldots)^{\mathrm{T}},
\end{aligned}$$

thus getting the order 1, 2, 4, 3, 7, 8, 6, 5 just as before.

5.4.2 Faure sequence

A natural question that arises after having seen the definition of the Sobol' sequence and the definition of the t-value is: Can we construct sequences for which $t = 0$? As mentioned before, this question was answered by Henri Faure in 1982 [112] when he presented a method to construct a sequence with $t = 0$ in any prime base b, with the dimension s satisfying $s \leq b$. In base b, the generating matrix C_j for the Faure sequence is given by the transpose of the Pascal matrix (with calculations done in $\mathbb{F}_b$) raised to the power $j - 1$ for $j = 1, \ldots, s$. Faure uses properties of Vandermonde matrices to show that the construction thus obtained has $t = 0$. As was the case for the Sobol' sequence, using a b-ary Gray code is helpful in implementing this sequence [440, 441]. Pseudocode is given in [441, p. 196].

As discussed in Prob. 5.9, it can be shown that, for each subset $I \subseteq \{1, \ldots, s\}$, the corresponding projection of the sequence over I is a $(0, d)$-sequence, where $d = |I|$. In particular, just as for the Sobol' sequence, each one-dimensional projection of the Faure sequence is a $(0, 1)$-sequence and thus the corresponding $(0, k, s)$-nets are fully projection-regular for $k \geq 0$.

The following example describes the Faure sequence in a very simple case.

Example 5.9. Suppose we take $s = b = 3$ and truncate the generating matrices to 3×3 matrices. Then

$$C_1 = \begin{pmatrix} 1 & 0 & 0 \\ 0 & 1 & 0 \\ 0 & 0 & 1 \end{pmatrix} \qquad C_2 = \begin{pmatrix} 1 & 1 & 1 \\ 0 & 1 & 2 \\ 0 & 0 & 1 \end{pmatrix} \qquad C_3 = \begin{pmatrix} 1 & 2 & 1 \\ 0 & 1 & 1 \\ 0 & 0 & 1 \end{pmatrix}.$$

Therefore (and not using a b-ary Gray code), the first five points are obtained as follows. As usual, we first have $\mathbf{u}_1 = (0, 0, 0)$, and then

$$C_1 \begin{pmatrix} 1 \\ 0 \\ 0 \end{pmatrix} = \begin{pmatrix} 1 \\ 0 \\ 0 \end{pmatrix}$$

so that $u_{2,1} = 1/3$. Similarly, since the first columns of C_2 and C_3 are given by $(1, 0, 0)^{\mathrm{T}}$, then $u_{2,2} = u_{2,3} = 1/3$. For the third point, we have

$$C_1 \begin{pmatrix} 2 \\ 0 \\ 0 \end{pmatrix} = \begin{pmatrix} 2 \\ 0 \\ 0 \end{pmatrix}$$

so that $u_{3,1} = 2/3$. Again, since C_2 and C_3 also have their first columns given by $(1, 0, 0)$, then $\mathbf{u}_3 = (2/3, 2/3, 2/3)$. For $\mathbf{u}_4$, we have that

$$C_1 \begin{pmatrix} 0 \\ 1 \\ 0 \end{pmatrix} = \begin{pmatrix} 0 \\ 1 \\ 0 \end{pmatrix},$$

and so $u_{4,1} = 1/9$. Similarly, since the second columns of C_2 and C_3 are $(1, 1, 0)$ and $(2, 1, 0)$, then $u_{4,2} = 1/3+1/9 = 4/9$ and $u_{4,3} = 2/3+1/9 = 7/9$, so that $\mathbf{u}_4 = (1/9, 4/9, 7/9)$. Continuing with $i = 5$, we get

$$C_1 \begin{pmatrix} 1 \\ 1 \\ 0 \end{pmatrix} = \begin{pmatrix} 1 \\ 1 \\ 0 \end{pmatrix},$$

so that $u_{5,1} = 1/3 + 1/9 = 4/9$. Similarly, since the sums of the first two columns of C_2 and C_3 are $(2, 1, 0)$ and $(0, 1, 0)$ respectively, this means $\mathbf{u}_5 = (4/9, 7/9, 1/9)$.

In addition to having a t-value equal to 0, another advantage of the Faure sequence over the Sobol' sequence is that the implied constant c_s in the star discrepancy bound (5.6) satisfies $\log c_s \in O(-s \log \log s)$ and thus goes to 0 exponentially fast with s. By contrast, for the Sobol' sequence, the best known bound is $\log c_s \in O(s \log \log s)$.

Although from the point of view of the t-value the Faure sequence is better than the Sobol' sequence, it is important to understand what exactly the t-value measures when looking at finite sets P_n. For instance, suppose $s = 360$ and we take $b = 367$ for the Faure sequence. If we look at its first 367^2 =134,689 points, then the fact that $t = 0$ means that all one- and two-dimensional projections have optimal equidistribution in base $b = 367$. For any u-dimensional projection with $u > 2$, we cannot say the corresponding t-value is 0 because even a $(1, 1, 1, 0, \ldots, 0)$-partition produces more boxes than points. Hence it is not clear that overall the Faure sequence is better in this setting, where s is large and n is not extremely big.

Also, while the Sobol' and Halton sequences are extensible in their dimension, the Faure sequence cannot be extended in this way since the base b must be at least as large as the dimension s. That is, for the Faure sequence, if we first construct the sequence in dimension s with a base $b \geq s$ and then decide to increase s to $s_1 > b$, then we need to choose a new base $\tilde{b} \geq s_1$ in order to define a new Faure sequence in dimension s_1. By contrast, with Sobol' and Halton, if we want to increase the dimension from s to s_1, we just need to compute additional parameters — new bases for Halton and new direction numbers for Sobol' — and can then simply extend the points from the previously constructed s-dimensional sequence by adding $s_1 - s$ new coordinates.

5.4.3 Niederreiter sequences

In addition to providing a general framework describing digital nets and sequences, Niederreiter proposed in [336] a digital sequence for arbitrary bases b that are a power of a prime that makes use of formal Laurent series. This construction includes the Faure sequence as a special case, but not the Sobol' sequence. The material presented in App. A may be useful for understanding what follows.

To define the Niederreiter sequence in base b, several "ingredients" are needed. First, we must choose s pairwise coprime polynomials $p_j(z)$ over $\mathbb{F}_b[z]$ for $j = 1, \ldots, s$. Let $e_j \geq 1$ be the degree of $p_j(z)$. Then, for each dimension j, we must also choose a sequence of polynomials $g_{j,m}(z)$ for $m \geq 1$ such that $\gcd(g_{j,m}(z), p_j(z)) = 1$ for all $1 \leq j \leq s$ and $m \geq 1$. Once we have those, we then build the generating matrices using the coefficients $a^j(m,k,r) \in \mathbb{F}_b$ from expansions of the form

$$\frac{z^k g_{j,m}(z)}{(p_j(z))^m} = \sum_{r=w}^{\infty} a^j(m,k,r) z^{-r} \tag{5.10}$$

for $0 \leq k < e_j$, $m \geq 1$, and $1 \leq j \leq s$. (The coefficient $w \leq 0$ may depend on j, m, k.) More precisely, the lth row of the jth generating matrix C_j is determined first by computing the pair (q, u) arising from

$$l - 1 = q e_j + u.$$

That is, $u = (l-1) \bmod e_j$ and $q = \lfloor (l-1)/e_j \rfloor$. Once we have the pair (q, u), we then construct the lth row of C_j as

$$(c_{j,l,1}, c_{j,l,2}, \ldots) = (a^j(q+1, u, 1), a^j(q+1, u, 2), \ldots). \tag{5.11}$$

That is, q determines which power of $p_j(z)$ and which polynomial $g_{j,m}(z)$ are used in the fraction shown on the left-hand side of (5.10), and u determines which power of z is used in the numerator of that same fraction. Then, the whole row contains the coefficients of the expansion of that fraction.

Now, using the fact that the coefficients a_r in a quotient of the form

$$\frac{p(z)}{P(z)} = \sum_{r=w} a_r z^{-r}$$

follow a recurrence whose characteristic polynomial is $P(z)$ and that the numerator $p(z)$ is used to initialize this recurrence, we can give the following intuitive description of a given generating matrix C_j. From (5.11), we have that its first group of e_j rows each have elements that follow a recurrence determined by $p_j(z)$, and each row is initialized differently as the powers z^k go from 0 to $e_j - 1$. The polynomial $g_{j,1}(z)$ is also used for the initialization of this first group. For the second group of e_j rows, the recurrence is now determined by $(p_j(z))^2$, and each row is again initialized differently through the increasing powers of z^k, each making use also of $g_{j,2}(z)$, and so on.

An important property of this construction is that it can be shown to be a (t, s)-sequence with $t = \sum_{j=1}^{s}(e_j - 1)$, just like we had for the Sobol' sequence. Unlike the Sobol' sequence, though, here we are not forcing $p_j(z)$ to be a primitive polynomial, and thus by choosing $p_j(z)$ in ascending order of degree within all irreducible polynomials in $\mathbb{F}_b[z]$, we obtain a smaller bound on the t-value than for the Sobol' sequence [337, p. 64]. Also, by choosing the base b appropriately — and thus possibly taking $b \geq s$, which might yield the Faure sequence — it is possible to show that the implied constant c_s in the star discrepancy of the Niederreiter sequence is such that $\log c_s \in O(-s \log\log s)$[335, p. 325]. Note that this is the same behavior as for the Faure sequence, which makes sense since the Faure sequence is one of the constructions we can choose when trying to minimize the behavior of c_s. On the other hand, if we only consider the base 2 Niederreiter sequences, then we get the same behavior for c_s as for the Sobol' sequence.

On a more practical note, it seems that, most of the time, implementations of this method take $g_{j,m}(z) = 1$ [44, 337]. Also, it is important to know that even if the bound on t given by $\sum_{j=1}^{s}(e_j - 1)$ is smaller for the Niederreiter sequence than for the Sobol' sequence, the one-dimensional projections of the Niederreiter sequence may not necessarily be (0,1)-sequences [397], which can be a disadvantage from a practical point of view.

5.4.4 Improvements to the original constructions of Halton, Sobol', Niederreiter, and Faure

Over the years, a lot of research has been done to try to improve the four constructions that we just described. Here we discuss these improvements, starting with the Halton sequence, then the Sobol' and Niederreiter sequences, and finally the Faure sequence. Numerical results illustrating the difference between the original and improved constructions are given at the end of Sect. 7.3.

Improvement to the Halton sequence

Although the Halton sequence suffers from severe space-filling problems in high dimensions, it continues to be a rather popular method because of its simplicity. An approach that has been studied by several researchers to try to improve its properties is to permute the digits $a_l(i)$ in each dimension before applying the radical-inverse function [18, 19, 41, 58, 113, 115, 222, 306, 454, 457]. More precisely, a *generalized Halton sequence* is defined by s sequences of permutations $\{\pi_{j,r}\}_{r\geq 1}$, $j = 1, \ldots, s$, where for each r the permutation $\pi_{j,r}$ acts on the integers $[0, \ldots, b_j - 1]$, and b_j is the jth base used (usually taken to be the jth smallest prime number). The ith point in this sequence is then

$$\mathbf{u}_i = \left(\sum_{r=1}^{\infty} \pi_{1,r}(a_r(i-1))b_1^{-r}, \ldots, \sum_{r=1}^{\infty} \pi_{s,r}(a_r(i-1))b_s^{-r} \right).$$

Figure 5.9 (left) shows the first 1000 points of a generalized Halton sequence where the permutations $\pi_{j,r} = \pi_j$ have been chosen according to a criterion that takes into account bounds obtained for the one-dimensional van der Corput sequence in base b [114] but that also measures the quality of two-dimensional projections (current work with Henri Faure [115]). The permutations used there are simply based on a multiplicative factor. That is, we choose for each dimension j a multiplier $f_j \in \{1, \ldots, b_j - 1\}$, and for all $r \geq 1$ we let

$$\pi_{j,r}(k) = f_j k \bmod b_j$$

for $k = 0, \ldots, b_j - 1$. The generalized Halton sequences described in [58, 306, 457] are also of this type.

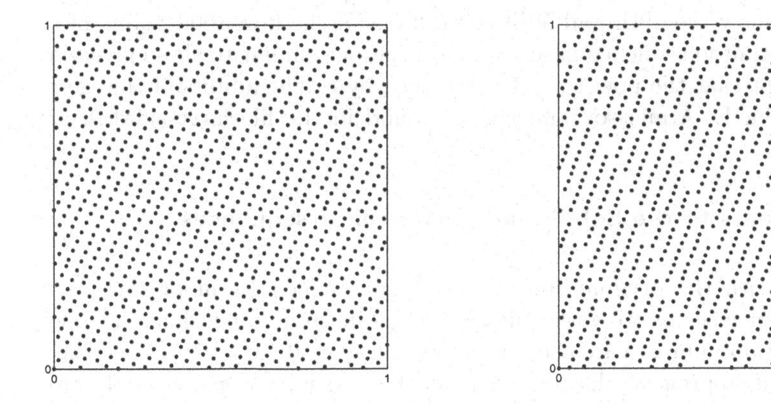

Fig. 5.9 First 1000 points of a generalized Halton sequence for the 49th and 50th coordinates based on the methods from [115] (left) and [19] (right).

A slightly more general type of permutation is used by Atanassov and Durchova [19], where a specific multiplier f_j is chosen for each dimension $j = 1, \ldots, s$, but then different permutations

$$\pi_{j,r} = f_j^{r-1} k \bmod b_j, \qquad r \geq 1, \tag{5.12}$$

are used for each digit. The multipliers in this case must be *admissible integers* satisfying certain properties. A description of these properties can be found in [18, 465], and they arise from a result by Atanassov [18], who showed that generalized Halton sequences built from such admissible integers had a better implied constant c_s for their star discrepancy. In addition, in the same paper, Atanassov was able to also improve the best implied constant c_s known so far [112] for the original Halton sequence, going from

$$c_s = \frac{1}{2^s} \prod_{j=1}^{s} \frac{(b_j - 1)}{\log b_j}$$

to

$$c_s = \frac{1}{s! 2^s} \prod_{j=1}^{s} \frac{(b_j - 1)}{\log b_j}.$$

This was an important result, as it proved that the implied constant for the star discrepancy of the Halton sequence was going to 0 with s rather than going to infinity. With a generalized Halton sequence based on admissible integers, the result is even better because then the implied constant is shown to be

$$\frac{1}{s!} \sum_{j=1}^{s} \log b_j \prod_{j=1}^{s} \frac{b_j(1 + \log b_j)}{(b_j - 1) \log b_j},$$

where the bases b_j are assumed to be the first s prime numbers. In Fig. 5.9 (right), we show the 49th and 50th coordinates of the first 1000 points of a generalized Halton sequence based on permutations similar to those given in (5.12), except that the power $r - 1$ is replaced by r. The admissible integers used are from E. Atanassov and can be found in the file `haltondat.h` at [491].

Improvements to the Sobol' and Niederreiter sequences

Compared with the original Halton and Faure sequences, in practice the Sobol' sequence seems to work quite well even in large dimensions, as long as the direction numbers are chosen appropriately [150, 202, 278]. Nevertheless, different approaches have been taken to find improvements and generalizations of the Sobol' sequence. An important class in that category is Tezuka's *generalized Sobol' sequence* [441], where instead of being restricted to primitive polynomials for the $p_j(z)$, the use of more general irreducible polynomials is allowed, just like for the Niederreiter sequence in base 2. Also, the generating matrices are obtained through a more general process with these sequences than with the original Sobol' sequences. In particular, the generating matrices for the generalized Sobol' sequences are not necessarily nonsingular upper-triangular.

In fact, in [439], Tezuka proposes a construction that generalizes not only the Sobol' sequence but also the Niederreiter sequences and therefore the Faure sequence. The main difference between the original and *generalized Niederreiter sequences* of Tezuka is the replacement of the numerator $z^k g_{j,m}(z)$ in (5.10) by a polynomial $y_{j,k}(z)$ such that each group of e_j polynomials of the form $y_{j,le_j}(z), y_{j,le_j+1}(z), \ldots, y_{j,(l+1)e_j-1}(z)$ has residues modulo $p_j(z)$ that are linearly independent over $\mathbb{F}_b$. Note that this condition was automatically met by the specific choice of polynomials made by Niederreiter.

Once we have these polynomials, then the lth row of C_j is determined by first computing $q = \lfloor (l-1)/e_j \rfloor$ and then letting

$$(c_{j,l,1}, c_{j,l,2}, \ldots) = (a^j(q+1,l,1), a^j(q+1,l,2), \ldots).$$

That is, as for the Niederreiter sequence, the first group of e_j rows contains coefficients that follow a recurrence whose characteristic polynomial is $p_j(z)$, and the lth row is initialized by the specific choice of polynomial $y_{j,l}(z)$. Then the second group has elements that follow a recurrence described by $(p_j(z))^2$, and so on. Hence a major difference between Niederreiter's sequence and the generalized Niederreiter sequence of Tezuka is in the way the recurrence determining each row of the generating matrices is initialized. Tezuka proves that generalized Niederreiter sequences are still low-discrepancy sequences and that their t-value is bounded above by $\sum_{j=1}^{s}(e_j - 1)$, just as for the original Niederreiter sequences. Conditions on the tightness of this upper bound are studied in [82].

As mentioned earlier, these generalized Niederreiter sequences also include the Sobol' and Faure sequences as special cases. In that setting, the *direction numbers* of the Sobol' sequence can be reformulated in terms of the polynomials $y_{j,l}(z)$ used above. More precisely, for the Sobol' sequence, we use e_j polynomials $\tilde{y}_{j,1}(z), \ldots, \tilde{y}_{j,e_j}(z)$ such that $\deg(\tilde{y}_{j,l}(z)) = e_j - l$, and then we set $y_{j,l}(z) = \tilde{y}_{j,(l-1) \bmod e_j+1}(z)$.

Furthermore, the framework of generalized Niederreiter sequences enabled Tezuka to propose a construction that he called a *polynomial arithmetic analogue* of the Halton sequence because it is constructed using principles similar to those of the Halton sequence but has better properties that can be proved using the fact that they are a special case of generalized Niederreiter sequences. The proposed construction turns out to be related to Faure sequences, as we will see shortly.

Another construction that can be thought of as improving on the Sobol' and Niederreiter sequences are the *Niederreiter-Xing sequences* [345, 346, 347, 482]. These sequences are not a special case of generalized Niederreiter sequences. They are based on *global function fields* and thus involve much deeper mathematical tools than what we have seen so far. We believe it would go beyond the scope of this text to explain this construction but still want to say a few words about it because of its theoretical importance. For the Sobol' and Niederreiter sequences, the bound on the t-value grows as $s \log s$ as the dimension s increases. By contrast, Niederreiter-Xing sequences are designed so that, for any base b that is a prime power, their t-value grows only linearly with s, which is the optimal rate that can be obtained.

An implementation of these sequences is discussed in [377]. They have been used in numerical experiments in [194], among others. Recent work on (t,s)-sequences based on global function fields can be found in [312, 342], where new improvements are presented. This active area of research is likely to continue to produce more improvements in the near future.

Improvements to the Faure sequence

As we mentioned previously, even if the Faure sequence is optimal from the point of view of the t-value, its space-filling properties are not always very good for small n and large s. Figure 5.10 shows the first 1000 points of the Faure sequence in base 53 over the 49th and 50th coordinates.

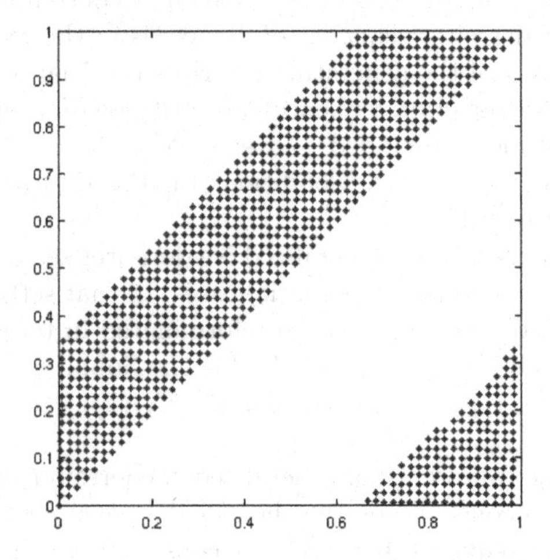

Fig. 5.10 First 1000 points of the Faure sequence in base 53 for the 49th and 50th coordinates.

A successful approach proposed by Tezuka [440] for improving the Faure sequence is to modify the generating matrices of the Faure sequence by multiplying them (from the left) by nonsingular lower-triangular matrices. More precisely, a generalized Faure sequence is obtained by taking C_j as $A_j(P^{j-1})^{\mathrm{T}}$, where A_j is some nonsingular lower-triangular matrix, for $j = 1, \ldots, s$, and P is the Pascal matrix in $\mathbb{F}_b$. It can be shown that the sequences obtained still have a t-value equal to 0 by using either the approach with Vandermonde matrices used by Faure in [112] or the framework of generalized Niederreiter sequences. Indeed, in this framework, generalized Faure sequences amount to setting the $y_{j,l}(z)$ to arbitrary polynomials in $\mathbb{F}_b[z]$, instead of fixing them to 1 as in the original Faure sequence. But just as in the original definition, the $p_j(z)$ are chosen to be $z - j + 1$, and so $t = \sum_{j=1}^{s}(e_j - 1) = 0$.

Figure 5.11 shows the 49th and 50th coordinates of the first 1000 points of a generalized Faure sequence. For this example, the matrices A_j have been chosen randomly. Experiments in finance done with a certain version of this generalized sequence are reported in [374]. Implementations are available in the software Finder from Columbia University [494].

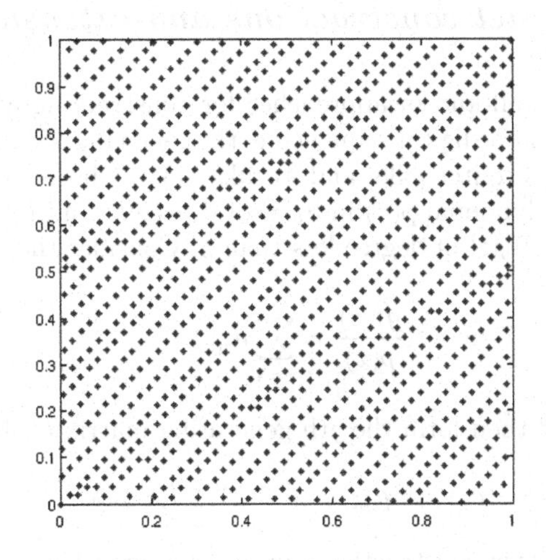

Fig. 5.11 First 1000 points of a generalized Faure sequence in base 53 for the 49th and 50th coordinates.

More work in this area has been done in [116, 445, 449]. Namely, in [449] Tezuka and Tokayama consider the generalized Faure sequence obtained by taking $A_j = P^{j-1}$. Interestingly, this particular choice corresponds to the polynomial arithmetic analogue of the Halton sequence that was discussed on p. 167 [439, 449]. In [116], Faure and Tezuka look at generating matrices of the form

$$(P^{j-1})^{\mathrm{T}}(\gamma_j U), \tag{5.13}$$

where U is some nonsingular upper-triangular matrix and the γ_j's are some constants in $\mathbb{Z}_b$. They show that the sequences obtained in this way still have $t = 0$. In the subsequent paper [445] by Tezuka and Faure, they refer to sequences obtained with $\gamma_j = 1$ in (5.13) as *reordered Faure sequences* because it can be shown that multiplying each generating matrix from the right by U simply amounts to reordering the points in the Faure sequence in a way that improves its space-filling properties. This is similar to the fact that using a Gray code amounts to reordering the points of a sequence and is equivalent to multiplying each generating matrix from the right by a certain matrix $\mathbf{G}$ as given in (5.9). However, in that case, it was motivated by making the implementation easier rather than improving the space-filling properties. It should be noted that Faure and Tezuka also prove in [116] the more general result that if a sequence based on the generating matrices $C_1, \ldots, C_s$ is a (t,s)-sequence, then the sequence based on $C_1(\gamma_1 U), \ldots, C_s(\gamma_s U)$ is also a (t,s)-sequence. That is, multiplication from the right by $\gamma_j U$ preserves the t-value.

5.4.5 Digital net constructions and extensions

One of the most well-known approaches for constructing a digital net that does not come from a digital sequence is the following idea, developed independently by Niederreiter [338] and Tezuka [438]. Choose a base b that is a prime power, a polynomial $p(z)$ in $\mathbb{F}_b[z]$ of degree k, and then s polynomials $g_1(z), \ldots, g_s(z)$ in $\mathbb{F}_b[z]$ of degree less than k. Consider the expansion

$$\frac{g_j(z)}{p(z)} = \sum_{r=1}^{\infty} a_{j,r} z^{-r}$$

for $1 \le j \le s$, and then form the jth generating matrix C_j by taking

$$c_{j,l,r} = a_{j,l+r-1} \qquad 1 \le l, r \le k. \tag{5.14}$$

It turns out that this construction can also be described as a lattice in the polynomial setup. To do so, we first need the function $\psi : \mathbb{F}_b((z^{-1})) \to \mathbb{R}$, which is an evaluation mapping defined as

$$\psi\left(\sum_{r=w}^{\infty} a_r z^{-r}\right) = \sum_{r=w}^{\infty} a_r b^{-r}. \tag{5.15}$$

Similarly, for a vector containing s components in $\mathbb{F}_b((z^{-1}))$, ψ evaluates each component using (5.15).

Definition 5.10. The digital net described by (5.14) is a *rank-1 polynomial lattice point set* of the form

$$P_n = \left\{\psi\left(q(z)\frac{g_1(z)}{p(z)}, \ldots, q(z)\frac{g_s(z)}{p(z)}\right) : q(z) \in \mathbb{F}_b[z]/(p(z))\right\},$$

where $p(z) \in \mathbb{F}_b[z]$ is a polynomial of degree k and the $g_j(z)$ are polynomials in $\mathbb{F}_b[z]$ of degree less than k, all multiplications are done modulo $p(z)$, and $n = b^k$.

The reason why $n = b^k$ is that there are b^k polynomials $q(z)$ in $\mathbb{F}_b[z]/(p(z))$. The analogy with the lattice construction is to view $(g_1(z), \ldots, g_s(z))$ as the generating vector, $p(z)$ plays the role of n, and letting $q(z)$ run over $\mathbb{F}_b[z]/(p(z))$ corresponds to multiplication by $i = 0, \ldots, n-1$.

The following example illustrates how to construct a small rank-1 polynomial lattice point set.

Example 5.11. Suppose $b = 2$ and $p(z) = z^3 + z + 1$. Then, for $s = 2$, take $g_1(z) = 1$ and $g_2(z) = z$. In this case, the polynomial $q(z)$ runs over $\{0, 1, z, z+1, z^2, z^2+1, z^2+z+1, z^2+z\}$. Also, we need the expansion

$$\frac{1}{z^3+z+1} = z^{-3} + z^{-5} + z^{-6} + z^{-7} + \ldots$$

(see App. A). Once we have that, we can easily compute quotients of the form

$$\frac{p(z)}{1+z+z^3}.$$

For instance, we get

$$\begin{aligned}\frac{z^2+1}{1+z+z^3} &= (z^{-3} + z^{-5} + z^{-6} + z^{-7} + \ldots) + \\ &\quad (z^{-1} + z^{-3} + z^{-4} + z^{-5} + \ldots) \\ &= z^{-1} + z^{-4} + z^{-6} + \ldots.\end{aligned}$$

Hence we have

$$\begin{aligned}\mathbf{u}_1 &= (0, 0), \\ \mathbf{u}_2 &= \psi(1/p(z), z/p(z)) \\ &= (2^{-3} + 2^{-5} + 2^{-6} + \ldots, 2^{-2} + 2^{-4} + 2^{-5} + 2^{-6} + \ldots) \\ &\approx (0.17, 0.36), \\ \mathbf{u}_3 &= \psi(z/p(z), z^2/p(z)) \\ &= (2^{-2} + 2^{-4} + 2^{-5} + \ldots, 2^{-1} + 2^{-3} + 2^{-4} + 2^{-5} + \ldots) \\ &\approx (0.36, 0.72), \\ \mathbf{u}_4 &= \psi((z+1)/p(z), (z^2+z)/p(z)) \\ &= (2^{-2} + 2^{-3} + 2^{-4} + \ldots, 2^{-1} + 2^{-2} + 2^{-3} + 2^{-6} + \ldots) \\ &\approx (0.44, 0.89), \\ \mathbf{u}_5 &= \psi(z^2/p(z), (z+1)/p(z)) \\ &= (2^{-1} + 2^{-3} + 2^{-4} + 2^{-5} + \ldots, 2^{-2} + 2^{-3} + 2^{-4} + \ldots) \\ &\approx (0.72, 0.44),\end{aligned}$$

$$\begin{aligned}
\mathbf{u}_6 &= \psi((z^2+1)/p(z), 1/p(z)) \\
&= (2^{-1}+2^{-4}+2^{-6}+\ldots, 2^{-3}+2^{-5}+2^{-6}+\ldots) \\
&\approx (0.58, 0.17), \\
\mathbf{u}_7 &= \psi((z^2+z+1)/p(z), (z^2+1)/p(z)) \\
&= (2^{-1}+2^{-2}+2^{-5}+\ldots, 2^{-1}+2^{-4}+2^{-6}+\ldots) \\
&\approx (0.78, 0.58), \\
\mathbf{u}_8 &= \psi((z^2+z)/p(z), (z^2+z+1)/p(z)) \\
&= (2^{-1}+2^{-2}+2^{-3}+2^{-6}+\ldots, 2^{-1}+2^{-2}+2^{-5}+\ldots \\
&\approx (0.89, 0.78).
\end{aligned}$$

This construction has been studied further in [80, 81, 83, 226, 238, 239, 240, 237, 339, 378, 396]. A special case of this construction consists in taking $g_j(z) = (g(z))^{j-1} \bmod p(z)$ for $j = 1, \ldots, s$, which in the lattice setting can be thought of as a polynomial equivalent of the Korobov point set. Parameters for $p(z)$ and $g(z)$ can be found in [378] in what is called the *Salzburg Tables.* These *polynomial Korobov* point sets will also be discussed in Section 5.5.

Generalized constructions called *polynomial integration lattices*, which are the polynomial version of the lattice point sets discussed in Sect. 5.3, have also been defined and studied [256, 285]. Here we choose a basis $\mathbf{v}_1(z), \ldots, \mathbf{v}_s(z)$, where each $\mathbf{v}_j(z)$ is in $\mathbb{F}_b^s((z^{-1}))$ and such that those vectors are independent over $\mathbb{F}_b((z^{-1}))$. We then have the following definition.

Definition 5.12. A *polynomial integration lattice* is a point set of the form

$$P_n = \{\psi(\mathbf{v}(z)) : \mathbf{v}(z) \in \mathcal{L}_s\} \cap [0,1)^s, \tag{5.16}$$

where $\mathcal{L}_s$ is the polynomial lattice defined by

$$\mathcal{L}_s = \left\{ \mathbf{v}(z) = \sum_{j=1}^{s} q_j(z)\mathbf{v}_j(z) : q_j(z) \in \mathbb{F}_b[z], j = 1, \ldots, s \right\}$$

and such that $\mathbb{F}_b^s[z] \subseteq \mathcal{L}_s$.

It can be shown that the number of points in (5.16) is b^k, where k is the degree of the polynomial given by $\det(\boldsymbol{V}^{-1})$ and $\boldsymbol{V}$ is the matrix whose rows are given by the $\mathbf{v}_j(z)$ [285].

This general construction can be used to draw interesting analogies between lattices and nets. However, as in the standard case, in practice, rank-1 polynomial lattices (including Korobov lattices) are mostly used and studied. It might be for that reason that most people refer to "polynomial lattices" to describe the rank-1 case [76, 80, 81, 226].

Just as for standard lattices, polynomial rank-1 lattices can also be made extensible in their number of points. This idea is briefly mentioned in [265] and discussed in much more detail in [340, 442]. The definition used in [340] is more general than the one in [442], but the discussion in [442] establishes a useful connection with the extensible construction for (standard) rank-1 lattices discussed on p. 150. We thus chose to present the construction given in [442], which goes as follows. Choose a base b and a vector $(g_1(z), \ldots, g_s(z))$ in $(\mathbb{F}_b[z])^s$. Then choose a polynomial $p(z) \in \mathbb{F}_b[z]$. For a given i, the polynomial version of the radical-inverse function is obtained through the following steps:

(1) First write

$$i = a_0(i) + a_1(i) \times b + a_2(i) \times b^2 + \ldots + a_m(i) \times b^m.$$

(2) Then construct the corresponding polynomial

$$v_i(z) = a_0(i) + a_1(i) \times z + a_2(i) \times z^2 + \ldots + a_m(i) \times z^m.$$

(3) We now want to expand this polynomial in base $p(z)$ rather than in base z. That is, we need to find the polynomials (which act as coefficients) $r_{i,0}(z), r_{i,1}(z), \ldots, r_{i,h}(z)$ such that

$$v_i(z) = r_{i,0}(z) + r_{i,1}(z) \times p(z) + \ldots + r_{i,h}(z) \times (p(z))^h$$

where h is given by

$$h = \lfloor m/e \rfloor$$

and where e is the degree of $p(z)$. It can be shown that

$$r_{i,l}(z) = \left[\frac{v_i(z)}{(p(z))^l} \right] \bmod p(z),$$

where for a formal Laurent series $g(z)$, the notation $[g(z)]$ represents the polynomial part of $g(z)$.

(4) Once we have these coefficients $r_{i,0}(z), \ldots, r_{i,h}(z)$ for a given i, the *polynomial analogue of the radical inverse function in base* $p(z)$ is defined as

$$\phi_{p(z)}(i) = \frac{r_{i,0}(z)}{p(z)} + \frac{r_{i,1}(z)}{(p(z))^2} + \ldots + \frac{r_{i,h}(z)}{(p(z))^h}.$$

An extensible polynomial rank-1 lattice (also called "polynomial version of Hickernell sequences" in [442]) can then be defined by the sequence

$$\mathbf{u}_i = \varphi\left(g_1(z)\phi_{p(z)}(i-1), \ldots, g_s(z)\phi_{p(z)}(i-1)\right), \qquad i \geq 1,$$

where

$$g_j(z)\phi_{p(z)}(i-1) = \frac{g_j(z) r_{i,0}(z) \bmod p(z)}{p(z)} + \frac{g_j(z) r_{i,1}(z) \bmod p(z)}{(p(z))^2} + \ldots + \frac{g_j(z) r_{i,h}(z) \bmod p(z)}{(p(z))^h}$$

and the evaluation function φ was defined in (5.15). More recent work on this topic can be found in [76], for instance, where some existence results are proved.

Another type of construction for digital nets that has been used to find nets with an improved t-value is a method called a *shift net*, which was introduced by Schmid [395]. The idea is as follows. A shift net with b^k points in dimension $s = k$ is built by first choosing a $k \times k$ matrix for C_1, the first generating matrix. Assume this matrix consists of the k column vectors $\mathbf{c}_1, \ldots, \mathbf{c}_k$. Then the $s-1$ remaining generating matrices are obtained by *shifting* the columns of C_1. That is,

$$C_j = (\mathbf{c}_j \mathbf{c}_{j+1} \ldots \mathbf{c}_s \mathbf{c}_1 \ldots \mathbf{c}_{j-1})$$

for $j = 2, \ldots, s$. Parameters describing shift nets that minimize the t-value are given in [395]. More recent work in this area can be found in [398]. Among others, one of the improvements of [398] compared with [395] is that exhaustive searches are performed for all dimensions considered, which enables the authors to improve on the shift nets that were found in [395]. It is worth mentioning that for some values of k and b (and recall that s must be equal to k here), shift nets provide digital nets with the smallest t-value known so far [489, 400].

Several constructions for digital nets that come from *linear codes* and *ordered orthogonal arrays* have also been found to provide optimal values for t, but we will not discuss these particular constructions here. Some references on this topic are [31, 244] and the MinT database [400, 489].

Finally, another line of research that has been pursued recently is to construct digital nets and sequences that work well for integrands belonging to certain classes of smooth functions [77, 78]. These nets are characterized not only by their t-value and are referred to as $(t, \alpha, \beta, n \times k, s)$-*nets*, which in the case $\alpha = \beta = 1$ and $n = k$ are the same as a (t, k, s)-net. We do not pursue this topic further here, as the applicability of these nets and sequences to practical problems has not yet been studied.

5.5 Recurrence-based point sets

In this section, we discuss a framework studied in [265] that describes a class of low-discrepancy point sets based on the same kind of recurrence-based constructions as those used to build pseudorandom number generators. This connection between pseudorandom number generators and constructions for quasi–Monte Carlo goes back at least to [334]. The point sets obtained are either lattices or digital nets. The framework that we are about to describe simply provides another way of defining them.

Definition 5.13. A *recurrence-based point set* P_n is obtained as follows. First choose a finite ring B of cardinality n, a *transition function* T, assumed to be a bijection over B, and then an output function $\eta : B \to [0,1)$, assumed to be one-to-one. Let $x_i = T(x_{i-1})$ for $i \geq 1$. Then

$$P_n = \{(\eta(x_0), \eta(x_1), \ldots, \eta(x_{s-1})) : x_0 \in B\}.$$

In other words, a recurrence-based point set is obtained by looking at all possible initial states x_0 for the recurrence $T(\cdot)$, and in each case by forming a point $\mathbf{u}_i$ by running this recurrence $s-1$ steps and applying $\eta(\cdot)$ to each element in B thus obtained.

For instance, with an LCG with modulus n and multiplier a, we have that $B = \mathbb{Z}_n$, $T(x) = ax$ (where operations are performed in $\mathbb{Z}_n$), and $\eta(x) = x/n$. Thus P_n is the same as the set Ψ_s that was defined in (3.10) in Chap. 3. More precisely,

$$P_n = \left\{ \frac{1}{n}(i, ai \bmod n, a^2 i \bmod n, \ldots, a^{s-1} i \bmod n), i = 0, \ldots, n-1 \right\},$$

which is the same as a Korobov point set with generator a in dimension s.

In the special case where the LCG has maximal period (which happens if n is prime and a is a primitive element modulo n), the connection above provides a very effective way of constructing a Korobov point set P_n using the fact that $x_0, x_1, \ldots, x_{n-2}$ runs over all numbers in $\{1, \ldots, n-1\}$, as long as $x_0 \neq 0$. More precisely, in this case we have that

$$P_n = \Psi_s = \{(u_i, u_{i+1}, \ldots, u_{i+1-s}), i = 0, \ldots, n-2\} \cup \{\mathbf{0}\}.$$

Thus P_n can be obtained by choosing a nonzero seed, running the LCG, forming one point from each overlapping s-dimensional vector output by the LCG, and adding the origin $\mathbf{0}$. Figure 5.12 gives pseudocode for generating P_n using this idea.

The connection between Korobov point sets and LCGs also allows us to discuss an important point. Since the generating vector has the form $(1, a, a^2 \bmod n, \ldots)$, it is clear that eventually its components will start repeating themselves, just like the output of an LCG does. In the best case — when n is prime and a is a primitive element modulo n — the cycle will be of length $n-1$, corresponding to an LCG with maximal period. This implies that if $s \geq n$, each point of the Korobov point set contains repeated coordinates. That is,

$$u_{i,j+l(n-1)} = u_{i,j}$$

for each $i = 1, \ldots, n$, $j = 1, \ldots, s$ and $l \geq 1$. While this might be considered problematic — especially in cases where s is very large — it is important to note that this problem disappears once the point set is *randomized.* For instance, if we add (modulo 1) a random shift $\mathbf{v}$ uniformly distributed in $[0,1)^s$ to each point in the Korobov point set — the same $\mathbf{v}$ being added to each point — then since the coordinates $v_1, v_2, \ldots$ of the random shift $\mathbf{v}$ do

```
for j ← 1 to s
    u[j] ← 0
// The first point is taken to be the origin
x ← 1
u[1] ← x/n
for j ← 2 to s
    x ← ax mod n
    u[j] ← x/n
    // we now have the first nonzero point
for i ← 1 to n − 2
    for j ← 1 to s − 1
        u[j] ← u[j + 1]
    x ← ax mod n
    u[s] ← x/n
    // we now have the (i + 1)th nonzero point
```

Fig. 5.12 Code to generate a Korobov point set based on maximum-period LCG.

not cycle, the cycle in the original deterministic points is broken after the shift has been added. We will come back to this point in Sect. 6.2.

Going back to the general setup for recurrence-based point sets, one of the properties of these point sets that can be very useful in practice is that they can handle problems with an unbounded dimension. This is because once B, T, and η are chosen, the dimension s of P_n can be taken to be arbitrarily large *with no additional parameters that need to be chosen.* Concretely, this can be implemented by generating the coordinates of each point $\mathbf{u}_i$ as needed until some condition (that depends on the coordinates generated so far) is satisfied. For instance, in the bank example from Chap. 1, one would generate coordinates $u_{i,j}$ for $j \geq 1$ until an arrival time is obtained that is after the bank's closing time of 3 pm. In practice, though, and as discussed in the LCG case in the previous paragraph, since problems of unbounded dimension might result in $s > n$, recurrence-based point sets should be *randomized* to handle such problems. For that reason, we will postpone our discussion about implementation issues for problems with an unbounded dimension until Sect. 6.2 in the chapter on randomized quasi–Monte Carlo methods.

Another interesting property of recurrence-based point sets is that, as shown in [284], they are fully projection-regular, and *dimension-stationary*, a concept that we now define.

Definition 5.14. A point set P_n is *dimension-stationary* if, for any set $I = \{i_1, \ldots, i_d\}$ of positive integers and integer $j \geq 1$ such that $i_d + j \leq s$, we have that $P_n(I) = P_n(\{I + j\})$.

In other words, for a dimension-stationary point set, projections over indices that have the same *spacings* are equal. So, for instance, a dimension-stationary point set with $s = 100$ is such that

$$P_n(\{1,3,4\}) = P_n(\{2,4,5\}) = P_n(\{3,5,6\}) = \ldots = P_n(\{97,99,100\}).$$

Since Korobov point sets with $\gcd(a,n) = 1$ are a special case of recurrence-based point sets, it means they are dimension-stationary. Also, it is not too hard to prove that the Faure sequence is dimension-stationary (see Prob. 5.9).

This brings us to the observation about Korobov point sets, which also holds for any recurrence-based point sets, that although the criteria used when searching for good Korobov generators a are typically restricted to some dimension s_0, once we have a generator a judged to be good for a given number of points n, we can use it to construct a point set in *any* dimension s. The only concern is that if s is larger than the dimension s_0 used in the computer search's criterion, then the quality of the r-dimensional point set for $s_0 < r \leq s$ is unknown and could potentially be bad. However, even if we only have information about the quality of projections $P_n(I)$ for some sets $I \in \mathcal{I}$ all such that, say, their largest index $i_d \leq s_0$, the dimension-stationarity of P_n implies that the same quality properties hold for any projection of the form $P_n(I+j)$ with $I \in \mathcal{I}$ and $j \geq 0$.

For example, suppose a Korobov generator a has been chosen so that it is the best with respect to a criterion that considers the projections

$$P_n(\{1,2\}),\ldots,P_n(\{1,8\}),P_n(\{1,2,3\}),P_n(\{1,2,4\}),\ldots,P_n(\{1,2,8\}),\ldots,$$
$$P_n(\{1,3,4\}),\ldots,P_n(\{1,7,8\}).$$

Then we know that any projection of the form $P_n(\{i,i+j_1\})$ with $i \geq 1$ and $1 \leq j_1 \leq 7$ and $P_n(\{i,i+j_2,i+j_2+j_3\})$ with $i \geq 1, j_2 \leq 6, j_3 \leq 7-j_2$ will also be good. In other words, even if the search was done based on $s_0 = 8$, because of the dimension-stationarity we know that, for instance, the projection $P_n(\{100,101,102\}) = P_n(\{1,2,3\})$ is also good. Hence it can be reasonable to use a generator a chosen in this way to construct point sets with dimension $s > s_0$.

We end this section by discussing another example of a recurrence-based point set, which is the *polynomial Korobov lattice point set* that was discussed on p. 172. In that case, the corresponding generator is a *polynomial LCG* [436, 437, 446, 441]. That is, here B is the ring $\mathbb{F}_2[z]/(p(z))$, where $p(z)$ is a polynomial in $\mathbb{F}_2[z]$ of degree k. For the transition function, choose $a(z) \in B$ and take

$$T(x(z)) = a(z)x(z) \bmod p(z).$$

For the output function, first consider the formal Laurent series

$$\frac{x(z)}{p(z)} = \sum_{l=0}^{\infty} x_l z^{-l}$$

and then let

$$\eta(x(z)) = \varphi(x(z)/p(z)) = \sum_{l=0}^{\infty} x_l 2^{-l}, \tag{5.17}$$

where the evaluation function $\varphi(\cdot)$ was defined in (5.15). One can prove that the recurrence-based point set thus obtained can also be described as follows. First form the set

$$P_n(z) = \{q(z)(1, a(z), \ldots, a^{s-1}(z))/p(z) : q(z) \in \mathbb{F}_2[z]/(p(z))\},$$

and then take $P_n = \{\varphi(\mathbf{v}(z)) : \mathbf{v}(z) \in P_n(z)\}$, which is just the polynomial Korobov lattice point set that we described in Sect. 5.4, p. 172. As we mentioned there, a polynomial Korobov lattice is a special case of a rank-1 polynomial lattice, which is itself a special case of a digital net.

Furthermore, a special case of a polynomial Korobov lattice is to take $a(z) = z^{\nu}$ for some $\nu \geq 1$. The point set P_n thus obtained corresponds to the s-dimensional output space Ψ_s of a Tausworthe generator.

Recurrence-based point sets based on the more general class of $\mathbb{F}_2$-linear generators are discussed in [371], where specific constructions are given. Constructions based on combined Tausworthe generators were used in [285] and are discussed in the next example.

Example 5.15. A combined Tausworthe generator can be described using the polynomial LCG formulation as follows [446]. We consider J polynomial LCGs based on the recurrences

$$x_{j,i}(z) = (z^{\nu_j} \bmod p_j(z)) x_{j,i-1}(z) \bmod p_j(z), i \geq 1$$

for $j = 1, \ldots, J$, where $p_j(z)$ is a primitive polynomial of degree k_j over $\mathbb{F}_b$ and $\gcd(\nu_j, b^{k_j} - 1) = 1$. The J generators are then combined to produce an output

$$\eta(x_{1,i}(z)) + \ldots + \eta(x_{J,i}(z)),$$

with η as in (5.17) and where + is taken to be a digitwise addition in $\mathbb{Z}_b$.

It can be shown that the output thus obtained is equivalent to the one obtained from a polynomial LCG based on the recurrence [446]

$$y_i(z) = g(z) y_{i-1}(z) \bmod p(z), \tag{5.18}$$

where

$$\begin{aligned} p(z) &= \prod_{j=1}^{J} p_j(z), \\ g(z) &= \sum_{j=1}^{J} g_j(z) h_j(z) p_{-j}(z), \qquad (5.19) \\ p_{-j}(z) &= p(z)/p_j(z), \end{aligned}$$

and $h_j(z)$ is such that

$$h_j(z)p_{-j}(z) = 1 \bmod p_j(z).$$

Also, if the polynomials $p_j(z)$ are pairwise relatively prime, then the period of this polynomial LCG is equal to the least common multiple of $(b^{k_1} - 1, \ldots, b^{k_J} - 1)$. This equivalent form can be useful to study theoretical properties of these generators.

In addition, using the fact that $g(z)$ and $p(z)$ are relatively prime, we can show that the recurrence (5.18) defines a bijection over the ring $\mathbb{F}_b[z]/(p(z))$, which we identify as the set of polynomials in $\mathbb{F}_b[z]$ of degree less than $k = \sum_{j=1}^{J} k_j$ (Prob. 5.11 asks you to prove this). This implies that the combined generator can be used to define a recurrence-based point set that corresponds to a polynomial Korobov lattice based on the generator $g(z)$ described by (5.19) and containing b^k elements.

In practice, however, polynomial Korobov lattices based on combined Tausworthe generators can be obtained by implementing each component separately and running the combined generator over all its cycles. More precisely, under the conditions that each component has maximal period and that those periods are relatively prime, we can construct P_n as in Fig. 5.13.

For example, suppose we take $b = 2$ and use $J = 2$ generators described respectively by the recurrences

$$x_{1,i}(z) = zx_{1,i-1}(z) \bmod (z^4 + z + 1),$$
$$x_{2,i}(z) = z^4 x_{2,i-1}(z) \bmod (z^7 + z^3 + 1).$$

In this case, the combined generator has three nontrivial cycles, the first one of length $(2^4 - 1)(2^7 - 1) = 15 \times 127 = 1905$ corresponding to using the seed $1 \in \mathbb{F}_2[z]$ for both components. Then we have the cycle of length 15 corresponding to only using the first component and the third cycle of length 127 corresponding to only using the second component. Adding the zero vector, we get $1905 + 15 + 127 + 1 = 2048$ points, as required.

We can use this idea to build the polynomial Korobov point set in an alternative way, where we initially construct the $2^J - 1$ cycles and then form the s-dimensional points by taking overlapping s-tuples over these cycles, as shown in Fig. 5.14. More details on this type of implementation are given in [74, 266].

5.6 Quality measures

So far in this chapter, the quality measures that we have mostly discussed are the star discrepancy and the t-value. Several other measures are described in this section. We start by giving more information on the star discrepancy

```
PolyCombTaus()
    u_1 ← 0
    i ← 2
    for l = 1 to 2^J − 1
        seed ← bin(l)
        length ← 1
        for j = 1 to J
            if seed[j] = 1
                length ← length ×(2^{k_j} − 1)
                InitTaus(j, 1)
        u_i ← 0
        for n ← 1 to length
            if n = 1
                for j = 1 to J
                    if seed[j] =1
                        for k = 1 to s
                            u_{i,k} ← u_{i,k} ⊕ Taus(j)
            else
                for k = 1 to s − 1
                    u_{i,k} ← u_{i,k+1}
                u_{i,s} ← 0
                for j = 1 to J
                    u_{i,s} ← u_{i,s} ⊕ Taus(j)
            i ← i + 1
```

Fig. 5.13 Code to generate all n points of a polynomial Korobov point set defined by a combined Tausworthe generator when $b = 2$. We assume `InitTaus`$(j, 1)$ initializes the jth generator to the seed $1 \in \mathbb{F}_2[z]$, `bin`(l) returns the binary representation of l, `Taus`(j) returns the next output of the jth Tausworthe generator, and $\oplus$ is a bitwise exclusive-or (addition in $\mathbb{Z}_2$).

and other variations of that measure and discuss their use for providing error bounds. We then use Fourier and Walsh expansions to study the integration error associated with lattices and nets, respectively. This allows us to derive more quality measures and to establish interesting connections between lattices and nets. To conclude the section, we briefly discuss why it is useful to look at alternative approaches to deterministic error bounds, which will lead us into the next chapter, on randomized quasi–Monte Carlo methods.

5.6.1 Discrepancy and related measures

Recall that the star discrepancy of a point set P_n is given by

```
PolyCombTausCycle()
// Initialization
    for l = 1 to 2^J - 1
        seed ← bin(l)
        length[l] ← 1
        for j = 1 to J
            if seed[j] = 1
                length[l] ← length[l] ×(2^{k_j} - 1)
                InitTaus(j, 1)
        for n ← 1 to length[l]
            v_{l,n} ← 0
            for j = 1 to J
                if seed[j] =1
                    v_{l,n} ← v_{l,n} ⊕ Taus(j)
// Generating the points
    u_1 ← 0
    i ← 2
    for l = 1 to 2^J - 1
        a ← length [l]
        for n = 1 to length(l)
            for k = 1 to s
                u_{i,k} ← v_{l,n+k-1 mod a}
            i ← i + 1
```

Fig. 5.14 Code to generate all n points of a polynomial Korobov point set defined by a combined Tausworthe generator when $b = 2$. We assume `InitTaus(j, 1)` initializes the jth generator to 1 and `Taus(j)` returns the next output of the jth Tausworthe generator.

$$D^*(P_n) = \sup_{\mathbf{v}\in[0,1)^s} |v_1 \dots v_s - \alpha(P_n, \mathbf{v})/n|,$$

where $\alpha(P_n, \mathbf{v})$ is the number of points from P_n that are in

$$\prod_{j=1}^{s} [0, v_j).$$

A first obvious variation to this measure is to not restrict one corner to be at the origin. This corresponds to the concept of *extreme discrepancy* $D(P_n)$, which for $\mathcal{J} = \{\mathbf{w}, \mathbf{v} \in [0,1)^s : 0 \le w_j \le v_j < 1, 1 \le j \le s\}$ is given by

$$D(P_n) = \sup_{\mathcal{J}} |R_n(J(\mathbf{w}, \mathbf{v}))|,$$

where

$$R_n(J(\mathbf{w}, \mathbf{v})) = \prod_{j=1}^{s} (v_j - w_j) - \frac{1}{n}\alpha(P_n, \mathbf{w}, \mathbf{v})$$

and $\alpha(P_n, \mathbf{w}, \mathbf{v})$ is the number of points in P_n that are in

$$\prod_{j=1}^{s} [w_j, v_j).$$

(Note that with this notation we can write the star discrepancy as

$$D^*(P_n) = \sup_{\mathbf{v}\in[0,1)^s} |R_n(J(\mathbf{0}, \mathbf{v}))|.)$$

Since the supremum in $D(P_n)$ is taken over more intervals than in $D^*(P_n)$, it is clear that $D(P_n) \geq D^*(P_n)$. One can actually show that [228]

$$D^*(P_n) \leq D(P_n) \leq 2^s D^*(P_n).$$

This can be generalized further by replacing $\mathcal{J}$ in the definition of the extreme discrepancy by the set of all convex sets in $[0, 1)^s$, thereby obtaining the *isotropic discrepancy.* This can be useful for domains that are more general than $[0, 1)^s$, but we will not discuss this further here.

In one dimension, there are simple formulas for computing the discrepancy of finite point sets. Namely, we have that if $0 \leq u_1 \leq u_2 \leq \ldots \leq u_n \leq 1$, then [339, Theorems 2.6 and 2.7]

$$\begin{aligned} D^*(u_1, \ldots, u_n) &= \frac{1}{2n} + \max_{1\leq i\leq n} \left| u_i - \frac{2i-1}{2n} \right|, \\ D(u_1, \ldots, u_n) &= \frac{1}{n} + \max_{1\leq i\leq n} \left(\frac{i}{n} - u_i \right) - \min_{1\leq i\leq n} \left(\frac{i}{n} - u_i \right). \end{aligned}$$

The case $s = 2$ can also lead to explicit formulas [339, p. 22], but beyond that it is very difficult to compute the star and extreme discrepancies.

On the other hand, if we consider yet another way to generalize the definition of discrepancy, which is to use a norm other than the L_∞ (or sup) norm, then it is possible to get discrepancy measures that can be computed relatively easily. Most notably, the L_2 *discrepancy* and L_2 *star discrepancy* are defined respectively as

$$\begin{aligned} T(P_n) &= \left(\int_{\mathcal{J}} (R_n(J(\mathbf{w}, \mathbf{v})))^2 \, d\mathbf{w} d\mathbf{v} \right)^{1/2}, \\ T^*(P_n) &= \left(\int_{[0,1)^s} (R_n(J(\mathbf{0}, \mathbf{v})))^2 \, d\mathbf{v} \right)^{1/2}. \end{aligned} \tag{5.20}$$

The L_2 star discrepancy $T^*(P_n)$ is discussed in [333], among others, but the *unanchored* version is more recent and was proposed by Morokoff and Caflisch in [326]. One motivation for defining it is that the L_2 star discrepancy is known to put a strong emphasis on points near $\mathbf{0}$, which can sometimes lead to misleading results [307, 326].

In contrast with the star discrepancy $D^*(P_n)$ and extreme discrepancy $D(P_n)$, their L_2 counterpart can be effectively computed. Namely, Warnock [468] showed that

$$(T^*(P_n))^2 = \frac{1}{n^2}\sum_{i=1}^{n}\sum_{j=1}^{n}\prod_{k=1}^{s}(1-\max(u_{i,k},u_{j,k})) - \frac{2^{-s+1}}{n}\sum_{i=1}^{n}\prod_{k=1}^{s}(1-u_{i,k}^2)+3^{-s}.$$

A faster algorithm that runs in $O(n(\log n)^s)$ is given by Heinrich in [170]. For the unanchored version, it is shown in [326] that

$$\begin{aligned} T^2(P_n) &= \frac{1}{n^2}\sum_{i=1}^{n}\sum_{j=1}^{n}\prod_{k=1}^{s}(1-\max(u_{i,k},u_{j,k}))\min(u_{i,k},u_{j,k}) \\ &\quad - \frac{2^{-s+1}}{n}\sum_{i=1}^{n}\prod_{k=1}^{s}u_{i,k}(1-u_{i,k}) + 12^{-s}. \end{aligned} \tag{5.21}$$

It is also useful to know that, for a random point set P_n,

$$\begin{aligned} E((T^*(P_n))^2) &= (2^{-s}-3^{-s})/n, \\ E((T(P_n))^2) &= 6^{-s}(1-2^{-s})/n. \end{aligned}$$

Going back to the star discrepancy and extreme discrepancy, the fact that they cannot be easily computed for a given P_n unless $s \le 2$ might suggest that these measures are useless. This would be wrong, as these measures are mostly used to understand the asymptotic behavior of different constructions. First, as we mentioned before, the concept of a low-discrepancy sequence makes use of the star discrepancy, namely by referring to sequences for which $D^*(P_n)$ is in $O(n^{-1}(\log n)^s)$. Second, more generally the concept of star (and extreme) discrepancy can be related to the concept of a *uniformly distributed sequence*, which means the sequence $\mathbf{u}_1, \mathbf{u}_2, \ldots$ is such that

$$\lim_{n\to\infty}\frac{1}{n}\sum_{i=1}^{n}\mathbf{1}_J(\mathbf{u}_i) = \lambda_s(J)$$

for any subinterval $J \in [0,1)^s$, where

$$\mathbf{1}_J(\mathbf{u}) = \begin{cases} 1 & \text{if } \mathbf{u} \in J \\ 0 & \text{otherwise} \end{cases}$$

and $\lambda_s(\cdot)$ is the s-dimensional Lebesgue measure. A sequence that is uniformly distributed has the useful property of providing an approximation Q_n for $I(f)$ that converges to $I(f)$ as n goes to infinity. The connection with discrepancy is that a sequence is uniformly distributed if and only if

$$\lim_{n \to \infty} D^*(P_n) = 0$$

(the equivalence holds for the extreme discrepancy as well) [339, p.17]. Third, as we mentioned in Sect. 5.4, in addition to the asymptotic order $O(n^{-1}(\log n)^s)$ of $D^*(P_n)$, looking at the hidden constant c_s in the O notation is frequently used to compare the quality of different low-discrepancy sequences as s increases.

More importantly, the concept of discrepancy can be used to derive deterministic upper bounds on the integration error. For example, a widely cited result is the *Koksma-Hlawka theorem* [191] (Koksma proved the one-dimensional version [223]), which states that for any function with a *variation in the sense of Hardy and Krause* $V(f)$ that is finite, we have the upper bound

$$E_n \leq D^*(P_n)V(f) \tag{5.22}$$

on the absolute error of integration E_n given by

$$E_n = |Q_n - I(f)|.$$

Thus, when $V(f) < \infty$ and $P_n = \{\mathbf{u}_1, \ldots, \mathbf{u}_n\}$ is based on a low-discrepancy sequence, the integration error is in $O(\log^s n/n)$. Comparing this with the probabilistic Monte Carlo error that is in $O(1/\sqrt{n})$, one can argue that for a fixed dimension s, the quasi–Monte Carlo error converges faster than with Monte Carlo. This result is often used to motivate the use of quasi–Monte Carlo by saying something like "for functions that are smooth enough and if you are willing to take n sufficiently large, you will obtain a smaller error with quasi–Monte Carlo than with Monte Carlo".

The variation of f in the sense of Hardy and Krause is a multidimensional version of the notion of variation in one dimension, defined for functions f over $[0, 1)$ as

$$V(f) = \sup_{P \in \mathcal{P}} \sum_{i=0}^{n_P - 1} |f(u_{i+1}) - f(u_i)|,$$

where $\mathcal{P}$ is the set of all partitions P of $[0, 1)$ of the form $P = \{u_0 = 0, u_1, \ldots, u_{n_P} = 1\}$, with $u_i < u_{i+1}$, and for some $n_P \geq 1$. When the function f is continuously differentiable, then

$$V(f) = \int_0^1 \left| \frac{\partial f(u)}{\partial u} \right| du,$$

so, for instance, the function $f(u) = (1-2u)^2$ has a total variation of $4(1/2 - 1/4) - 4(1/2 - (1 - 1/4)) = 2$.

To extend this concept in higher dimensions, we first need to define the *variation of* f *on* $[0, 1)^s$ *in the sense of Vitali*, given by

$$V^{(s)}(f) = \sup_{P \in \mathcal{P}} \sum_{J \in P} |\Delta(f;J)|,$$

where $\mathcal{P}$ is the set of all partitions P of $[0,1)^s$. That is, a partition P is defined by s sets of the form $\{u_{0,j} = 0, u_{1,j}, \dots, u_{n_{P,j}}\}$, for $j = 1, \dots, s$, and the sum over $J \in P$ means we sum over all intervals J of the form

$$\prod_{j=1}^{s} [u_{l_j,j}, u_{l_j+1,j})$$

for some $0 \leq l_j < n_{P,j}$, $j = 1, \dots, s$. The notation $\Delta(f;J)$ represents the alternating sum of the values of f at the vertices of J. That is, for $J = \prod_{k=1}^{s} [a_k, b_k)$, we have [213, p. 20]

$$\Delta(f;J) = \sum_{j_1=0}^{1} \dots \sum_{j_s=0}^{1} (-1)^{\sum_{k=1}^{s} j_k} f(j_1 a_1 + (1-j_1)b_1, \dots, j_s a_s + (1-j_s)b_s).$$

Here again, if f has continuous partial derivatives, we have the more convenient formula

$$V^{(s)}(f) = \int_{[0,1)^s} \left| \frac{\partial^s f}{\partial u_1 \dots \partial u_s} \right| du_1 \dots du_s.$$

The last ingredient is to look at what we could call *projections* of the variation in the sense of Vitali. That is, for a subset $I = \{i_1, \dots, i_d\} \subseteq \{1, \dots, s\}$, we let $V^{(d)}(f;I)$ be the value of $V^{(d)}$ for the function $f^{(I)}(u_{i_1}, \dots, u_{i_d}) = f(\tilde{u}_1, \dots, \tilde{u}_s)$, where

$$\tilde{u}_j = \begin{cases} u_j & \text{if } j \in I \\ 1 & \text{else.} \end{cases}$$

That is, $f^{(I)}$ is obtained by fixing to 1 the variables with indices that are not in I. So, for instance, if $f(\mathbf{u}) = u_1^2 + u_1 u_2 + 3u_3$, then $f^{(\{1,2\})}(u_1, u_2) = u_1^2 + u_1 u_2 + 3$.

We then have

$$V(f) = \sum_{d=1}^{s} \sum_{I:|I|=d} V^{(d)}(f;I).$$

Before going further, it should be noted that, for this definition of variation, several simple functions can be shown to have $V(f) = \infty$. For instance, as pointed out in [213], consider the two-dimensional function

$$f(u_1, u_2) = \begin{cases} 0 & \text{if } u_1 \leq u_2 \\ 1 & \text{otherwise.} \end{cases}$$

That is, f is 0 below the diagonal line that joins (0,0) and (1,1) and 1 above it. It is easy to see that $V(f) = \infty$ for this function. More precisely, we can find partitions of the unit square containing an arbitrarily large number of intervals J along the main diagonal such that $|\Delta(f,J)| = 1$.

Going back to the Koksma-Hlawka inequality (5.22), we can see that it gives a bound on the integration error to which two distinct quantities contribute: $D^*(P_n)$ measures the quality of the point set, and $V(f)$ measures how difficult it is to integrate the function f. Also, the particular choice of norm $V(\cdot)$ used to measure the variability of f is specifically related to the star discrepancy. A different choice of norm $\|\cdot\|_{\mathcal{F}}$ for f would thus lead to a different error bound of the form

$$E_n \leq \|P_n\|_{\mathcal{P}} \|f\|_{\mathcal{F}},$$

where $\|P_n\|_{\mathcal{P}}$ is a certain discrepancy measure associated to $\|\cdot\|_{\mathcal{F}}$ [180, 181, 182]. For instance, in [181], Hickernell shows that a bound similar to (5.22) exists if we replace the star discrepancy by an L_2 version *different* from the more common $T^*(P_n)$ that was defined in (5.20). Indeed, the L_2 star discrepancy $T^*(P_n)$ can be shown to give the *expected* squared error for a certain class of functions, where the expectation is taken over a *Brownian sheet measure* over this set of functions [479].

To get an analogue of (5.22), we must instead use a *generalized* L_2 *discrepancy* defined by

$$D_2(P_n) = \left[\sum_I \int_{[0,1)^d} |\alpha(P_n(I), \mathbf{v}_I) - v_{i_1} \ldots v_{i_d}|^2 \, d\mathbf{v}_I \right]^{1/2}, \tag{5.23}$$

where the sum over $I = \{i_1, \ldots, i_d\}$ runs over all nonempty subsets $I \subseteq \{1, \ldots, s\}$, $\mathbf{v}_I$ represents the d-dimensional vector $(v_{i_1}, \ldots, v_{i_d})$, and the quantity $\alpha(P_n(I), \mathbf{v}_I)$ is defined as the number of points in the d-dimensional projection $P_n(I)$ that fall in

$$\prod_{j=1}^{d} [0, v_{i_j}).$$

Just as was the case for $T^*(P_n)$, a formula for $D_2(P_n)$ exists and is given by [181, Eq. (5.1c)]

$$(D_2(P_n))^2 = \left(\frac{4}{3}\right)^s - \frac{2}{N} \sum_{i=1}^{n} \prod_{j=1}^{s} \left(\frac{3 - u_{i,j}^2}{2} \right) + \frac{1}{n^2} \sum_{i,i'=1}^{n} \prod_{j=1}^{s} [2 - \max(u_{i,j}, u_{i',j})] .$$

The class of functions for which $D_2(P_n)$ can provide an upper bound is one for which the (generalized) L_2 norm $V_2(f)$ is finite, where

$$V_2(f) = \left[\sum_{\emptyset \neq I \subseteq \{1,\dots,s\}} \int_{[0,1)^d} \left| \frac{\partial^d f}{\partial \mathbf{u}_I} \bigg|_{\mathbf{u}_{-I}=(1,\dots,1)} \right|^2 d\mathbf{u}_I \right]^{1/2},$$

and where for $I = \{i_1, \dots, i_d\}$, we define $\mathbf{u}_{-I} = (u_j : j \notin I)$. Then, for any function f such that $V_2(f)$ is finite, the upper bound

$$E_n \leq D_2(P_n)V_2(f)$$

holds for the absolute integration error E_n.

One way to generalize this result is to use an L_p norm to measure the discrepancy and an L_q norm to measure the functions, where p and q are such that $1/p + 1/q = 1$. Several other discrepancy measures are discussed in [181], including some that use weights. We will discuss weighted measures in more detail in the appendix at the end of Chap. 6.

We conclude this subsection with Table 5.1, which summarizes the different discrepancy measures that we discussed in this section. More detailed information on the concept of discrepancy can be found in books such as [27, 89, 228, 308, 339, 429].

Table 5.1 Summary of discrepancy measures discussed in this text

notation	name	anchored?	norm	comments
$D^*(P_n)$	star discrepancy	yes	sup	Used in Koksma-Hlawka inequality.
$D(P_n)$	extreme discrepancy	no	sup	Closely related to $D^*(P_n)$.
$T^*(P_n)$	L_2 star discrepancy	yes	L_2	Can be computed. Used for average-case error analysis.
$T(P_n)$	L_2 discrepancy	no	L_2	Can be computed.
$D_2(P_n)$	generalized L_2 discrepancy	yes	L_2	Used in generalized Koksma-Hlawka inequality. Can be computed.

5.6.2 Criteria based on Fourier and Walsh decompositions

We already described how the concept of $(q_1, \dots, q_s)$-equidistribution relates to the t-value, which is often used to measure the quality of digital nets and sequences. The equidistribution concept can be used to assess the quality of these constructions in several other ways, as we will see in this section.

Also, although the nature of the high uniformity that characterizes lattice point sets is unrelated to this concept of equidistribution, several interesting connections can be drawn between quality measures that are used to assess the uniformity of lattice and net constructions. To do so, it is useful to look at the equidistribution properties of nets from a functional point of view. More precisely, if a point set is $(q_1, \ldots, q_s)$-equidistributed in base b, then a function that is constant on each b-adic elementary interval (or cell) $J(\mathbf{r})$ of the form (5.5) will be integrated with zero error by this point set. This holds because the $(q_1, \ldots, q_s)$-equidistribution property implies that each b-adic elementary interval $J(\mathbf{r})$ contains b^{k-q} points from P_n. Thus a function of the form

$$f(\mathbf{u}) = \sum_{r_1=0}^{b^{q_1}-1} \cdots \sum_{r_s=0}^{b^{q_s}-1} c_{\mathbf{r}} \mathbf{1}_{\mathbf{u} \in J(\mathbf{r})}$$

has its integral approximated by

$$Q_n = \frac{1}{n} \sum_{i=1}^{n} f(\mathbf{u}_i) = \sum_{\mathbf{r}} c_{\mathbf{r}} \frac{b^{m-q}}{b^m} = \sum_{\mathbf{r}} c_{\mathbf{r}} b^{-q} = I(f)$$

since the volume of $J(\mathbf{r})$ is equal to b^{-q}. Therefore the corresponding integration error is 0.

As building blocks for functions like that, we can use *Walsh functions* in base b [26, 171, 172, 240] of the form

$$\xi_{\mathbf{h}}(\mathbf{u}) = e^{2\pi \mathrm{i} \langle \mathbf{h}, \mathbf{u} \rangle_b},$$

where $\mathrm{i} = \sqrt{-1}$, $\mathbf{h} \in \mathbb{N}_0^s$, and the product $\langle \mathbf{h}, \mathbf{u} \rangle_b$ is computed as follows. For each j, write the base b expansion of h_j and u_j

$$h_j = \sum_{l=0}^{\infty} h_{j,l} b^l,$$

$$u_j = \sum_{l=1}^{\infty} u_{j,l} b^{-l}.$$

Then

$$\langle \mathbf{h}, \mathbf{u} \rangle_b = \frac{1}{b} \sum_{j=1}^{s} \sum_{l=0}^{\infty} h_{j,l} u_{j,l+1}, \tag{5.24}$$

where all operations are done in $\mathbb{Z}_b$. (As before, we are assuming b is prime.) For instance, if $b = 2$, $\mathbf{h} = (3, 1)$, and $\mathbf{u} = (0.375, 0.875)$, then

$$\langle \mathbf{h}, \mathbf{u} \rangle_b = \frac{1}{2}(1 \times 0 + 1 \times 1 + 0 \times 1) + (1 \times 1 + 0 \times 1 + 0 \times 1) = 0.$$

Note that

$$\int_{[0,1)^s} \xi_{\mathbf{h}}(\mathbf{u}) d\mathbf{u} = 0.$$

To see how the functions $\xi_{\mathbf{h}}(\mathbf{u})$ can be used to study the $(q_1, \ldots, q_s)$-equidistribution of a point set, observe that the digits

$$u_{1,1}, \ldots, u_{1,q_1}; \ldots; u_{s,1}, \ldots, u_{s,q_s}$$

can be used as labels to identify which cell $J(\mathbf{r})$ of the $(q_1, \ldots, q_s)$-partition $\mathbf{u}$ is falling in. Hence, if $\mathbf{u}$ differs from $\mathbf{v}$ only through digits of the form $u_{j,l}$ with $l > q_j$, then the two points are in the same cell $J(\mathbf{r})$. Now pick an $\mathbf{h}$ such that $h_{j,l} = 0$ for all $l \geq q_j, j = 1, \ldots, s$. Its dot product with $\mathbf{u}$ and $\mathbf{v}$ will be the same. Hence a Walsh function $\xi_{\mathbf{h}}(\mathbf{u})$ with an $\mathbf{h}$ of this form is constant over the b-adic boxes induced by a $(q_1, \ldots, q_s)$-partition and is therefore integrated with zero error by such point sets. More precisely, we have the following lemma.

Lemma 5.16. *Let* $\mathbf{h}$ *be a vector and for each* h_j *consider its expansion*

$$h_j = \sum_{l=0}^{\infty} h_{j,l} b^l$$

in base b*. Let* $d_b(h_j)$ *be the smallest integer* l *such that* $h_{j,l} = 0$ *for all* $l > d_b(h_j)$*. If a point set* $P_n = \{\mathbf{u}_1, \ldots, \mathbf{u}_n\}$ *in base* b *is* $(q_1, \ldots, q_s)$*-equidistributed for values* q_j *given by*

$$q_j = d_b(h_j) + 1$$

for each $j = 1, \ldots, s$*, then*

$$\frac{1}{n} \sum_{i=1}^{n} \xi_{\mathbf{h}}(\mathbf{u}_i) = 0.$$

Note that in the case $b = 2$ we can write $\xi_{\mathbf{h}}(\mathbf{u}) = (-1)^{\langle \mathbf{h}, \mathbf{u} \rangle_2}$ because

$$\begin{aligned} e^{2\pi \mathrm{i} \langle \mathbf{h}, \mathbf{u} \rangle_2} &= \cos 2\pi \langle \mathbf{h}, \mathbf{u} \rangle_2 + \mathrm{i} \sin 2\pi \langle \mathbf{h}, \mathbf{u} \rangle_2 \\ &= \cos 2\pi \langle \mathbf{h}, \mathbf{u} \rangle_2 = \begin{cases} 1 & \text{if } \langle \mathbf{h}, \mathbf{u} \rangle_2 = 0 \\ -1 & \text{if } \langle \mathbf{h}, \mathbf{u} \rangle_2 = 1/2. \end{cases} \end{aligned}$$

If we look at the vector $(q_1, \ldots, q_s)$ for which $q_j = d_2(h_j) + 1$, the function $\xi_{\mathbf{h}}(\mathbf{u})$ alternates between 1 and -1 over all dyadic boxes $J(\mathbf{r})$ in the corresponding $(q_1, \ldots, q_s)$-partition. Figure 5.15 illustrates the sign change pattern for $s = 2$ and $\mathbf{h} = (3, 1)$ over the corresponding dyadic $(d_2(h_1) + 1, d_2(h_2) + 1) = (2, 1)$-partition.

Alternatively, the vectors $\mathbf{h}$ can be represented as polynomials. More precisely, for a given $\mathbf{h} = (h_1, \ldots, h_s)$ where each h_j has the decomposition

$$h_j = \sum_{l=0}^{\infty} h_{j,l} b^l,$$

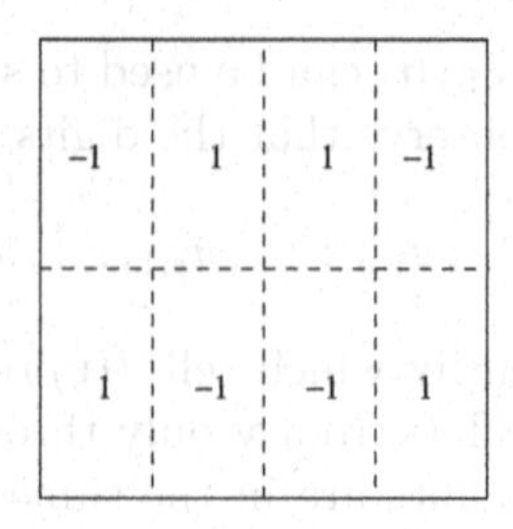

Fig. 5.15 Function $(-1)^{\langle \mathbf{h},\mathbf{u}\rangle_2}$ over $[0,1)^2$ for $\mathbf{h} = (3,1)$.

we associate the vector $\mathbf{h}(z) = (h_1(z), \ldots, h_s(z))$ of polynomials where each $h_j(z) \in \mathbb{F}_b[z]$ is given by

$$h_j(z) = \sum_{l=0}^{\infty} h_{j,l} z^l.$$

From our discussion above, it is easy to see that the quantity $d_b(h_j)$ introduced in Lemma 5.16 simply corresponds to the degree of the polynomial $h_j(z)$ associated to h_j, where we assume that $\deg(0) = -1$. Hence we can reformulate the result of this lemma by saying that if a point set is $(q_1, \ldots, q_s)$-equidistributed with $q_j = \deg(h_j(z)) + 1$, then $\xi_{\mathbf{h}}(\mathbf{u})$ is integrated with zero error by this point set. We can also look at

$$\tau(\mathbf{h}(z)) = \sum_{j=1}^{s} (\deg(h_j(z)) + 1), \tag{5.25}$$

which turns out to be related to the t-value. More precisely, for a given (t,k,s)-net, if $\tau(\mathbf{h}(z)) \leq k - t$, then $\xi_{\mathbf{h}}(\mathbf{u})$ is integrated with zero error by the net. (Problem 5.12 asks you to prove this.)

So far we only gave *sufficient* conditions for $\xi_{\mathbf{h}}(\mathbf{u})$ to be integrated with zero error. In order to find necessary *and* sufficient conditions, we need to talk about the *dual space* of a digital net. Before we do that, let us discuss lattices first as for these we already introduced the concept of a dual lattice in Chap. 3, but we will recall it here for the sake of completeness.

For lattices, the functions that can be used as building blocks to understand which functions are well integrated by a lattice point set are those from a Fourier basis. More precisely, consider the function

$$\nu_{\mathbf{h}}(\mathbf{u}) = e^{2\pi i \mathbf{h}\cdot\mathbf{u}},$$

where the dot product $\mathbf{h} \cdot \mathbf{u}$ is now simply the usual

$$\mathbf{h} \cdot \mathbf{u} = \sum_{j=1}^{s} h_j u_j. \tag{5.26}$$

For any nonzero $\mathbf{h}$, the integral of $\nu_{\mathbf{h}}(\mathbf{u})$ over $[0,1)^s$ is zero.

It can be shown that as long as $\mathbf{h}$ is not in the *dual lattice* corresponding to a lattice point set P_n, then $\nu_{\mathbf{h}}(\mathbf{u})$ is integrated with zero error [408]. Recall that the dual lattice is the set

$$L_s^* = \{\mathbf{h} \in \mathbb{R}^s : \mathbf{h} \cdot \mathbf{u}_i \in \mathbb{Z} \text{ for all } \mathbf{u}_i \in P_n\}.$$

More precisely, we have [408, Theorem 1] the following lemma.

Lemma 5.17. *If P_n is a lattice point set, then*

$$\frac{1}{n}\sum_{i=1}^{n} e^{2\pi i \mathbf{h}\cdot\mathbf{u}_i} = \begin{cases} 1 & \text{if } \mathbf{h} \in L_s^* \\ 0 & \text{otherwise.} \end{cases}$$

For digital nets, one can also define a *dual space* $\mathcal{C}_s^*$ based on the generating matrices $C_1, \ldots, C_s$. More precisely, let $C_1, \ldots, C_s$ be the $\infty \times k$ generating matrices associated with a digital net in base b with $n = b^k$ points, and let C be the $k \times \infty$ matrix obtained by concatenating the transpose of each C_j; that is,

$$C = (C_1^{\mathrm{T}}|\ldots|C_s^{\mathrm{T}}).$$

Let $\mathcal{C}_s^*$ be the null space of the row space of C,

$$\mathcal{C}_s^* = \{\mathbf{h} \in \underbrace{\mathbb{F}_b^\infty \times \ldots \times \mathbb{F}_b^\infty}_{s \text{ times}} : C \cdot \mathbf{h} = \mathbf{0}\}, \tag{5.27}$$

where the product $C \cdot \mathbf{h}$ is given by the k-dimensional vector

$$\begin{pmatrix} \sum_{j=1}^{s}\sum_{l=1}^{\infty} c_{j,l,1}h_{j,l-1} \\ \vdots \\ \sum_{j=1}^{s}\sum_{l=1}^{\infty} c_{j,l,k}h_{j,l-1} \end{pmatrix}.$$

The following result [265, Lemma 2] is the equivalent of Lemma 5.17 for digital nets.

Lemma 5.18. *Let P_n be a digital net in base b with b^k points, and let $\mathcal{C}_s^*$ be defined as in (5.27). Then*

$$\frac{1}{n}\sum_{I=1}^{n} \xi_{\mathbf{h}}(\mathbf{u}_i) = \begin{cases} 1 & \text{if } \mathbf{h} \in \mathcal{C}_s^* \\ 0 & \text{otherwise.} \end{cases}$$

With this in mind, one way to draw a parallel between lattice point sets and digital nets is to observe that they both perfectly integrate basis functions of the form $e^{2\pi i \mathbf{h}\cdot\mathbf{u}}$ and $e^{2\pi i \langle \mathbf{h},\mathbf{u}\rangle_b}$ — for lattices and nets, respectively — where $\mathbf{h}$ is a vector that is not in the dual space corresponding to the point set.

Furthermore, let us assume, for the sake of argument, that for typical functions arising in practice, the most important terms in their Walsh or

Fourier decomposition are those associated with wave functions $\nu_{\mathbf{h}}(\mathbf{u})$ or $\xi_{\mathbf{h}}(\mathbf{u})$ with a "small" $\mathbf{h}$. If the "shortest" $\mathbf{h}$ in the dual space is "big" enough, it means several functions with a small $\mathbf{h}$ are perfectly integrated by P_n, and thus a large part of f is correctly integrated by P_n. From this point of view, it makes sense to choose P_n so that the smallest $\mathbf{h}$ in the dual space is as large as possible.

Several connections between nets and lattices can be done by looking at quality measures based on the property above. The t-value can be written as the "length" of the shortest vector in the dual space of the digital net by using a certain measure of distance or weight (see [343, 405] and prior to that [438] for $b = 2$). That is, we can write the t-value as [343]

$$t = k + 1 - \min_{0 \neq \mathbf{h} \in C_s^*} \tau(\mathbf{h}(z)), \tag{5.28}$$

where we use the representation of $\mathbf{h}$ as a polynomial when writing $\tau(\mathbf{h}(z))$, which was defined in (5.25). Similarly, the *resolution* of the digital net — which is the largest ℓ_s such that the net is $(\ell_s, \ldots, \ell_s)$-equidistributed — is given by [265]

$$\ell_s = -1 + \min_{0 \neq \mathbf{h} \in C_s^*} \|\mathbf{h}\|_\infty,$$

where

$$\|\mathbf{h}\|_\infty = \max_{1 \leq j \leq s} (d_b(h_j) + 1)$$

and $d_b(h)$ was defined in Lemma 5.16. This result has been widely studied in the case where the net comes from a recurrence-based point set derived from different types of $\mathbb{F}_2$-linear generators in [66, 67, 436, 437].

Similarly, a measure sometimes used to assess the quality of lattices is the *Babenko-Zaremba index* [25, 24], defined as

$$\rho = \min_{0 \neq \mathbf{h} \in L^*} \|\mathbf{h}\|_\pi,$$

where

$$\|\mathbf{h}\|_\pi = \prod_{j=1}^{s} \max(1, |h_j|).$$

This is closely related to the quantity

$$l_s = \min_{0 \neq \mathbf{h} \in L_s^*} \|\mathbf{h}\|_2,$$

where

$$\|\mathbf{h}\|_2 = \left(\sum_{j=1}^{s} h_j^2 \right)^{1/2},$$

which is computed in the spectral test that was discussed in Chap. 3 as a way of measuring the quality of the lattice Ψ_s induced by an MRG. The only difference is that for ρ we use the sup norm to compute the shortest vector in the dual lattice, while for l_s we use the usual L_2 norm. Bounds relating ρ and l_s can be found in [107] along with parameters for Korobov point sets based on a quality measure that depends on the spectral test.

Each of the measures t, ℓ_s, and l_s can be used within more global criteria that evaluate several projections, such as the quantities $M_{\mathcal{I}}$ and $\Delta_{\mathcal{I}}$ defined in Sect. 3.5.1. We will come back to this in Chap. 6. Also, while these three measures enjoy nice geometric interpretations, it is not the case for ρ, which also turns out to be quite difficult to compute. For this reason, tables giving good parameters with respect to ρ are usually limited to small values of s, like $s \leq 10$ [300].

For lattice point sets, a more popular measure based on the product norm $\|\cdot\|_\pi$ is to use the *weighted* P_α [182], given by

$$\tilde{\mathcal{P}}_\alpha = \sum_{\mathbf{0} \neq \mathbf{h} \in L_s^*} \beta_I \|\mathbf{h}\|_\pi^{-\alpha}, \tag{5.29}$$

where β_I is a nonnegative weight that depends on the set of indices

$$I = I(\mathbf{h}) := \{j : h_j \neq 0\}.$$

These weights can be used to give more or less importance to the different projections $P_n(I)$. Compared with the measures ρ and ℓ_s, here we compute a weighted sum of the inverse of the length of the vectors in the dual lattice rather than focusing on the shortest one. Based on this interpretation, it is clear that a *smaller* $\tilde{\mathcal{P}}_\alpha$ is preferred.

This measure generalizes the $\mathcal{P}_\alpha$ studied in [407] and the references therein, in which the weights are set to 1. The weighted P_α can also be used as the "discrepancy" component of an error bound of the type (5.22) but for a weighted space of periodic functions [182, Eq. (4.8c)]. From this point of view, as before, we conclude that a smaller $\tilde{\mathcal{P}}_\alpha$ is preferred.

If the weights β_I are given by a product of the form

$$\beta_I = \beta_0 \prod_{j \in I} \beta_j^\alpha,$$

then the infinite sum defining $\tilde{\mathcal{P}}_\alpha$ can be shown to be equal to a sum over the n points in P_n. More precisely, for α a positive even integer, we have [182, Eq. (4.15)]

$$\tilde{\mathcal{P}}_\alpha = \beta_0 \left\{ -1 + \frac{1}{n} \sum_{i=0}^{n-1} \prod_{j=1}^{s} \left[1 - (-\beta_j^2)^{\alpha/2} \frac{(2\pi)^\alpha}{\alpha!} B_\alpha(u_{ij}) \right] \right\}, \tag{5.30}$$

where $B_\alpha(\cdot)$ is the Bernoulli polynomial of degree α. The first few Bernoulli polynomials are given by [1]

$$\begin{aligned}
B_0(x) &= 1,\\
B_1(x) &= x - 1/2,\\
B_2(x) &= x^2 - x + 1/6,\\
B_3(x) &= x^3 - 3x^2/2 + x/2,\\
B_4(x) &= x^4 - 2x^3 + x^2 - 1/30.
\end{aligned}$$

The compact formulation (5.30) is used when this criterion is computed in practice. Several tables of good parameters for lattice point sets are based on searches made using criteria related to $\tilde{\mathcal{P}}_\alpha$. Examples can be found in [86, 160, 412, 407] and more recently in [186, 410].

Related to the weighted P_α is the concept of *diaphony*, introduced by Zinterhof [484]. More precisely, the diaphony of a point set P_n is given by

$$F(P_n) = \left(\sum_{\mathbf{h} \neq \mathbf{0}} \|\mathbf{h}\|_\pi^2 S_n^2(\mathbf{h}) \right)^{1/2}, \tag{5.31}$$

where

$$S_n(\mathbf{h}) = \frac{1}{n} \sum_{i=1}^{n} e^{2\pi i \mathbf{h}\cdot\mathbf{u}_i}.$$

Note that, for a lattice, the diaphony $F(P_n)$ and $\mathcal{P}_2$ are equal. (Problem 5.18 asks you to prove this.) For general point sets, Zinterhof [484] proved the important identity

$$F^2(P_n) = \frac{1}{n^2} \sum_{i=1}^{n} \sum_{i'=1}^{n} g((\mathbf{u}_i - \mathbf{u}_{i'}) \bmod 1), \tag{5.32}$$

where

$$g(\mathbf{u}) = -1 + \prod_{j=1}^{s} \left(1 - \frac{\pi^2}{6} + \frac{\pi^2}{2}(1 - 2u_j)^2 \right).$$

Another connection between nets and lattices can be established through the general concept of a *weighted spectral test* [174]. This quantity, denoted $F_r(P_n)$, generalizes the diaphony by replacing the term $\|\mathbf{h}\|_\pi^2$ by a weight $r(\mathbf{h})$ in the sum (5.31) and also by generalizing the quantity $S_n(\mathbf{h})$ to be based on either Fourier or Walsh basis functions. That is, the weighted spectral test is defined as

$$F_r(P_n) = \left(\sum_{\mathbf{h} \neq \mathbf{0}} r(\mathbf{h}) \tilde{S}_n^2(\mathbf{h}) \right)^{1/2},$$

where

$$\tilde{S}_n(\mathbf{h}) = \begin{cases} \frac{1}{n}\sum_{i=1}^{n} e^{2\pi \mathrm{i} \mathbf{h}\cdot\mathbf{u}_i} & \text{for Fourier} \\ \frac{1}{n}\sum_{i=1}^{n} e^{2\pi \mathrm{i} \langle \mathbf{h},\mathbf{u}_i\rangle_b} & \text{for Walsh,} \end{cases}$$

and where the products $\mathbf{h}\cdot\mathbf{u}$ and $\langle \mathbf{h},\mathbf{u}\rangle_b$ were defined in (5.26) and (5.24), respectively. The weight function $r(\cdot)$ must satisfy the following three conditions: (i) $r(\mathbf{h}) > 0$ for all $\mathbf{h}$; (ii) $r(\mathbf{0}) = 1$; and (iii) $\sum_{\mathbf{h}} r(\mathbf{h}) < \infty$.

In addition to the classical diaphony, another special case of the weighted spectral test is the *dyadic diaphony* introduced by Hellekalek and Leeb [173], which can be viewed as the digital, base 2, version of the diaphony. More precisely, it is obtained by taking the Walsh functions in base 2 and the weight function

$$r(\mathbf{h}) = \frac{1}{3^s - 1}\prod_{j=1}^{s} \rho(h_j), \tag{5.33}$$

where

$$\rho(h_j) = \begin{cases} 2^{-2g} & \text{if } 2^g \le h_j < 2^{g+1} \\ 1 & \text{if } h_j = 0. \end{cases}$$

Formulated using the polynomial setup, this means that for nonzero h_j we have $\rho(h_j) = 2^{-2d_j}$, where d_j is the degree of $h_j(z)$. It is shown in [173] that the dyadic diaphony can be computed as

$$\left(\frac{1}{3^s-1}\frac{1}{n^2}\sum_{i=1}^{n}\sum_{i'=1}^{n} \zeta(\mathbf{u}_i \oplus \mathbf{u}_{i'})\right)^{1/2},$$

where

$$\zeta(\mathbf{u}) = -1 + 3^s \prod_{j=1}^{s}\left(\mathbf{1}_{u_j=0} + \mathbf{1}_{0<u_j<1}(1 - 2^{1+\lfloor \log_2 u_j \rfloor})\right).$$

A weighted version of the dyadic diaphony was introduced in [285] and defined by replacing the weight function (5.33) by

$$r(\mathbf{h}) = \frac{\beta_0}{3^s - 1}\prod_{j=1}^{s} \beta_j \rho(h_j)$$

for some weights $\beta_0, \ldots, \beta_s > 0$. It is shown in [285] that, for polynomial integration lattices in base 2, this weighted dyadic diaphony can be computed as

$$\left(\frac{1}{3^s-1}\frac{\beta_0}{n}\sum_{i=0}^{n-1} \tilde{\psi}(\mathbf{u}_i)\right)^{1/2},$$

where

$$\tilde{\psi}(\mathbf{u}) = -1 + \prod_{j=1}^{s}\left[1 + 2\beta_j^2\left(1 - 3\cdot 2^{\lfloor \log_2 u_j \rfloor}\right)\right].$$

The weighted dyadic diaphony is also studied in [227], where an additional parameter $\alpha > 1$ similar to the one used to define $\tilde{\mathcal{P}}_\alpha$ is introduced.

Our discussion of which types of functions are integrated perfectly by lattice point sets leads us to the observation that the periodicity of the Fourier basis functions suggests that lattices work best with periodic functions. As a matter of fact, the integration error based on a lattice P_n can be shown to be [408]

$$Q_n - I(f) = \sum_{\mathbf{0} \neq \mathbf{h} \in L_s^*} \hat{f}(\mathbf{h}) \tag{5.34}$$

for functions whose Fourier expansion is absolutely convergent and where $\hat{f}(\mathbf{h})$ is the Fourier coefficient of f evaluated at $\mathbf{h}$. An error bound of the Koksma-Hlawka type can then be obtained, where $D^*(P_n)$ is replaced by the weighted P_α [180, 182, 407].

Since f needs to be one-periodic with respect to each u_j in order for (5.34) to hold, several ways of periodizing f have been proposed [197, 407, 483]. Typically, and as explained in [407, Sect. 2.12], the idea is to choose a transformation $\eta : [0,1] \to [0,1]$ that is smooth, increasing, and such that $\eta'(0) = \eta'(1) = 0$. If we transform f into

$$\tilde{f}(u_1, \ldots, u_s) = f(\eta(u_1), \ldots, \eta(u_s))\eta'(u_1) \ldots \eta'(u_s), \tag{5.35}$$

then $\tilde{f}$ is periodic since $\tilde{f}(u_1, \ldots, u_s) = 0$ whenever $u_j \in \{0,1\}$ for some j, and also, from calculus, we know that

$$\int_{[0,1)^s} \tilde{f}(\mathbf{u}) d\mathbf{u} = \int_{[0,1)^s} f(\mathbf{u}) d\mathbf{u}.$$

For instance, Sidi proposed [403]

$$\eta(t) = t - \frac{1}{2\pi} \sin(2\pi t),$$

which was used with success in [40] for option pricing in finance.

Transformations can also be applied to the lattice points themselves. One such example is the *baker transformation* of Hickernell [183], where each coordinate $u_{i,j}$ is replaced by $2u_{i,j}$ if $u_{i,j} < 0.5$ and by $2(1 - u_{i,j})$ otherwise. Applying this transformation to a lattice point set makes it possible to get error bounds for nonperiodic integrands that are similar to those obtained for periodic integrands (and original lattice point sets). This transformation has been shown to be useful for practical problems in [69, 263]. We can also think of the baker transformation as periodizing the integrand based on the function

$$\eta(u) = 1 - |2u - 1|,$$

and since the absolute value of this derivative (where it exists) is one, $f(\eta(u_1), \ldots, \eta(u_s))$ integrates to the same thing as f, which is why we can think of this method as simply changing the point set without affecting the integrand. This is in contrast with periodizations based on an increasing η, where the corresponding integrand $\tilde{f}$ given in (5.35) is different from f. It should also be pointed out that recent work by Kuo et al. on periodization indicates these transformations may fail in high dimensions [232].

5.6.3 Motivation for going beyond error bounds

Going back to the Koksma-Hlawka inequality (5.22), it is important to mention that there are serious limitations that prevent this result from being really useful in practical settings. First, even for moderate values of s, n must be very large in order for $\log^s n/n$ to be smaller than $1/\sqrt{n}$. For instance, for $s = 10$, n must be at least about 10^{39} for the inequality to hold. Second, the condition $V(f) < \infty$ often is not met for functions arising in practical applications. In computer graphics, f often includes an indicator function for sets whose boundaries are not parallel to the axes [213], which results in an unbounded $V(f)$. Functions with infinite variation are also often encountered in derivative pricing in finance [366]. Finally, even when $V(f) < \infty$, the inequality (5.22) cannot reliably be used to give an idea of the error since it only provides an upper bound, which turns out to be very hard to compute anyway since both D_n^* and $V(f)$ are hard to compute.

In practical settings where the user wants an estimate of the integration error $|Q_n - I(f)|$ for a given n, point set P_n, and function f, something other than the Koksma-Hlawka inequality must be used. At this point, it might be tempting to revert to Monte Carlo, for which an easy way of estimating the error is available. If we remind ourselves why it is so — Monte Carlo is based on independent random samples that allow simple variance estimates of $\hat{\mu}$ to be made, and the central limit theorem then provides a way to construct confidence intervals — a natural next question is "How could we allow error estimation through random sampling in the context of quasi–Monte Carlo"? Randomized quasi–Monte Carlo is an answer to that question and will be discussed in the next chapter.

Problems

5.1. Show that a rank-1 lattice point set with n points based on the generating vector $(z_1, \ldots, z_s)$ is fully projection-regular if and only if $\gcd(z_j, n) = 1$ for all $j = 1, \ldots, s$.

5.2. We briefly mentioned copy rules on p. 148. Following [277], we define a ν^r *copy rule* point set as being of the form

$$P_n = \bigcup_{m_1=0}^{\nu-1} \cdots \bigcup_{m_r=0}^{\nu-1} \bigcup_{i=1}^{p} \{((m_1/\nu, \ldots, m_r/\nu, \underbrace{0, \ldots, 0}_{s-r \text{ times}}) + \mathbf{u}_i) \bmod 1\}, \quad (5.36)$$

where $\{\mathbf{u}_i,\ i = 1, \ldots, p\}$ is a rank-1 lattice point set and $\nu \geq 2$ is an integer such that $\gcd(p, \nu) = 1$. Hence the number of points is $n = \nu^r p$. Show that P_n as given in (5.36) is not fully projection-regular for $s \geq 2$.

5.3. Write a program that, given an integer s, a generating vector $\mathbf{z} = (z_1, \ldots, z_s)$ in $\mathbb{Z}^s$, and an integer $i \geq 1$, returns the ith point from an extensible lattice sequence in base 2 based on $\mathbf{z}$.

5.4. (a) Show that the first b^k points of an extensible lattice sequence form a lattice point set. (b) Show that, for $k \geq 2$, the set of the first 2^k points from an extensible lattice sequence in base 2 can be obtained by taking the union of the first 2^{k-1} points of the sequence and the set

$$\left\{\frac{i}{2^k}(z_1, \ldots, z_s) \bmod 1, i = 1, 3, \ldots, 2^k - 1\right\}.$$

5.5. (a) Write a program that, given an integer s and an integer $i \geq 1$, outputs the ith point of the Halton sequence. (b) Repeat (a) but for a generalized Halton sequence that uses permutations based on multiplicative factors f_j for $j = 1, \ldots, s$. An example of good factors can be found at [498]. (c) For each of (a) and (b), write a program that returns the ith point to which a random digital $(b_1, \ldots, b_s)$-shift is added (see App. B).

5.6. (a) Consider the first 2^k points of the Sobol' sequence in $[0, 1)^s$, where $k < d_s$ and d_s is the degree of the sth smallest primitive polynomial in base 2. Show that the t-value of the point set obtained might depend on the choice of direction numbers. (b) Consider a projection of the form $P_n(\{j, j+1\})$ with $n = 2^k$ for the Sobol' sequence. For $j = 10, 20, 30, 40, 50, 100$, find the smallest value of k such that the t-value for $P_n(\{j, j+1\})$, denoted $t_{\{j,j+1\}}$, is independent of the direction numbers chosen in dimension j and $j+1$.

5.7. Show that the Niederreiter sequence in base 2 does not include the Sobol' sequence as a special case.

5.8. Compare the value of $T(P_n)$ — using the formula (5.21) — for $n = 1000, 5000, 10000$ and $s = 10, 20, 40$ for the Sobol' sequence with (i) the direction numbers as in [501] and (ii) setting the direction numbers $v_{j,k} = m_{j,k}/2^k$ by choosing $m_{j,k} = 1$ for $k = 1, \ldots, d_j$. (Code to implement the Sobol' sequence is available on the Web. For instance, a widely used source is [501] from the paper [43].)

5.9. (a) Show that for each $I \subseteq \{1, \ldots, s\}$ and each $k \geq 0$, for $n = b^k$, the projection $P_n(I)$ of the Faure sequence in base b has a t-value equal to 0. (b) Show that the $n = b^k$ first points of the Faure sequence in base $b \geq s$ form a dimension-stationary point set. Show that this is not necessarily true for a generalized Faure sequence based on nonsingular lower-triangular matrices $A_1, \ldots, A_s$. (c) Show that the first n points of the Sobol' sequence do not form a dimension-stationary point set.

5.10. (a) Consider a Tausworthe generator specified by a trinomial of the form $P(z) = z^k + z^r + 1$ and parameters ν and L, where $\gcd(\nu, 2^k - 1) = 1$. Write a program that, given an integer s, generates the corresponding s-dimensional recurrence-based point set. (b) Suppose now that you have two Tausworthe generators as above. Repeat (a) for the combined generator based on these two Tausworthe generators. (c) Repeat (b) with three components.

5.11. Show that the recurrence (5.18) defines a bijection over the ring $\mathbb{F}_b[z]/(p(z))$, which we identify with the set of polynomials in $\mathbb{F}_b[z]$ of degree less than $k = \sum_{j=1}^{J} k_j$.

5.12. Prove the statement on p. 190 saying that if $\tau(\mathbf{h}(z)) \leq k - t$, then $\xi_{\mathbf{h}}(\mathbf{u})$ is integrated with a zero error by a (t, k, s)-net, where $t \leq k$.

5.13. Show the *propagation rule* that says that if P_n is a (t, k, s)-net, then for $u < k$ the first b^u points of P_n form a (t, u, s)-net.

5.14. Define t_I to be the t-value of the projection $P_n(I)$ of a digital net P_n. Show that $t = \max_{\emptyset \neq I \subseteq \{1,\ldots,s\}} t_I$.

5.15. (a) Compute the value of $T^*(P_n)$, $T(P_n)$, and $D_2(P_n)$ for $n = 1000, 5000, 10000, 20000, 50000$ and P_n obtained as (i) first n points of the Halton sequence; (ii) first n points of the generalized Halton sequence implemented in Prob. 5.5; and (iii) an extensible Korobov lattice sequence in base 2 based on the generator $a = 14471$. Use $s = 5, 10, 20, 50$. (b) Repeat (a) but only for the two-dimensional projection $P_n(\{39, 40\})$ and the values of n listed in (a).

5.16. Show that (5.30) is a valid formula for $\tilde{\mathcal{P}}_\alpha$ using the fact that for the Bernoulli polynomial of degree α — where α is even — we have the Fourier expansion

$$B_\alpha(u) = -\alpha! \sum_{h \neq 0} \frac{e^{2\pi i h u}}{(2\pi i h)^\alpha}, \quad 0 < u < 1.$$

5.17. Show that applying the baker transformation to a point set is equivalent to using the original point set to the function $f(\eta(u_1), \ldots, \eta(u_s))$, where $\eta(u) = 1 - |2u - 1|$.

5.18. Prove that the formula (5.32) for the diaphony is equivalent to (5.30) when $\alpha = 2$ and the underlying point set is a lattice point set.

Chapter 6
Using Quasi–Monte Carlo in Practice

6.1 Introduction

In the preceding chapter, we presented several constructions that can be used for quasi–Monte Carlo sampling and discussed how to assess their quality. In this chapter, we focus on issues that arise when applying quasi–Monte Carlo methods in practice. We first discuss randomized quasi–Monte Carlo, which, as we mentioned at the end of the previous chapter, is an essential tool to make low-discrepancy sampling applicable in practice. In Sect. 6.3, we discuss ANOVA decompositions, which have been very useful for understanding the success of quasi–Monte Carlo methods in practice. We discuss in Sect. 6.4 the use of quasi–Monte Carlo sampling in simulation studies and how it can be combined with other variance reduction techniques. We conclude in Sect. 6.5 with a short discussion of different issues and suggestions that might be helpful to practitioners.

We include an appendix to this chapter, where we briefly discuss the concept of *tractability* and related results that have had a great impact on the construction of low-discrepancy point sets over the last few years. This area of study has connections with ANOVA decompositions, which is why we chose to present it in this chapter rather than the previous one, but it does not exactly fit with the more simulation-oriented issues discussed in the rest of the chapter, which is why we put it in an appendix.

This chapter does not focus on specific applications. The next chapter will discuss the use of quasi–Monte Carlo sampling in finance, which is probably the most well-known application for these methods. Another area where quasi–Monte Carlo has been quite successful is computer graphics [213, 460]. The survey [364] by Owen describes quasi–Monte Carlo sampling for people working in that area.

C. Lemieux, *Monte Carlo and Quasi–Monte Carlo Sampling*,
Springer Series in Statistics 692, DOI: 10.1007/978-0-387-78165-5_6,

6.2 Randomized quasi–Monte Carlo

The fact that the Monte Carlo method is based on an i.i.d. sample of points makes it easy to get error estimates when applying this method. Since low-discrepancy point sets do not have this property, we cannot directly estimate the error in the same fashion. However, we can create a *random sample of quasi-random estimators*, each based on a low-discrepancy point set of size n. More precisely, randomized quasi–Monte Carlo consists in choosing a deterministic low-discrepancy point set P_n and applying a randomization such that (i) each point $\tilde{\mathbf{u}}_i$ in the randomized point set $\tilde{P}_n$ is $U([0,1)^s)$ and (ii) the low discrepancy of P_n is preserved (in some sense) after the randomization.

Condition (i) guarantees that the estimator based on $\tilde{P}_n$ is unbiased. This is because

$$\mathrm{E}\left(\frac{1}{n}\sum_{i=0}^{n-1} f(\mathbf{u}_i)\right) = \frac{1}{n}\sum_{i=0}^{n-1}\mathrm{E}(f(\mathbf{u}_i)) = \frac{1}{n}\sum_{i=0}^{n-1}\int_{[0,1)^s} f(\mathbf{u}_i)d\mathbf{u}_i = I(f),$$

where the second equality comes from the fact that each $\mathbf{u}_i \sim U([0,1)^s)$. Condition (ii) is a natural one to ask for because our main motivation for using quasi-random sampling is that we expect the low discrepancy of the underlying point set P_n to produce a more accurate estimator than Monte Carlo. We do not want this advantage to be lost by using a randomization that would destroy this low discrepancy and take us back to random sampling.

In general, randomizations are designed for a certain class of constructions so that at least one of the characterizations of the construction's low discrepancy is preserved. For example, the random shift mentioned briefly on p. 175 is designed for lattice point sets, in which points have the property of lying on parallel equidistant lines, and this property is preserved after the shift is applied.

Once a randomization method is chosen, we can create a sample of m i.i.d. estimators of the form

$$\hat{\mu}_{\mathrm{rqmc},l} = \frac{1}{n}\sum_{i=0}^{n-1} f(\tilde{\mathbf{u}}_{i,l}),$$

where $\{\tilde{\mathbf{u}}_{i,1}, i = 0,\ldots,n-1\},\ldots, \{\tilde{\mathbf{u}}_{i,m}, i = 0,\ldots,n-1\}$ are m independent randomized copies of P_n. For instance, with the random shift method, we have

$$\tilde{\mathbf{u}}_{i,l} = (\mathbf{u}_i + \mathbf{v}_l) \bmod 1,$$

where $\mathbf{v}_1,\ldots,\mathbf{v}_m$ are m i.i.d. uniform vectors over $[0,1)^s$. With these m i.i.d. estimators, we can construct the unbiased estimator

$$\hat{\mu}_{\mathrm{rqmc}} = \frac{1}{m}\sum_{l=1}^{m}\hat{\mu}_{\mathrm{rqmc},l}$$

for $I(f)$ and estimate its variance, denoted $\mathrm{Var}(\hat{\mu}_{\mathrm{rqmc}})$, by the unbiased estimator

$$\hat{\sigma}^2_{m,\mathrm{rqmc}} = \frac{1}{m}\hat{\sigma}^2_{\mathrm{rqmc}},$$

where

$$\hat{\sigma}^2_{\mathrm{rqmc}} = \frac{1}{m-1}\sum_{l=1}^{m}(\hat{\mu}_{\mathrm{rqmc},l} - \hat{\mu}_{\mathrm{rqmc}})^2 \tag{6.1}$$

is an unbiased estimator of $\mathrm{Var}(\hat{\mu}_{\mathrm{rqmc},l})$. The empirical variance $\hat{\sigma}^2_{m,\mathrm{rqmc}}$ can then be compared with the one obtained from a Monte Carlo estimator based on a total of nm sample points or with other randomized quasi–Monte Carlo estimators. In addition, if m is large enough, one can construct confidence intervals for $I(f)$ based on the randomized quasi–Monte Carlo estimator.

This brings us to an important question that is often raised when presenting the randomized quasi–Monte Carlo approach: For a fixed computing budget, how should we choose the number of points n and the number of randomizations m relative to each other? There is no obvious answer to this question. A large n has the benefit of getting an increased quality/uniformity from the low-discrepancy point set, possibly with a faster error reduction than when m is increased. This is because in some settings (to be discussed shortly) it can be shown that the variance of $\hat{\mu}_{\mathrm{rqmc},l}$ is in $O(n^{-3}\log^{s-1} n)$, while in terms of m we only have that the variance of $\hat{\mu}_{\mathrm{rqmc}}$ is the usual $O(1/m)$ that we get with Monte Carlo. In other words, in good scenarios we have $\mathrm{Var}(\hat{\mu}_{\mathrm{rqmc}}) \in O(\log^{s-1} n/(mn^3))$, so if we want to do better than $\mathrm{Var}(\hat{\mu}_{\mathrm{mc}}) \in O(1/nm)$, it seems like we should take n as large as possible. On the other hand, m must be taken large enough — say $m \geq 10$ — so that the variance estimate (6.1) is sufficiently reliable.

Randomization can also be used to improve the quality of a point set or sequence. For example, we saw in Sect. 5.4.4 that one way of improving the quality of the Halton sequence was to use permutations, and that Faure sequences could be improved by using nonsingular lower-triangular (NLT) matrices multiplying the generating matrices. While these improvements can be chosen in a deterministic way, they can also be chosen randomly. Sometimes this is done only once and the resulting (randomized) point set is then used just as in the deterministic quasi–Monte Carlo framework [348, 440]. That is, no independent repetitions of this process are done in order to estimate the error or variance.

We now describe the most common approaches used to randomize low-discrepancy point sets.

6.2.1 Random shift (or rotation sampling)

A very simple randomization method is to use a *random shift* [72], also called a *Cranley-Patterson rotation* or *rotation sampling*, as shown on Fig. 6.1.

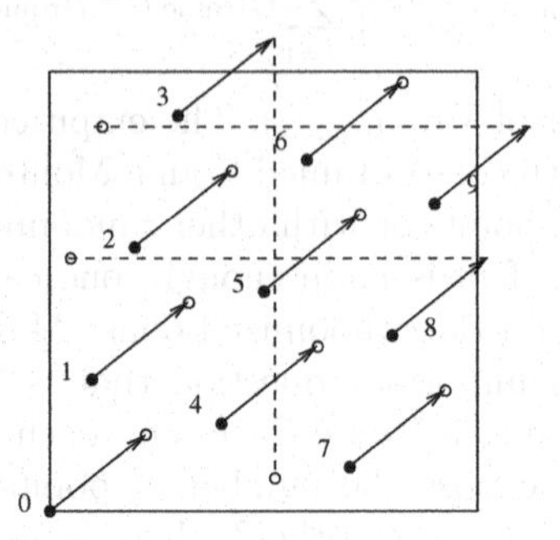

Fig. 6.1 A shifted point set. Original points are marked with filled circles and shifted points are marked with white circles. Dotted lines indicate the effect of the mod1 operation.

As discussed before, the idea here is to generate a uniform random vector $\mathbf{v} \sim U([0,1)^s)$ and then let

$$\tilde{\mathbf{u}}_i = (\mathbf{u}_i + \mathbf{v}) \bmod 1,$$

for $i = 1, \ldots, n$, where the modulo 1 operation is taken coordinatewise. Since $\mathbf{v}$ is uniform, each point $\tilde{\mathbf{u}}_i$ in the randomized point set is also uniformly distributed. For the sake of completeness, we prove this in the following proposition.

Proposition 6.1. *Let* $\mathbf{u} \in [0,1)^s$, $\mathbf{v} \sim U([0,1)^s)$, *and* $\mathbf{w} = (\mathbf{u}+\mathbf{v}) \bmod 1$. *Then* $\mathbf{w} \sim U([0,1)^s)$.

Proof. It is sufficient to show that, for any $\mathbf{x} \in [0,1]^s$, $P(w_j \le x_j, j = 1,\ldots,s) = x_1 \ldots x_s$. First, by independence of the coordinates v_j, we have

$$P(w_j \le x_j, j = 1, \ldots, s) = P(w_1 \le x_1) \ldots P(w_s \le x_s).$$

This means we just need to prove $P(w_j \le x_j)$ for each j. We consider two cases: (1) if $u_j \le x_j$, then $P(w_j \le x_j) = P(v_j \le x_j - u_j$ or $1 - u_j \le v_j \le 1) = x_j - u_j + u_j = x_j$; (2) if $u_j > x_j$, then $P(w_j \le x_j) = P(1 - u_j \le v_j \le 1 - u_j + x_j) = 1 - u_j + x_j - (1 - u_j) = x_j$.

This proves that Condition (i) is satisfied. To see in what sense Condition (ii) is satisfied, we show in Fig. 6.1 an example of a small lattice point set and the effect of the shift on it. For this example, the shift preserves the structure of the point set in the sense that the original points lie on parallel equidistant lines — in fact, on an infinite number of families of lines — and this remains true after the shift. Also, the distance between these lines remains the same for each family of parallel equidistant lines.

As we mentioned on p. 175, a major incentive for using a random shift with Korobov point sets is that it breaks the cycles that would otherwise appear in the coordinates of each point when $s \geq n$. In particular, if the dimension is unbounded, then applying a random shift becomes crucial. It can be done rather easily if the random number generator used to generate the shift can be reset to a given state. For instance, one can initially choose an upper bound s_0 on the maximum dimension, generate an s_0-dimensional random shift $\mathbf{v}$, and save the current state x_0 of the generator. Then, for a given point $\mathbf{u}_i = (u_{i,1}, \ldots, u_{i,s_0})$, if it turns out that f cannot be evaluated only with the first s_0 coordinates of the randomized point

$$\tilde{\mathbf{u}}_i = (\mathbf{u}_i + \mathbf{v}) \bmod 1$$

(for instance, in the bank example, this would happen if, say, $s_0 = 600$ and it turns out that the 300th customer arrives *before* 3 pm, and thus at least one more client needs to be simulated), then additional coordinates can be obtained as

$$\tilde{u}_{i,j} = (u_{i,j} + \texttt{Rand01}()) \bmod 1, j > s_0, \tag{6.2}$$

where `Rand01()` represents a call to the random number generator, and $u_{i,j}$ should be easily obtainable from the construction chosen. For instance, if the underlying point set is a Korobov point set enumerated in the order

$$\mathbf{u}_i = \frac{i-1}{n}\big((1, a, a^2 \bmod n, \ldots, a^{s_0-1} \bmod n) \bmod 1\big),$$

then

$$u_{i,j} = \frac{1}{n}\big((i-1)a^{j-1} \bmod n\big).$$

Once enough additional coordinates of the form $\tilde{u}_{i,j}$ with $j = s_0+1, s_0+2, \ldots$ have been generated, then the generator should be reset to the state x_0 so that if another point $\tilde{\mathbf{u}}_{i'}$ with $i' > i$ also needs more than s_0 coordinates, then the same shift is added to that point when calling `Rand01()` in (6.2).

Summing up, the random shift is a very simple randomization that is easy to apply. Although it is designed for lattice point sets, it can also be applied to digital nets and sequences [325, 455], but in those cases it does not exactly preserve the low-discrepancy properties of those point sets. In particular, if P_n is $(q_1, \ldots, q_s)$-equidistributed, then the randomly shifted version of P_n is not necesssarily $(q_1, \ldots, q_s)$-equidistributed.

The random shift method is discussed further in [264, 325, 453, 455].

6.2.2 Digital shift

This method is to digital nets the equivalent of what the random shift modulo 1 is to lattices. It adds a random, uniform shift to the points of a digital net P_n but using operations in $\mathbb{Z}_b$ rather than ordinary real addition, where b is the base in which the net is defined.

More precisely, for a digital net P_n in base b, generate a random vector $\mathbf{v} = (v_1, \ldots, v_s)$ uniformly in $[0,1)^s$ and consider the base b expansion of its coordinates. That is, write

$$v_j = \sum_{l=0}^{\infty} v_{j,l} b^{-l}.$$

Then the *digitally shifted* version of P_n — denoted $\tilde{P}_n$ — has points $\tilde{\mathbf{u}}_i$ such that

$$\tilde{u}_{i,j} = \sum_{l=0}^{\infty} (u_{i,j,l} + v_{j,l}) b^{-l},$$

where the addition is performed in $\mathbb{Z}_b$ and the digits $u_{i,j,l}$ come from the base b expansion of $u_{i,j}$. That is,

$$u_{i,j} = \sum_{l=0}^{\infty} u_{i,j,l} b^{-l}.$$

This can also be used for randomizing Halton sequences, but in this case each coordinate is defined with a different base b_j [19, 464] (see App. B).

It is easy to see that this randomization preserves $(q_1, \ldots, q_s)$-equidistribution properties because performing a digital shift simply amounts to a relabeling of the b-ary boxes for a given $(q_1, \ldots, q_s)$-partition.

6.2.3 Scrambling and permutations

These randomization methods are also designed for digital nets, but they perturb their structure more deeply than a simple random digital shift. We start by describing a *random linear scrambling*, which can be thought of as a randomized version of the transformations suggested by Tezuka to improve the Faure sequence [440]. Its use as a randomization technique is studied in detail in [194, 307, 365].

A random linear scrambling is applied by choosing s lower-triangular nonsingular matrices $R_1, \ldots, R_s$ with elements in $\mathbb{Z}_b$ and multiplying them with the generating matrices of a digital net in base b. That is, this method amounts to using randomized generating matrices of the form $R_j C_j$, $j = 1, \ldots, s$, where the C_j are the original generating matrices. In addition, a digital shift can be performed and has the advantage of simplifying the analysis of the point set obtained [307, p. 540], in particular because it ensures that each randomized point is uniformly distributed over $[0,1)^s$ [365, Remark 3.2]. As with the random digital shift, this method preserves $(q_1,\ldots,q_s)$-equidistribution properties. In fact, Matousěk shows in [308] that the t-value of the scrambled net is not larger than the original net's t-value, so the scrambling can potentially improve the quality of the net.

Another way to think about linear scrambling is to write [307]

$$R_j C_j \mathbf{a} = R_j \tilde{\mathbf{a}}_j = \begin{pmatrix} R_{j,1,1} & 0 & \ldots\ldots & 0 \\ R_{j,2,1} & R_{j,2,2} & 0 & \ldots & 0 \\ \vdots & \vdots & & \ddots & \vdots & \vdots \end{pmatrix} \tilde{\mathbf{a}} = \begin{pmatrix} \pi_0^j(\tilde{a}_{j,0}) \\ \pi_{0,\tilde{a}_{j,0}}^j(\tilde{a}_{j,1}) \\ \vdots \end{pmatrix},$$

where

$$\mathbf{a} = (a_0(i), a_1(i), a_2(i), \ldots)^{\mathrm{T}}$$

is the vector containing the base b expansion of i, and

$$\tilde{\mathbf{a}}_j = C_j \mathbf{a} = \begin{pmatrix} \tilde{a}_{j,0} \\ \tilde{a}_{j,1} \\ \vdots \end{pmatrix}.$$

(We should really use the notation $\tilde{a}_{j,l}(i)$ instead of $\tilde{a}_{j,l}$, but we choose to omit the i to make the notation less heavy.) In other words, we can think of R_j as a way of defining *nested permutations* $\pi_0^j, \pi_{0,\tilde{a}_{j,0}}^j, \pi_{0,\tilde{a}_{j,0},\tilde{a}_{j,1}}^j, \ldots$ that are applied to the digits of $\tilde{\mathbf{a}}_j$. For instance,

$$\begin{aligned} \pi_0^j(\tilde{a}_{j,0}) &= R_{j,1,1} \times \tilde{a}_{j,0}, \\ \pi_{0,\tilde{a}_{j,0}}^j(\tilde{a}_{j,1}) &= R_{j,2,1}\tilde{a}_{j,0} + R_{j,2,2}\tilde{a}_{j,1}, \end{aligned}$$

and so on, where all operations are performed in $\mathbb{Z}_b$. Note that the types of permutations used above are linear. That is, they are restricted to be of the form

$$\pi(\tilde{a}_{j,r}) = x\tilde{a}_{j,r} + y$$

for some constants x and y in $\mathbb{Z}_b$, where y depends on the previous digits $\tilde{a}_{j,r-1},\ldots,\tilde{a}_{j,0}$.

This formulation allows us to establish a parallel with the original scrambling technique proposed by Owen in 1994 [357], which amounts to the approach above, but where general permutations — not necessarily linear — are allowed. The process by which nested permutations are used to scramble a point set is referred to as *nested scrambling* in [365]. In the terminology of [365], random linear scrambling is called *affine matrix scrambling.*

Other forms of scrambling are discussed and compared in [194, 307, 365, 445]. For instance, we have:

(1) *Random digit scrambling* [307]. Choose s random independent permutations $\pi_1, \ldots, \pi_s$ and apply the same π_j to each digit in the expansion of $\tilde{\mathbf{a}}_j$. Hence, more general permutations (not necessarily linear) are applied than with random linear scrambling, but the same permutation is applied to each digit in dimension j. This is also called *positional uniform scrambling* in [365].
(2) *Random linear digit scrambling* [307]. This is a special case of the random digit scrambling where permutations are of the form $\pi_j(a) = h_j a + g_j$ for some (random) $h_j \in \{1, \ldots, b-1\}$ and $g_j \in \{0, \ldots, b-1\}$. This is also called *positional linear scrambling* in [365].
(3) *Fully random scrambling* [307]. This is used to refer to Owen's scrambling, which is also called *nested uniform scrambling* in [365].
(4) *I-binomial scrambling* [445]. This corresponds to using a random lower-triangular scrambling matrix of the form

$$R_j = \begin{pmatrix} h_1 & 0 & 0 & 0 & \ldots \\ g_2 & h_1 & 0 & 0 & \ldots \\ g_3 & g_2 & h_1 & 0 & \ldots \\ g_4 & g_3 & g_2 & h_1 & \ldots \\ \vdots & \ddots & \ddots & \ddots & \ddots \end{pmatrix}$$

for $j = 1, \ldots, s$, where the integers h_l are nonzero elements of $\mathbb{Z}_b$, while $g_l \in \mathbb{Z}_b$. One interesting aspect of the I-binomial scrambling is that it can be categorized as nested scrambling, although it requires only $O(k)$ integers to scramble k digits (for each dimension), while the two other forms of nested scrambling — random linear scrambling and fully random scrambling — require $O(k^2)$ and $O(b^k)$ integers, respectively [365].
(5) *Affine striped matrix scrambling* [365]. This is a special case of random linear (or affine matrix) scrambling where we use

$$R_j = \begin{pmatrix} h_1 & 0 & 0 & 0 & \ldots \\ h_1 & h_2 & 0 & 0 & \ldots \\ h_1 & h_2 & h_3 & 0 & \ldots \\ h_1 & h_2 & h_3 & h_4 & \ldots \\ \vdots & \ddots & \ddots & \ddots & \ddots \end{pmatrix}$$

for $j = 1, \ldots, s$, and here the integers h_l are nonzero elements of $\mathbb{Z}_b$.

These different scramblings will be discussed again later on, in Sect. 6.2.6.

Remark on Latin hypercube sampling

A well-known technique in simulation studies that bears some resemblance to the scrambling and permutations that we just described is the *Latin hypercube sampling* approach [313]. In Latin hypercube sampling, a point set P_n is constructed so that each one-dimensional projection contains exactly one point in each interval of the form $[(j-1)/n, j/n)$, for $j = 1, \ldots, n$. This is done by generating s random uniform permutations π_j over $[0, \ldots, n-1]$ and then defining

$$P_n = \left\{ \left(\frac{\pi_1(i)}{n} + v_{i1}, \ldots, \frac{\pi_s(i)}{n} + v_{is} \right), i = 0, \ldots, n-1 \right\}, \tag{6.3}$$

where the $v_{i,j}$ are independent and uniformly distributed over $[0, 1/n)$. Sometimes, a variant where each v_{ij} is replaced by $1/2n$ is used. To see the connection with a scrambled digital net, we can think of the Latin hypercube sampling point set (6.3) as being a scrambled $(0, 1, s)$-net in base n. We will talk about Latin hypercube sampling in more detail in Chap. 8.

6.2.4 Partitions and Latin supercube sampling

For problems with a very large dimension s, one possible approach is to split the set of variables into two sets, $\{u_1, \ldots, u_d\}$ and $\{u_{d+1}, \ldots, u_s\}$, and apply quasi–Monte Carlo to the first set and Monte Carlo to the second set [351, 352, 353]. The idea behind this *hybrid* approach is that if the problem is formulated so that the first d variables are the most important, then applying quasi–Monte Carlo to this first subset should help, while for the second set we simply rely on random sampling. By choosing d appropriately, one thus hopes to improve on pure Monte Carlo.

In [360], Owen develops a generalization of this idea that also has similarities with Latin hypercube sampling. He called this method *Latin supercube sampling*, and it works as follows:

(1) Split $\{1, \ldots, s\}$ into r groups $\{1, \ldots, d_1\}, \{d_1 + 1, \ldots, d_1 + d_2\}, \ldots, \{d_1 + \ldots + d_{r-1} + 1, \ldots, s\}$, where $\sum_{l=1}^{r} d_l = s$.
(2) Choose r low-discrepancy point sets (randomized or not) of dimension $d_1, \ldots, d_r$, denoted $P_n^1, \ldots, P_n^r$.
(3) Choose r random uniform permutations $\pi_1, \ldots, \pi_r$ of $[1, \ldots, n]$.
(4) The Latin supercube sampling method then uses as its ith point

$$\left(\mathbf{u}^1_{\pi_1(i)}, \mathbf{u}^2_{\pi_2(i)}, \ldots, \mathbf{u}^r_{\pi_r(i)} \right).$$

That is, the first d_1 coordinates of the ith point are obtained from the $\pi_1(i)$th point of P_n^1, the next d_2 are obtained from the $\pi_2(i)$th point of P_n^2, and so on. Latin hypercube sampling is a special case of this method where $r = s$, $d_l = 1$ for all l, and $P_n^l = \{v_{1,l}, 1/n + v_{2,l}, \ldots, (n-1)/n + v_{n,l}\}$, where the variables $v_{i,l}$ are either independent and uniformly distributed in $[0, 1/n)$ or simply set to $1/2n$.

This method is useful for problems where variables can be partitioned into groups within which there is a lot of correlation but variables from different groups interact only mildly. By applying Latin supercube sampling over these corresponding subgroups and using the *ANOVA decomposition* framework, one can expect that improvement over the Monte Carlo method should be obtained [360].

6.2.5 Array-RQMC

We now describe an approach proposed by L'Ecuyer, Lécot, and Tuffin to simulate Markov chains defined over an ordered space that relies on randomized low-discrepancy point sets [263]. This method — called *array-RQMC* — works as follows.

Suppose that we want to generate N steps of a Markov chain X defined over an ordered space and such that d uniform numbers are required to generate the next state of the chain given the current state. Instead of using an $s = Nd$-dimensional point set to perform this type of simulation — assigning each Nd-dimensional point to one path of the chain — the idea of array-RQMC is to use N i.i.d. randomized copies of a d-dimensional point set for each step of the chain. Furthermore, at each step, the order in which the points are assigned to the chain paths is determined by the current state. That is, we can think of array-RQMC as using $r = N$ underlying point sets in the same way as Latin supercube sampling but where the permutations π_l for $l = 1, \ldots, r$ are determined in the following way.

First, $\pi_1(i) = i$ for $i = 1, \ldots, n$. Then, for $l \geq 2$, let $x_{1,l-1}, \ldots, x_{n,l-1}$ be the sample of the Markov chain obtained at step $l-1$. Rearrange this sample according to the order defined on the state space of the Markov chain, thereby obtaining

$$x_{(1),l-1} < \ldots < x_{(n),l-1}.$$

Then let π_l be defined so that $\pi_l(j) = k$, where k is such that $x_{j,l-1} = x_{(k),l-1}$. That is, the permutation used at step l is such that points from the lth copy of the underlying d-dimensional point set are assigned to the Markov chain paths according to the ordering obtained at the previous step $l-1$.

Hence an important difference with Latin supercube sampling is that the permutations used to reorder the point sets at each step are determined in a systematic way from the definition of the Markov chain rather than being generated randomly. In some settings — including common problems in finance — the array-RQMC method can provide much more accurate results than the standard randomized quasi–Monte Carlo approach based on an Nd-dimensional point set, where each point is assigned to a path [263].

The idea of reordering low-discrepancy point sets for Markov chain simulation had been studied previously in a deterministic context in [245, 246], among others.

6.2.6 Studying the variance

When a deterministic low-discrepancy point set is randomized, we can look at the corresponding estimator and try to analyze its variance. At the beginning of this section, we mentioned how to *estimate* that variance. Here, we give expressions for its theoretical value. Although these expressions cannot be evaluated exactly in general, they can provide useful insight on the different randomization methods mentioned above and how they perform compared with Monte Carlo. Let us start with the variance for a randomly shifted lattice point set [264].

Proposition 6.2. *If f is square-integrable and $\tilde{P}_n = \{\tilde{\mathbf{u}}_i, \ldots, \tilde{\mathbf{u}}_n\}$ is a randomly shifted lattice point set, then the corresponding estimator $\hat{\mu}$ defined by*

$$\hat{\mu} = \frac{1}{n}\sum_{i=1}^{n} f(\tilde{\mathbf{u}}_i)$$

has variance

$$\mathrm{Var}(\hat{\mu}) = \sum_{\mathbf{0}\neq\mathbf{h}\in L_s^*} |\hat{f}(\mathbf{h})|^2, \tag{6.4}$$

where $\hat{f}(\mathbf{h})$ is the Fourier coefficient of f evaluated in $\mathbf{h}$, given by

$$\hat{f}(\mathbf{h}) = \int_{[0,1)^s} f(\mathbf{u})e^{-2\pi i\mathbf{h}\cdot\mathbf{u}}d\mathbf{u},$$

and L_s^* is the dual lattice associated with the lattice L_s such that the unshifted point set $P_n \subseteq L_s$.

Similarly, for a digitally shifted net, we have the following result [265], which makes use of the dual space $\mathcal{C}_s^*$ of the net that was defined in (5.34) and the product $\langle\mathbf{h},\mathbf{u}\rangle_b$ that was defined on p. 188.

Proposition 6.3. *If f is square-integrable and $\tilde{P}_n$ is a digitally shifted net in base b, then the corresponding estimator $\hat{\mu}$ has variance*

$$\mathrm{Var}(\hat{\mu}) = \sum_{\mathbf{0}\neq\mathbf{h}\in\mathcal{C}_s^*} |\tilde{f}(\mathbf{h})|^2, \tag{6.5}$$

where $\tilde{f}(\mathbf{h})$ is the b-ary Walsh coefficient of f evaluated in $\mathbf{h}$, given by

$$\tilde{f}(\mathbf{h}) = \int_{[0,1)^s} f(\mathbf{u}) e^{-2\pi i \langle \mathbf{h},\mathbf{u}\rangle_b} d\mathbf{u}.$$

A similar result can be obtained for point sets based on the Halton sequence, with the multibase digital shift discussed on p. 206, and is discussed in App. B.

A few words about these two results are in order. First, the variance expressions given in (6.4) and (6.5) are closely related to formulas for the deterministic error of the corresponding point set that can be found in [408] for lattices and [265] for nets. More precisely, for digital nets, we have the expression

$$Q_n - I(f) = \sum_{0 \neq \mathbf{h} \in \mathcal{C}_s^*} \tilde{f}(\mathbf{h}), \tag{6.6}$$

similar to the error bound (5.34) for lattice point sets, and this holds as long as

$$\sum_{0 \neq \mathbf{h} \in \mathcal{C}_s^*} |\tilde{f}(\mathbf{h})| < \infty.$$

Note that for the variance expressions (6.4) and (6.5) to hold, we only need f to be square-integrable, or equivalently to have

$$\sum_{\mathbf{h}} |\hat{f}(\mathbf{h})|^2 < \infty \text{ (Fourier/lattice) or } \sum_{\mathbf{h}} |\tilde{f}(\mathbf{h})|^2 < \infty \text{ (Walsh/nets).}$$

By contrast, for the error bounds (5.34) and (6.6), we needed the (much) stronger condition of absolute convergence

$$\sum_{\mathbf{h}} |\hat{f}(\mathbf{h})| < \infty \text{ (Fourier/lattice) or } \sum_{\mathbf{h}} |\tilde{f}(\mathbf{h})| < \infty \text{ (Walsh/net).}$$

Recall that in Sect. 5.6 we argued informally that, under the assumption that the most "important" basis functions (Fourier or Walsh) were the ones associated with "small" vectors $\mathbf{h}$, it made sense to try to make sure the dual space (or lattice) had no "short vectors". In terms of the variance expressions above, doing that will avoid large contributions in (6.4) and (6.5).

Note that, for Monte Carlo, the variance of an estimator based on n points is given by

$$\frac{1}{n} \sum_{0 \neq \mathbf{h} \in \mathbb{Z}^s} |\hat{f}(\mathbf{h})|^2$$

(where we can also replace $\hat{f}$ by a b-ary Walsh coefficient $\tilde{f}$ and sum over $\mathbb{N}_0^s$ instead). The difference from the shifted randomized quasi–Monte Carlo methods is that here we sum over *all* $\mathbf{h}$ but each term is divided by n. Since the dual space (or dual lattice) corresponding to a point set of cardinality n contains n times less vectors than the whole set of vectors $\mathbf{h}$ [407], this means that the shifted randomized quasi–Monte Carlo estimators have smaller variances than the Monte Carlo estimator if, on average, the Fourier (Walsh) coefficients are smaller over the dual lattice (space).

This also means that, based on the expressions (6.4) and (6.5), we can easily construct "bad" functions for which the variance of the shifted randomized quasi–Monte Carlo estimator will be larger than for Monte Carlo. For example, a nonconstant function whose Fourier coefficients are 0 for all nonzero $\mathbf{h}$ that are not in the dual space will have a variance n times larger than Monte Carlo. Although it is important to be aware of these worst cases, functions like this are not necessarily likely to arise in practice. In a way, this potential problem comes from the fact that randomizations based on a shift are "too simple" and do not sufficiently "shuffle" or "scramble" the point set in order to prevent the existence of functions that interact in a destructive way with the deterministic point set on which the estimator is based.

The scrambling approach proposed by Owen in 1995 to randomize nets does not suffer from this drawback. It inputs enough randomness in the deterministic construction to prevent the occurrence of bad functions for which the scrambled net estimator performs significantly worse than Monte Carlo. More precisely, we have the following proposition [361].

Proposition 6.4. *Let $\hat{\mu}_{\rm scr}$ be the estimator constructed from a (fully random) scrambled (t, k, s)-net in base b. For any square-integrable function f with variance σ^2,*

$$\mathrm{Var}(\hat{\mu}_{\rm scr}) \leq \frac{b^t}{n}\left(\frac{b+1}{b-1}\right)^s \sigma^2, \tag{6.7}$$

where $n = b^k$.

It should be pointed out that, as discussed in [444], the price to pay for using fully random scrambling is that in some cases where the integrand happens to be approximated with small error by a deterministic net, the scrambled version of the net might have an increased error compared with its deterministic counterpart. Here again, while it is important to be aware of this possible disadvantage, one could argue that it is outweighed by the ability of the scrambled net to provide an error estimate and destroy potential bad interactions between its underlying deterministic point set and the function to be integrated.

Another possible disadvantage of the full scrambling approach compared with the random digital shift is that its implementation requires significantly more space and time. An alternative is to use a random linear scrambling or I-binomial scrambling with a digital shift. As shown in [194, 307], Prop. 6.4 also holds for these two randomization techniques. However, their advantage over scrambled nets is that their implementation is about as simple as for the random digital shift since the generating matrices $R_j C_j$ can be recomputed at the beginning and then each point is generated as in the random digital shift approach. More generally, Prop. 6.4 holds for any randomization approach that satisfies the following properties [194, 307]:

(1) Each point $\tilde{\mathbf{u}}_i$, $i = 1, \ldots, n$ in the randomized point set $\tilde{P}_n$ is uniformly distributed over $[0,1)^s$.
(2) For $1 \leq i, i' \leq n$ and $1 \leq j \leq s$, if $u_{i,j,l} = u_{i',j,l}$ for $l = 1, \ldots, r$ but $u_{i,j,r+1} \neq u_{i',j,r+1}$, then
    a. $\tilde{u}_{i,j,l} = \tilde{u}_{i',j,l}$ for $l = 1, \ldots, r$;
    b. $(\tilde{u}_{i,j,r+1}, \tilde{u}_{i',j,r+1})$ is uniformly distributed over $\{(a_1, a_2) \in \mathbb{F}_b^2 : a_1 \neq a_2\}$; and
    c. $(\tilde{u}_{i,j,p}, \tilde{u}_{i,j,q})$ are uncorrelated for any $p, q > r+1$.

As mentioned in [365], it is crucial that the scrambling approach be nested in order for Property 2(b) to be satisfied. Simple digital shifts and positional scrambling do not satisfy this property.

Note that although Prop. 6.4 suggests that scrambled nets cannot do much worse than Monte Carlo, the convergence rate for their variance is still $O(1/n)$. This might lead to the conclusion that randomized quasi–Monte Carlo does not capture the advantage of quasi–Monte Carlo over Monte Carlo deduced from the Koksma-Hlawka inequality and similar results. Recall, however, that these results required f to be of bounded variation, while Prop. 6.4 only assumes f is square-integrable. It turns out that if we make further assumptions on f, the convergence rate of the variance can be improved to $O(n^{-3+\epsilon})$ [359]. More precisely, for any scrambling satisfying the two properties above, we have the following theorem.

Theorem 6.5. *[359, Theorem 2] If f is a "smooth" function (that is, there exists $A \geq 0$ and $\beta \in (0,1]$ such that*

$$\left| \frac{\partial^s}{\partial u_1 \ldots \partial u_s} f(\mathbf{u}) - \frac{\partial^s}{\partial u_1 \ldots \partial u_s} f(\mathbf{u}^*) \right| \leq A \|\mathbf{u} - \mathbf{u}^*\|^\beta$$

for all $\mathbf{u}, \mathbf{u}^* \in [0,1)^s$), then for a scrambled digital net we have that the corresponding estimator $\hat{\mu}_{\mathrm{scr}}$ is such that

$$\mathrm{Var}(\hat{\mu}_{\mathrm{scr}}) \in O(n^{-3} \log^{s-1} n).$$

6.3 ANOVA decomposition and effective dimension

In Chap. 1, we illustrated the advantage of Monte Carlo methods over rectangular grids by considering the simple function $f(\mathbf{u}) = \sum_{j=1}^{s} \sqrt{u_j}$, which is a sum of s univariate functions. Similarly, to understand the behavior of quasi–Monte Carlo methods for numerical integration, it is useful to decompose an s-dimensional integrand as a sum of 2^s components based on each possible subset $u_I = (u_{i_1}, \ldots, u_{i_d})$ of variables, where $I = \{i_1, \ldots, i_d\} \subseteq \{1, \ldots, s\}$.

More precisely, we can use a *functional ANOVA decomposition* [99, 193, 416]

$$f(\mathbf{u}) = \sum_{I \subseteq \{1,\dots,s\}} f_I(\mathbf{u}),$$

where, for nonempty subsets I, we have

$$f_I(\mathbf{u}) = \int_{[0,1)^{s-d}} f_I(\mathbf{u}) d\mathbf{u}_{-I} - \sum_{J \subset I} f_J(\mathbf{u}),$$

where $d = |I|$ and $-I = \{1,\dots,s\} \backslash I$ is the complement of I in $\{1,\dots,s\}$. The ANOVA component $f_\emptyset(\mathbf{u})$ is simply the integral

$$I(f) = \int_{[0,1)^s} f(\mathbf{u}) d\mathbf{u}.$$

We also have that $\int_{[0,1)^s} f_I(\mathbf{u}) = 0$ for all nonempty I and that

$$\int_{[0,1)^s} f_I(\mathbf{u}) f_J(\mathbf{u}) d\mathbf{u} = 0$$

for all $I \neq J$. That is, the ANOVA components are orthogonal. Here is an example to illustrate these definitions.

Example 6.6. Suppose $s = 2$ and $f(\mathbf{u}) = u_1 + 2u_1u_2^2 + u_2^3$. Then

$$\begin{aligned}
f_\emptyset(\mathbf{u}) &= 13/12, \\
f_{\{1\}}(\mathbf{u}) &= \int_0^1 f(\mathbf{u}) du_2 - 13/12 = 5u_1/3 - 5/6, \\
f_{\{2\}}(\mathbf{u}) &= \int_0^1 f(\mathbf{u}) du_1 - 13/12 = u_2^2 + u_2^3 - 7/12, \\
f_{\{1,2\}}(\mathbf{u}) &= 2u_1u_2^2 - 2u_1/3 - u_2^2 + 1/3.
\end{aligned}$$

The usefulness of this decomposition in the context of quasi–Monte Carlo was first noticed by Sobol' [416, 417] and developed further in [185, 360, 421]. One way to use it is to look at the components' variance

$$\sigma_I^2 = \int_{[0,1)^s} f_I^2(\mathbf{u}) d\mathbf{u}$$

and then write $\mathrm{Var}(f) = \mathrm{Var}(f(\boldsymbol{U})) = \sigma^2 = \sum_I \sigma_I^2$. Therefore

$$S_I = \frac{\sigma_I^2}{\sigma^2} \in [0,1]$$

can be interpreted as a measure of the relative importance of f_I and is called the *global sensitivity index* in [419]. If we know — or can guess — which subsets I correspond to important components f_I, then, informally speaking, we can say that quasi–Monte Carlo approximations based on point sets for which the corresponding projections $P_n(I)$ are of high quality should be accurate. Going further, information on global sensitivity indices can be used as a guide for constructing or choosing a low-discrepancy point set for that problem. For instance, in the weighted P_α criterion (5.29), one could try to choose the weights β_I proportionally to the indices S_I. Similar ideas for other criteria will be discussed briefly in Sect. 6.3.4.

It should be noted that finding a closed-form expression for the ANOVA components f_I is typically not possible since, among other things, it requires knowing the value of $I(f)$. For the same reason, the variance σ_I^2 is usually not known exactly. It is, however, possible to estimate σ_I^2 and approximate $f_I(\mathbf{u})$ [9, 286, 419, 421], as we will see in Sect. 6.3.3.

6.3.1 Effective dimension

Studying the ANOVA decomposition of a function can help in assessing the difficulty level of the corresponding integration problem. One way to summarize that assessment is through the concept of *effective dimension*, which was first introduced in [375] to explain the success of quasi–Monte Carlo methods on a 360-dimensional problem in finance. This concept was used as a way of measuring the number of "important" variables in this problem, and the fact that it was much smaller than 360 was used as an argument to explain why quasi–Monte Carlo methods could be successful on such problems.

More precisely, we have the following definitions [51, 185].

Definition 6.7. The *effective dimension of f in the superposition sense* (and in proportion p) is the smallest integer d_S such that

$$\frac{1}{\sigma^2} \sum_{I:|I| \le d_S} \sigma_I^2 \ge p.$$

The *effective dimension of f in the truncation sense* (and in proportion p) is the smallest integer d_T such that

$$\frac{1}{\sigma^2} \sum_{I:|I| \le d_T} \sigma_I^2 \ge p.$$

What this definition says is that a function with an effective dimension d can be well approximated (from a least-squares point of view) by a sum of functions of at most d variables each (for the superposition sense) or a sum of functions involving only the first d variables $u_1, \ldots, u_d$ (for the truncation sense version).

For example, in [51], a 360-dimensional problem involving the pricing of a *mortgage-backed security* is shown to have a dimension of 1 in the superposition sense, with p very close to 1; in [286], an Asian option pricing problem with $s = 32$ is shown to have an effective dimension of 2 in the superposition sense in proportion $p = 0.97$. More examples in finance are studied in [463, 466].

Having a small effective dimension in the truncation sense is believed to be especially important for functions integrated with the Sobol' sequence, since the upper bound on the quality parameter t_I of its projections $P_n(I)$ increases as the indices in I increase [422]. Indeed, we have that

$$t_I \leq \sum_{j \in I} (d_j - 1),$$

where d_j is the degree of the primitive polynomial $p_j(z)$ used in dimension j. Since these degrees form a nondecreasing sequence, it is clear that the bound on t_I increases with the value of the indices in I. This fact was noted explicitly by Sobol' and his collaborators back in 1992 [422]. Shortly after, researchers in finance noticed this fact as well, and this led to the development of techniques that can be used to modify f so that this type of effective dimension can be reduced, as was discussed in Sect. 6.3.2.

The following example studies the effective dimension in the superposition and truncation senses for simple functions. As we mentioned before, in practice it is usually impossible to compute these quantities exactly, but they can at least be approximated, as we will see in Sect. 6.3.3.

Example 6.8. We consider three functions. To simplify things, assume we are interested in computing effective dimensions for a proportion $p = 0.99$.

(1) As in [417], consider a linear function of the form

$$f(\mathbf{u}) = f_0 + \sum_{j=1}^{s} c_j(u_j - 1/2), \qquad c_j \in \mathbb{R},$$

which is already written in its ANOVA form. That is, for this function, we have

$$f_{\{j\}} = c_j(u_j - 1/2)$$

for $j = 1, \ldots, s$, and $f_I(\mathbf{u}) = 0$ for all subsets I containing more than one index. It is easy to see that

$$\sigma^2 = \frac{1}{12} \sum_{j=1}^{s} c_j^2,$$

$$\sigma^2_{\{j\}} = \frac{c_j^2}{12},$$

and therefore the global sensitivity indices are given by

$$S_{\{j\}} = \frac{c_j^2}{\sum_{j=1}^s c_j^2}$$

for $j = 1, \ldots, s$. This function has an effective dimension of 1 in the superposition sense for any constants c_j, but the truncation sense version has a value that depends on these constants. For instance, if they are all equal, then $d_T = \lceil 0.99s \rceil$. If we have $c_j = c^j$ for some $0 < c < 1$, then d_T is the smallest integer d such that

$$\sum_{I \subseteq \{1,\ldots,d\}} \sigma_I^2 = \frac{c^2(1-c^{2d})}{1-c^2} \geq 0.99\sigma^2 = 0.99\frac{c^2(1-c^{2s})}{1-c^2}.$$

After rearranging, we get that

$$d_T = \left\lceil \frac{\log(1-0.99(1-c^{2s}))}{2\log c} \right\rceil.$$

Table 6.1 gives values of d_T for different combinations of c and s.

Table 6.1 Effective dimension in the truncation sense for linear function $f(\mathbf{u}) = f_0 + \sum_{j=1}^s c^j(u_j - 1/2)$.

$c \backslash s$	5	10	20	50	100
0.99	5	10	20	50	97
0.95	1	2	4	10	19
0.9	1	1	2	5	10
0.5	1	1	1	1	2
0.1	1	1	1	1	1

(2) One way of constructing a function with bivariate components is to take the previous function and raise it to the power two. That is, consider

$$f(\mathbf{u}) = \left(c_0 + \sum_{j=1}^s c_j(u_j - 1/2)\right)^2$$
$$= c_0^2 + \sum_{j=1}^s c_j^2(u_j - 1/2)^2 + 2\sum_{i<j} c_i c_j (u_i - 1/2)(u_j - 1/2).$$

In that case,

$$I(f) = c_0^2 + \frac{1}{12}\sum_{j=1}^{s} c_j^2,$$

$$f_{\{j\}} = c_j^2(u_j - 1/2)^2 - \frac{c_j^2}{12},$$
$$f_{\{i,j\}} = 2c_i c_j (u_i - 1/2)(u_j - 1/2),$$

so that

$$\sigma^2_{\{j\}} = \frac{c_j^4}{80} - \frac{c_j^4}{144} = \frac{c_j^4}{180},\ j = 1, \ldots,$$
$$\sigma^2_{\{i,j\}} = \frac{c_i^2 c_j^2}{36},\ 1 \le i < j \le s.$$

Therefore, if the constants c_j are all equal to some constant c, then we have

$$\sigma^2 = \frac{sc^4}{180} + \frac{s(s-1)c^4}{2 \times 36},$$

and the global sensitivity indices are given by

$$S_{\{j\}} = \left(s + \frac{5s(s-1)}{2}\right)^{-1}, j = 1, \ldots, s,$$
$$S_{\{i,j\}} = \left(\frac{s}{5} + \frac{s(s-1)}{2}\right)^{-1}, 1 \le i < j \le s.$$

Thus, as expected, all components of the same size have the same global sensitivity index value. On the other hand, if $c_j = c^j$ for some $0 < c < 1$, then we have

$$\sigma^2 = \frac{c^4(1-c^{4s})}{180(1-c^4)} + \frac{1}{18(1-c^2)}\left[c^6\left(\frac{1-c^{4(s-1)}}{1-c^4}\right) - (s-1)c^{2s}\right],$$

and similarly we get

$$\sum_{I \subseteq \{1,\ldots,d\}} \sigma_I^2 = \frac{c^4(1-c^{4d})}{180(1-c^4)} + \frac{1}{18(1-c^2)}\left[c^6\left(\frac{1-c^{4(d-1)}}{1-c^4}\right) - (d-1)c^{2d}\right].$$

Using this, we can compute the effective dimension in the truncation sense, as shown in Table 6.2.

(3) Consider now a multiplicative function of the form

$$f(\mathbf{u}) = \prod_{j=1}^{s} \frac{|4u_j - 2| + a_j}{a_j + 1}$$

Table 6.2 Effective dimension in the truncation sense for $f(\mathbf{u}) = (f_0 + \sum_{j=1}^{s} c^j (u_j - 1/2))^2$.

c\s	5	10	20	50	100
0.99	5	10	20	50	97
0.95	5	10	20	49	72
0.9	5	10	20	36	36
0.5	5	7	7	7	7
0.1	3	3	3	3	3

for some nonnegative constants a_j, $j = 1, \ldots, s$ [363, 419]. The case $a_j = 0$ is often used as a test function for comparing different integration methods [19, 43, 58, 73, 127]. As shown in [363], we have that

$$\sigma_I^2 = \prod_{j \in I} \frac{1}{3(1 + a_j)^2}.$$

For this type of function, the smaller a_j is, the more important the variable j is. For instance, in [419], Sobol' shows that, for the choice $a_1 = a_2 = 0$ and $a_3 = \ldots = a_8 = 3$ in dimension 8, we have that $S_{\{1\}} = S_{\{2\}} = 0.329$, while $S_{\{j\}} = 0.021$ for all other values $j \geq 3$. In addition, $S_{\{1,2\}} = 0.110$, $S_{\{i,j\}} = 0.007$ if one of the indices i or j is 1 or 2, and $S_{\{i,j\}} = 0.0004$ otherwise. Based on his calculations, we get that $d_S = 3$ for this example, but $d_T = 8$.

Another way of defining the effective dimension that is especially relevant in the context of simulation is as follows [284].

Definition 6.9. The *effective dimension of f in the successive-dimensions sense* (and in proportion p) is the smallest integer d_U such that

$$\frac{1}{\sigma^2} \sum_{I : I \in \mathcal{I}_{s,d_u}} \sigma_I^2 \geq p,$$

where $\mathcal{I}_{s,d_u} = \{\{i, \ldots, i + d_U - 1\}, 1 \leq i \leq s - d_U + 1\}$.

Here we not only restrict the subsets I to contain no more than d_U indices but also restrict the *range* of I (the largest index in I minus the smallest one) to be no larger than $d_U - 1$. So, for instance, the subset $I = \{1, 2, 100\}$ would be considered when summing up the variance contributions for $d_S = 3$ but not for $d_U = 3$ since the range of I in this case is $99 > 2$. This added restriction is relevant in settings where f is defined so that variables u_j with indices that are not too far apart interact more than those with indices that are far apart.

The question of whether or not problems need to have a small effective dimension in order for quasi–Monte Carlo to work well might appear as a controversial issue based on recently published papers [362, 443, 444]. More precisely, what is shown by Owen in [362] is that, for scrambled nets, high-dimensional square-integrable functions must have a low effective dimension in order for the corresponding estimator to have a variance much smaller than Monte Carlo for practical sample sizes. By contrast, in [443], what is shown is that it is possible to construct a class of functions with maximal effective dimension (both in the truncation and superposition senses) for which generalized Sobol' sequences — defined specifically for this class of functions — achieve an error rate of $O(n^{-1})$, which is much better than the $O(1/\sqrt{n})$ associated with Monte Carlo. Hence the result in [362] is for randomized nets and looks at a wide class of functions, while in [443] the result is for deterministic constructions whose defining parameters are allowed to depend on the specific (small) class of functions under study.

From our point of view, the practical implication of these two different results is that if one needs to work with a wide class of functions and decides to use scrambled nets, then he or she should know that, for reasonable values of n, improvement on the Monte Carlo method will be significant only if the functions to be integrated have a small effective dimension. If one is interested in a very specific class of functions, then it is possible (at least theoretically) to construct a deterministic point set that will provide a very good approximation, even if the function has a high effective dimension.

Finally, the ANOVA decomposition framework can be used to characterize functions further by using the concept of *dimension distribution* introduced in [365] and studied also in [17, 297]. A dimension distribution for a function f is a probability distribution on the values $\{1, \ldots, s\}$. The effective dimension then becomes a certain quantile of that distribution. More precisely, following [365], we have the following definition.

Definition 6.10. For a given function $f : [0,1)^s \to \mathbb{R}$, the *dimension distribution of d in the superposition sense*, denoted $p_S(\cdot)$, is such that

$$p_S(d) = \sum_{I:|I|=d} \sigma_I^2/\sigma^2$$

for $d = 1, \ldots, s$. The *dimension distribution of d in the truncation sense* is such that

$$p_T(d) = \sum_{I:m(I)=d} \sigma_I^2/\sigma^2$$

for $d = 1, \ldots, s$, where $m(I) = \max(j|j \in I)$ is the largest index in I.

Based on the dimension distribution, one can define the concept of *average dimension* given by

$$D_S = \frac{\sum_I \sigma_I^2 |I|}{\sum_I \sigma_I^2}$$

in the superposition sense and by

$$D_T = \frac{\sum_I \sigma_I^2 m(I)}{\sum_I \sigma_I^2}$$

in the truncation sense. The average dimension discussed in [17] is taken in the superposition sense. This concept provides an alternative way of characterizing functions that can be useful for understanding how successful quasi–Monte Carlo methods will be in integrating them.

6.3.2 Brownian bridge and related techniques

Recall the formulation for $\mu = \mathrm{E}(h(\mathbf{X}))$ discussed in Chap. 4,

$$\mu = \mathrm{E}(h(\mathbf{X})) = \int_{\Omega} h(\mathbf{x})\varphi(\mathbf{x})d\mathbf{x} = \int_{[0,1)^s} f(\mathbf{u})d\mathbf{u},$$

where we view $\mathbf{X}$ as the vector of random variables to be simulated, and $\varphi(\mathbf{x})$ represents the pdf of $\mathbf{X}$. As discussed before, we can write $f(\mathbf{u}) = h(g(\mathbf{u}))$, where $g(\cdot)$ is the transformation used to generate an observation $\mathbf{x}$ with joint density function $\varphi(\mathbf{x})$.

When $\mathbf{X}$ consists of observations $B(t_1), \ldots, B(t_s)$ from a standard Brownian motion $\{B(t), t \geq 0\}$, the most straightforward way to generate these observations is to take $g(u_1, \ldots, u_s) = (x_1, \ldots, x_s)$, where

$$x_j = x_{j-1} + \sqrt{t_j - t_{j-1}}\Phi^{-1}(u_j),$$

for $j = 1, \ldots, s$, and $x_0 = 0$. That is, each u_j is used to generate the increment of the Brownian motion between t_j and t_{j-1}.

Furthermore, in matrix notation, this approach can be described as follows [2, 327]. We have that

$$\begin{pmatrix} B(t_1) \\ \vdots \\ B(t_s) \end{pmatrix} = A \begin{pmatrix} \Phi^{-1}(u_1) \\ \vdots \\ \Phi^{-1}(u_s) \end{pmatrix}, \tag{6.8}$$

where

$$A = \begin{pmatrix} \sqrt{t_1} & 0 & 0 & \ldots & 0 \\ \sqrt{t_1} & \sqrt{t_2 - t_1} & 0 & \ldots & 0 \\ & & \vdots & \ddots & \\ \sqrt{t_1} & \sqrt{t_2 - t_1} & \sqrt{t_3 - t_2} & \ldots & \sqrt{t_s - t_{s-1}} \end{pmatrix}.$$

Note that AA^{T} equals the covariance matrix of $B(t_1), \ldots, B(t_s)$. That is,

$$\Sigma = AA^{\mathrm{T}} = \begin{pmatrix} t_1 & t_1 & \dots & t_1 \\ t_1 & t_2 & \dots & t_2 \\ \vdots & \vdots & & \vdots \\ t_1 & t_2 & \dots & t_s \end{pmatrix},$$

which holds because $\mathrm{Cov}(B(t), B(s)) = \min(s,t)$ for a Brownian motion (see Prob. 2.1).

For this type of problem, one approach that can be used to reduce the effective dimension is to exploit the Brownian bridge property of $B(\cdot)$ to generate the observations $B(t_1), \dots, B(t_s)$ in an arbitrary order. More precisely, for any $u < v < w$, this property tells us that $B(v)|(B(u) = a, B(w) = b)$ has a normal distribution with mean

$$\frac{w-v}{w-u}a + \frac{v-u}{w-u}b$$

and variance

$$\frac{(v-u)(w-v)}{w-u}.$$

This idea was first studied in [52], where it was suggested to use u_1 to generate the final observation $B(t_s)$, then u_2 to generate $B(t_{\lfloor s/2 \rfloor})$, then u_3 and u_4 to generate $B(t_{\lfloor s/4 \rfloor})$ and $B(t_{\lfloor 3s/4 \rfloor})$, respectively, and so on.

The technique above can be generalized as follows. In (6.8), replace the matrix A by any matrix B such that $BB^{\mathrm{T}} = AA^{\mathrm{T}} := \Sigma$, where Σ is the covariance matrix of $B(t_1), \dots, B(t_s)$. This approach is called the *generalized Brownian bridge technique* in [327]. For example, in [2], principal components analysis is used to define B. That is, B is defined as $B = PD^{1/2}$, where P's columns are formed by the eigenvectors of the covariance matrix Σ and D is a diagonal matrix containing the corresponding eigenvalues of Σ in decreasing order. This method was shown to numerically outperform the Brownian bridge technique in [2], but its computation time is longer since to simulate n Brownian motion paths, it runs in $O(ns^2)$ rather than the $O(ns)$ required for the standard and Brownian bridge methods. Following this work, different modifications were proposed in [4] to reduce the computation time for the principal components method.

Another approach that in some sense goes even further in that direction is one proposed by Imai and Tan [200], where a matrix V of the form $V = AH$ is used, with A the lower-triangular matrix obtained from the Cholesky decomposition of Σ and H an orthogonal matrix chosen so as to minimize the effective dimension of the problem in the truncation sense. A feature of this technique not present in the methods based on the Brownian bridge and principal components analysis is that the chosen matrix V depends on the problem. In the examples provided in [200], this technique results in a smaller error than the principal components approach. A generalization of this technique that seems quite promising is studied in [201].

A generalization of principal components analysis called the *Karhunen-Loève expansion* [3, p. 75],[382, p. 141] is also discussed in [2]. This method is used in the context of quantization-based option pricing in [369]. It rewrites the realization of a Gaussian process $X = \{X(t), t \geq 0\}$ into an infinite sum of the form

$$X(t,\omega) = \sum_{l \geq 1} \sqrt{\lambda_l}\xi_l(\omega)e_l(t),$$

where the ξ_l are i.i.d. standard normal variables, the $\{e_l(t), l \geq 1\}$ are eigenbasis functions that depend on the structure of the process X, and the constants λ_l are the corresponding eigenvalues, sorted in decreasing order. This decomposition thus separates the randomness in X — modeled via the dependence on ω — from its time dependence. Hence, once we have the terms λ_l and ξ_l (which must usually be determined numerically), we can approximate the whole process $\{X(t), t \geq 0\}$ by drawing a sufficiently large number of normal variables ξ_l. As for the principal components decomposition, it has the property that the approximation

$$\sum_{l=1}^{m} \sqrt{\lambda_l}\xi_l(\omega)e_l(t)$$

based on m terms maximizes the explained variability among all approximations based on m normal variables.

Although the methods above succeed in making the transformation g rely more heavily on the first few variables u_j, this does not necessarily mean that once we apply the transformation h to $\mathbf{x} = g(\mathbf{u})$ it will reduce the effective dimension of $f(\mathbf{u}) = h(g(\mathbf{u}))$. In some cases, it works quite well. For instance, for a 32-dimensional Asian option pricing problem in finance, using the Brownian bridge makes the one- and two-dimensional ANOVA components explain 99% of the variance instead of 80% with the standard method [286]. This translates into variance reduction factors (compared with the standard method) of about 9 for the Sobol' sequence and 6 for a Korobov point set, both with $n = 1024$. By contrast, Papageorgiou provides numerical results in [373] showing that for a certain type of *digital option* in finance, the Brownian bridge technique produces estimators with a larger error than the standard method does. Hence the Brownian bridge technique should not be applied blindly.

Similar ideas can be used to generate Poisson processes [128]. For instance, one can use u_1 to generate the total number N of arrivals over the simulation horizon and then generate the actual arrival times conditioned on N. Using the fact that the ordered arrival times conditioned on N have a beta distribution, they can be generated in an order that intuitively should reduce the effective dimension. That is, we can first generate the median arrival time, then the one corresponding to the 25th percentile, and so on, just as in the Brownian bridge technique. Other ideas for transforming f can be found in

[425], an early reference that contains several useful ideas for the successful application of quasi–Monte Carlo sampling in a practical setting.

Finally, using conditional Monte Carlo (discussed in Chap. 4) typically amounts to reducing the number of input variables that need to be generated, thereby resulting in an automatic reduction of the (nominal) dimension. For instance, the dimension decreased from 13 to 8 in the SAN example discussed in Sect. 4.6.

6.3.3 Methods for estimating σ_I^2 and approximating $f_I(\mathbf{u})$

In practice, it is usually not possible to compute the variance contributions σ_I^2 explicitly or to get exact expressions for the ANOVA components $f_I(\mathbf{u})$ since, among other things, they require knowing the value $I(f)$ of the integral of f. In this section, we discuss two approaches that have been proposed to approximate these quantities.

First, in [9, 419, 417], the authors directly write σ_I^2 as an integral and estimate it using either Monte Carlo or quasi–Monte Carlo methods. In particular, when $I = \{j\}$ contains only one index j, we have that

$$\begin{aligned}\sigma_{\{j\}}^2 &= \int_0^1 \left(\int_{[0,1)^{s-1}} f(\mathbf{u}) d\mathbf{u}_{-j} - I(f) \right)^2 du_j \\ &= \int_0^1 \left(\int_{[0,1)^{s-1}} f(\mathbf{u}) d\mathbf{u}_{-j} \right)^2 du_j - (I(f))^2,\end{aligned}$$

where $\mathbf{u}_{-j} = (u_1, \ldots, u_{j-1}, u_{j+1}, \ldots, u_s)$. Hence one can use the Monte Carlo estimator

$$\hat{\sigma}_{\{j\}}^2 = \frac{1}{n} \sum_{i=1}^{n} f(u_{i,j}, \mathbf{u}_{i,-j}^{(1)}) f(u_{i,j}, \mathbf{u}_{i,-j}^{(2)}) - \hat{\mu}^2, \tag{6.9}$$

where the superscripts (1) and (2) refer to two independent samples for $\mathbf{u}_{-j}$, and $\hat{\mu}$ is the Monte Carlo estimator for $I(f)$. Hence, to construct the s estimates $\hat{\sigma}_{\{1\}}, \ldots, \hat{\sigma}_{\{s\}}$, one needs two independent samples, $\{\mathbf{u}_1^{(1)}, \ldots, \mathbf{u}_n^{(1)}\}$ and $\{\mathbf{u}_1^{(2)}, \ldots, \mathbf{u}_n^{(2)}\}$. Confidence intervals for the sensitivity indices can then be constructed using bootstrapping, as discussed by Archer et al. [9]. Here, the bootstrap resampling is done over the sample of size n corresponding to the summands in (6.9). Archer et al. propose this approach because it does not require any additional function evaluations, which typically represent the most expensive part of the calculation. Based on this, they choose to use as many as B =10,000 resamples, arguing that 1000 would probably be sufficient for the application at hand.

For subsets I containing more than one index, Sobol' [417] suggests looking at the quantity

$$\gamma_I = \frac{1}{\sigma^2} \sum_{\emptyset \neq J \subseteq I} \sigma_J^2, \tag{6.10}$$

which can be estimated by

$$\hat{\gamma}_I = \frac{1}{\hat{\sigma}^2} \left(\frac{1}{n} \sum_{i=1}^{n} f(\mathbf{u}_i^{(1)}) f(\mathbf{u}_{i,I}^{(1)}, \mathbf{u}_{i,-I}^{(2)}) - \hat{\mu}^2 \right) \tag{6.11}$$

using the fact that

$$\gamma_I = \frac{1}{\sigma^2} \int \left(\int f(\mathbf{u}) d\mathbf{u}_{-I} \right)^2 d\mathbf{u}_I \tag{6.12}$$

(see Prob. 6.9). Here the notation $(\mathbf{u}_{i,I}^{(1)}, \mathbf{u}_{i,-I}^{(2)})$ represents a point whose coordinates $j \in I$ are taken from the point $\mathbf{u}_i^{(1)}$ and the coordinates $j \notin I$ are taken from the point $\mathbf{u}_i^{(2)}$. One of the reasons why the quantity γ_I is interesting is that it is closely connected to the effective dimension d_T in the truncation sense. That is, for a level p, one can determine d_T by computing $\hat{\gamma}_I$ for subsets I of the form $I = \{1, 2, \ldots, d\}$, increasing d until $\hat{\gamma}_I \geq p$. The smallest value of d for which this holds is thus an approximation for d_T. This approach is used in [463].

Rather than using the Monte Carlo method, it is also possible to use quasi–Monte Carlo to construct the estimators above. For instance, in [9] the authors choose a $2s$-dimensional low-discrepancy point set of size n and use its first s coordinates to define the first point set and the last s ones to define the second point set. That is, we let

$$\begin{aligned} \mathbf{u}_i^{(1)} &= (u_{i,1}, \ldots, u_{i,s}), \\ \mathbf{u}_i^{(2)} &= (u_{i,s+1}, \ldots, u_{i,2s}), \end{aligned}$$

for $i = 1, \ldots, n$, and can then construct the estimators (6.9) and (6.11). (Note that $\hat{\sigma}^2$ must be estimated differently when using quasi-random sampling; see Prob. 8.12.)

A different approach, discussed in [286] and based on ideas developed in [5] in the context of *quasi-regression*, is to use a complete orthonormal polynomial basis $\{v_l(\mathbf{u})\}_l$ to decompose f and then rewrite $f_I(\mathbf{u})$ and σ_I^2 in terms of the coefficients in this decomposition. Approximations for $f_I(\mathbf{u})$ and σ_I^2 can then be built by approximating a large enough number of those coefficients.

More precisely, we write

$$f(\mathbf{u}) = \sum_l \beta_l v_l(\mathbf{u}),$$

where

$$\beta_l = \int f(\mathbf{u}) v_l(\mathbf{u}) d\mathbf{u}.$$

Furthermore, assuming that $v_0(\mathbf{u}) = 1$, we have that $\beta_0 = I(f)$.

In what follows, we assume that the basis $\{v_l(\mathbf{u})\}_l$ is defined as a tensor product of a one-dimensional complete basis $\{w_r(u)\}_{r \geq 0}$, where $w_r(u)$ is a polynomial of degree r. In that case, the index l is a vector containing the degrees r_j of each polynomial in the product defining v_l. That is,

$$v_l := v_{\mathbf{r}} = \prod_{j=1}^{s} w_{r_j}(u_j),$$

where $\mathbf{r} = (r_1, \ldots, r_s)$. Since

$$f(\mathbf{u}) = \sum_l \beta_l v_l(\mathbf{u}),$$

it can then easily be proved that

$$\sigma_I^2 = \sum_{\mathbf{r} \in \mathcal{R}_I} \beta_{\mathbf{r}}^2, \tag{6.13}$$

where the set $\mathcal{R}_I$ in (6.13) consists of all vectors $\mathbf{r} \in \mathbb{N}_0^s$ satisfying $r_j = 0$ if and only if $j \notin I$. Similarly, we have

$$\gamma_I = \frac{1}{\sigma^2} \sum_{\emptyset \neq J \subseteq I} \sum_{\mathbf{r} \in \mathcal{R}_J} \beta_{\mathbf{r}}^2,$$

where γ_I was defined in (6.10). Based on these expressions, estimators for σ_I^2 and γ_I can be obtained as follows.

(1) Choose a finite set $\mathcal{R}$ of vectors $\mathbf{r}$ for which the corresponding coefficient $\beta_{\mathbf{r}}$ will be estimated.
(2) Replace $\beta_{\mathbf{r}}$ by their Monte Carlo (or quasi–Monte Carlo) estimators

$$\hat{\beta}_{\mathbf{r}} = \frac{1}{n} \sum_{i=1}^{n} f(\mathbf{u}_i) v_{\mathbf{r}}(\mathbf{u}_i).$$

(3) Make the adjustments required to take into account the fact that $\hat{\beta}_{\mathbf{r}}^2$ is not an unbiased estimator of $\beta_{\mathbf{r}}^2$. For instance, it can be proved that [286]

$$\hat{\beta}_{\mathbf{r},bc}^2 = \frac{n}{n-1} \left(\hat{\beta}_{\mathbf{r}}^2 - \frac{1}{n^2} \sum_{i=1}^{n} v_{\mathbf{r}}^2(\mathbf{u}_i) f^2(\mathbf{u}_i) \right)$$

is an unbiased estimator of $\beta_{\mathbf{r}}^2$.

(4) Build the estimator

$$\hat{\sigma}^2_{I,\mathcal{R}} = \sum_{\mathbf{r}\in\mathcal{R}_I\cap\mathcal{R}} \hat{\beta}^2_{\mathbf{r},bc}$$

for σ_I^2, and in turn use it to give estimated lower bounds of the form

$$\hat{\gamma}_I = \sum_{J\subseteq I} \hat{\sigma}^2_{J,\mathcal{R}}$$

for γ_I.

For this approach also, Monte Carlo methods can be replaced by quasi–Monte Carlo ones, as done in [286]. Further improvement can be obtained with more advanced quasi-regression tools based on wavelets, such as those presented in [205].

6.3.4 Using the ANOVA insight to find good constructions

In the previous chapter, we described several approaches that can be used for constructing low-discrepancy point sets. In each case, some parameters must be chosen: the generator a for Korobov point sets, the generating vector $\mathbf{z}$ for a rank-1 lattice, the direction numbers for the Sobol' sequence, the lower-triangular matrices for generalized Faure sequences, the permutations for the Halton sequence, etc. If we start with the assumption that, *generally*, quasi–Monte Carlo works better with functions having a low effective dimension in some sense, then we can use this as a guide in our search for "good" parameters. More precisely, if we assume that we are working with such functions — that either occur "naturally" or that have been "engineered" to have this property, for example by using the Brownian bridge technique — then a sensible approach for choosing these parameters is to try to define selection criteria that look closely at the low-dimensional projections of P_n. One such criterion was discussed briefly in Chap. 3 and is denoted $M_{t_1,\ldots,t_d}$ in [282], where it is used to find good generators a for Korobov point sets. This criterion looks at projections of the form $P_n(I)$ with $I = \{1, j_2, \ldots, j_l\}$, where $j_l \le t_l$ for $l = 2, \ldots, d$, and $I = \{1, 2, \ldots, t\}$ for $t \le t_1$. For instance, $M_{32,24,12,8}$ means we consider all projections of the form

$$\begin{aligned}
\{1, 2, \ldots, d\} &\quad \text{for } d \le 32, \\
\{1, j_1\} &\quad \text{for } j_1 \le 24, \\
\{1, j_1, j_2\} &\quad \text{for } 1 < j_1 < j_2 \le 12, \\
\{1, j_1, j_2, j_3\} &\quad \text{for } 1 < j_1 < j_2 < j_3 \le 8.
\end{aligned}$$

The quality of each projection is measured by the normalized spectral test $d^*_{|I|}/d_I$, and $M_{t_1,\ldots,t_d}$ returns the worst case — that is, the smallest value of $d^*_{|I|}/d_I$ — over all projections.

In [284], the performance of Korobov point sets based on generators chosen via the more traditional measure M_8 — used to find good LCGs in [123, 119, 253] — is compared with those chosen via $M_{8,8,8}$. Numerical results show that in some cases the M_8 generators perform significantly worse than the $M_{8,8,8}$ ones. Intuitively, what this means is that if we do not look carefully at some of the low-dimensional projections of a point set, then undetected defects might cause the corresponding estimator to be subperforming. In Table 1 of [264], for each value of n, three generators a are given. The first one is based on M_{32} and therefore fails to look at several important low-dimensional projections compared with the ones based on $M_{32,24,12,8}$ and $M_{32,24,16,12}$ given on the second and third rows, respectively.*

A criterion similar to $M_{t_1,\ldots,t_d}$ based on the *resolution gap* rather than the spectral test is used in [285] to find polynomial Korobov lattices based on combined Tausworthe generators, where the resolution gap is simply the difference $\ell^*_s - \ell_s$ between the best possible resolution ℓ^*_s for an s-dimensional point set and its actual resolution ℓ_s. More generally, Panneton and L'Ecuyer [371] use criteria like this based on either the resolution gap, the t-value, or a quantity called the *neighbor-free gap* to find recurrence-based point sets based on $\mathbb{F}_2$-linear generators. In [397], the value t_I of the quality parameter t is computed for several projections $P_n(I)$ of the Sobol' sequence and is investigated as a way of defining alternatives to the t-value, which corresponds to the largest t_I for all $I \subseteq \{1,\ldots,s\}$. Similar ideas are discussed briefly in [237].

Alternatively, this kind of reasoning can be used to define *weighted* spaces of functions in which variables u_j are assumed to have less and less importance as j increases. The introduction of such spaces has allowed important breakthroughs in the study of the *tractability* of integration, which in turn have led to several new ideas for constructing low-discrepancy point sets in high dimensions. This is what we discuss in the appendix to this chapter.

6.4 Using quasi–Monte Carlo sampling for simulation

We go back again to the formulation from Chap. 4, where we write

---

* We would like to use this opportunity to point out that the values of a used in [150] that come from [264] were not the ones that were recommended as being the best in [264]. The "good" a's in [264] are those given in the second or third row of each group of three generators given for each n in Table 1 of that paper, as pointed out at the beginning of p. 1226 in [264]. For instance, for $n = 1021$, $a = 76$ or $a = 306$ should be chosen over $a = 331$.

$$\mu = \mathrm{E}(Y) = \mathrm{E}(h(\mathbf{X})) = \int_{\Omega} h(\mathbf{x})\varphi(\mathbf{x})d\mathbf{x}, \tag{6.14}$$

and where $\varphi(\mathbf{x})$ is the joint density function of the vector $\mathbf{X}$ of random variables to be simulated. From this point of view, Monte Carlo and randomized quasi–Monte Carlo amount to estimating μ by

$$\hat{\mu} = \frac{1}{n}\sum_{i=1}^{n} h(\mathbf{x}_i),$$

where each $\mathbf{x}_i$ has density $\varphi(\cdot)$.

In the Monte Carlo case, the $\mathbf{x}_i$'s are independent, while in the randomized quasi–Monte Carlo case, they are correlated. In Fig. 6.2, we show a sample of 1024 bivariate standard normal random variates with correlation 0 (top) generated (pseudo)randomly (left) or based on a two-dimensional randomly digitally shifted Sobol' sequence (right). The lower figures show the same types of samples but with a correlation of 0.5 for the bivariate normal. In both cases, inversion of the normal CDF has been used to generate the standard normal variates.

For the specific examples investigated in these figures, it looks like the low discrepancy of the Sobol' point set managed to produce a bivariate sample that in some sense deviates less from the true bivariate normal density than in the random case. If in turn the function h of interest is able to capture this improved behavior, we can hope that the Sobol' sequence will give rise to a more precise estimator. But if, for example, the function h is zero on most of its domain and nonzero only in a small region far away from (0,0), then randomized quasi–Monte Carlo, just like Monte Carlo, will suffer from having too few sample points in the region of interest. In this case, the improved empirical distribution of the sample based on randomized quasi–Monte Carlo might be of little help. For that reason, randomized quasi–Monte Carlo, just like Monte Carlo, will benefit from importance sampling in such cases. The use of importance sampling within quasi–Monte Carlo is studied in [195, 233, 264]. Moreover, recent work studying the combination of quasi–Monte Carlo methods with splitting techniques — which are related to importance sampling — can be found in [260].

More generally, the formulation (6.14) is helpful in understanding the interaction between randomized quasi–Monte Carlo and common variance reduction techniques. Importance sampling affects the transformation of $\mathbf{u}$ into $\mathbf{x}$ and also h by multiplying the original function h by the likelihood ratio derived from the chosen change of measure. Control variates only affect the definition of h. Note, however, that h should be modified differently for randomized quasi–Monte Carlo than for Monte Carlo because the optimal coefficient β depends on the distribution of the estimators for μ and μ_c — where, as in Chap. 4, μ_c denotes the expectation of the control variable — which is modified by the use of randomized quasi–Monte Carlo methods [187].

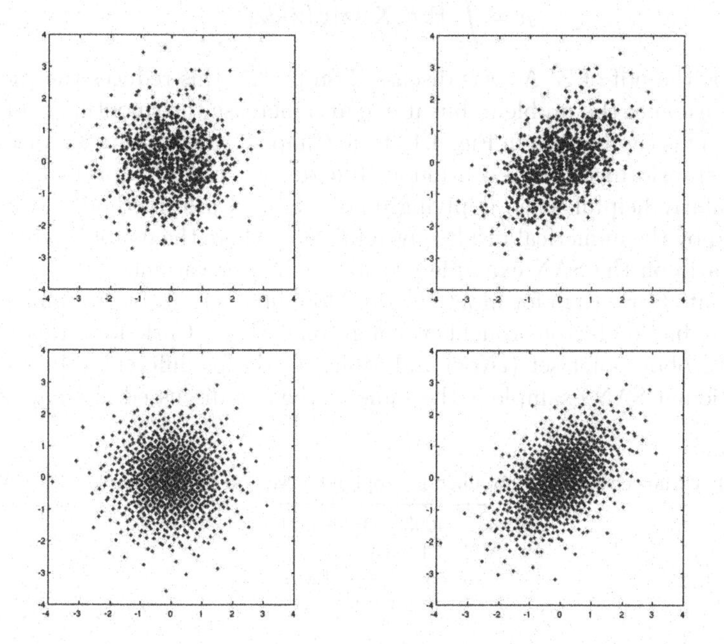

Fig. 6.2 Sample of 1024 bivariate normal variates with $\rho = 0$ (top) and $\rho = 0.5$ (bottom), based on random sampling (left) or quasi–Monte Carlo sampling (right).

That is, one should use

$$\beta^{\mathrm{rqmc}} = \frac{\mathrm{Cov}(\hat{Y}_{\mathrm{rqmc}}, \hat{C}_{\mathrm{rqmc}})}{\mathrm{Var}(\hat{C}_{\mathrm{rqmc}})},$$

where

$$\hat{Y}_{\mathrm{rqmc}} = \frac{1}{n}\sum_{i=1}^{n} Y_i \text{ and } \hat{C}_{\mathrm{rqmc}} = \frac{1}{n}\sum_{i=1}^{n} C_i$$

are the two estimators for $\mu = \mathrm{E}(Y)$ and $\mu_c = \mathrm{E}(C)$ based on a randomized low-discrepancy point set with n points. This optimal β^{rqmc} can be estimated by constructing m i.i.d. copies of the estimators $\hat{Y}_{\mathrm{rqmc}}$ and $\hat{C}_{\mathrm{rqmc}}$ in a manner similar to how $\mathrm{Var}(\hat{\mu}_{\mathrm{rqmc}})$ is estimated, as discussed on p. 202.

Conditional Monte Carlo completely changes the formulation (6.14), as it amounts to having

$$\mu = \int_{\mathcal{Z}} \mathrm{E}(h(\mathbf{X})|\mathbf{z})\psi(\mathbf{z})d\mathbf{z},$$

where ψ is the pdf of $\mathbf{Z}$. As was discussed on p. 225, this reduces the (nominal) dimension of the problem, but it can also increase the smoothness of the function. This can be seen in Fig. 4.13 from Chap. 4, where an indicator function was transformed into a continuous function for the SAN problem. This is particularly helpful when applying quasi–Monte Carlo methods, which is confirmed by the numerical results given in [285], where the use of conditional Monte Carlo on the SAN example provides a greater variance reduction for quasi–Monte Carlo than for Monte Carlo. More precisely, Table 6.3 shows the variance reduction factor brought by conditional Monte Carlo for a randomly shifted Korobov point set (rKor) and Monte Carlo for different values of n. The particular SAN example is the same as the one discussed in Chap. 4.

Table 6.3 Variance reduction obtained by applying CMC on SAN example with $s = 13$.

n	4093	16381	65521
MC	4.1	4.1	4.1
rKor	43	200	126

Summing up, quasi-random sampling based on randomized low-discrepancy point sets can be thought of as a general variance reduction technique in the sense that it can be applied to a wide class of problems without necessarily any specific information on the problem at hand. However, its success clearly depends on certain properties of the function to be integrated that have to do with its interaction with the low-discrepancy point set used. For example, if the point set has good uniformity properties for the projections that correspond to important ANOVA components, then the randomized quasi–Monte Carlo estimator should have a lower variance than the Monte Carlo estimator. Also, the gains are usually larger in terms of efficiency since several randomized low-discrepancy constructions can be generated rather quickly — faster than constructing a random point set by repeatedly calling a pseudorandom number generator. When using techniques meant to reduce the effective dimension — such as the Brownian bridge technique — the computation time usually increases, but the greater variance reduction that can be obtained may compensate for this drawback, thus yielding efficiency gains as well.

Estimating quantiles

So far, we have focused on the use of quasi–Monte Carlo methods for estimating an integral or an expectation. With randomized quasi–Monte Carlo, it is possible to go beyond this type of problem, just as was the case for Monte Carlo, as discussed in Sect. 1.5. In particular, one can construct an empirical CDF and estimate quantiles using the same ideas as in [23], where it is shown how to use Latin hypercube sampling for that purpose. In fact, quantile estimators based on randomized quasi–Monte Carlo methods have been used for *value-at-risk* estimation problems in [206, 235, 367].

More precisely, as in Sect. 1.5, suppose that, for a given value of $p \in (0,1)$ and a random variable $Y = h(\mathbf{X})$ representing the output of a simulation based on the vector of random variables $\mathbf{X}$, we want to find an estimate for the 100pth quantile q_p. Here we use the representation $Y = f(\boldsymbol{U})$, where $\boldsymbol{U}$ is uniformly distributed in $[0,1)^s$. Suppose $P_n = \{\mathbf{u}_1, \ldots, \mathbf{u}_n\}$ is a randomized quasi–Monte Carlo point set, and let $y_i = f(\mathbf{u}_i)$ for $i = 1, \ldots, n$. We can then define as the approximation for the CDF of Y

$$\hat{F}_{n,\mathrm{rqmc}}(y) = \frac{1}{n} \sum_{i=1}^{n} \mathbf{1}_{y_i \leq y}$$

and the corresponding quantile estimate

$$\hat{q}_{p,\mathrm{rqmc}} = \hat{F}^{-1}_{n,\mathrm{rqmc}}(p) = y_{(\lceil np \rceil)},$$

where $y_{(1)} \leq \ldots \leq y_{(n)}$ are the order statistics of the sample. Note that since the y_i's have a different multivariate distribution than when random sampling is used — in particular, they are not independent — the bias of the estimator $\hat{q}_{p,\mathrm{rqmc}}$ is different from the bias of the estimator $\hat{q}_p$ based on random sampling. We illustrate this with the following example, which parallels Example 1.4 discussed in Sect. 1.5.

Example 6.11. Suppose $n = 4$, $y = f(u) = u$, and we use the point set $P_4 = \{v, (0.25 + v) \bmod 1, (0.5 + v) \bmod 1, (0.75 + v) \bmod 1)\}$, where $v \sim U(0,1)$. This corresponds to a one-dimensional randomly shifted lattice point set with $n = 4$. Then

$$\hat{q}_{0.5,\mathrm{rqmc}} = y_{(3)} \sim U(0.5, 0.75),$$

and therefore $\mathrm{E}(\hat{q}_{0.5,\mathrm{rqmc}}) = 0.625$, with a corresponding bias of 0.125. Recall that for a random sample, we saw in Example 1.4 that the bias was 0.1.

More generally, for a sample of size n based on a randomly shifted lattice point set, we have

$$\hat{q}_{0.5,\mathrm{rqmc}} = \begin{cases} y_{\left(\frac{n+1}{2}\right)} \sim U\left(\frac{n-1}{2n}, \frac{n+1}{2n}\right) & \text{if } n \text{ is odd} \\ y_{\left(\frac{n}{2}+1\right)} \sim U\left(\frac{n}{2n}, \frac{n+2}{2n}\right) & \text{if } n \text{ is even.} \end{cases}$$

Therefore, when n is odd, we have $\mathrm{E}(\hat{q}_{0.5,\mathrm{rqmc}}) = 1/2$ and thus the randomized quasi–Monte Carlo quantile estimator has no bias, but when n is even, $\mathrm{E}(\hat{q}_{0.5,\mathrm{rqmc}}) = 1/2 + (1/2n)$, and therefore in this case the bias is $1/2n$. Recall that, for the corresponding quantile estimator described in Example 1.4, the bias was instead $1/(2(n+1))$. However, if we compare the variances of the two estimators when n is even, then we have that

$$\mathrm{Var}(\hat{q}_{0.5,\mathrm{rqmc}}) = \frac{1}{12n^2}$$

since $\hat{q}_{0.5,\mathrm{rqmc}} \sim U(n/2n, (n+2)/2n)$, while with the Monte Carlo method we have

$$\mathrm{Var}(\hat{q}_{0.5}) = \frac{n}{4(n+1)^2}$$

since, as seen in Example 1.4, $\hat{q}_{0.5} = y_{(n/2+1)}$ has a beta distribution with parameters $(n/2+1, n/2)$. Hence the randomized quasi–Monte Carlo estimator has a mean-square error in $O(1/n^2)$, while for the Monte Carlo estimator it is in $O(1/n)$ because the variance is in $O(1/n)$.

This example demonstrates that quantile estimators based on randomized quasi–Monte Carlo point sets have different properties than estimators based on Monte Carlo. In general settings, it might be difficult to assess the bias and variance as we did in the example above, but the hope is that if the variables Y_i are sampled so that the corresponding approximation $\hat{F}_{n,\mathrm{rqmc}}$ for the CDF of Y is more accurate than the one obtained by random sampling, then the resulting quantile estimator $\hat{q}_{p,\mathrm{rqmc}}$ extracted from that better approximation should also be more accurate. In [23, Sect. 2], the authors show that under mild conditions this is the case when Latin hypercube sampling is used. That is, they show that as $n \to \infty$, the estimator $\hat{q}_{p,\mathrm{rqmc}}$ obtained from an n-point Latin hypercube sampling estimator has a bias that goes to 0 and a variance no larger than that of the Monte Carlo estimator.

Numerical results confirm the superiority of this estimator, not only to Monte Carlo but also to estimators of the form

$$\frac{1}{n}\sum_{i=1}^{n} \hat{q}_{p,i}, \tag{6.15}$$

where $\hat{q}_{p,i}$ is a quantile estimator based on a random sample $\{Y_i^{(1)}, \ldots, Y_i^{(m)}\}$ but where there might be a dependence within the sample $\{Y_i^{(l)}, i = 1, \ldots, n\}$ for a given l. In other words, here we compute n different estimators $\hat{q}_{p,i}$, $i = 1, \ldots, n$ of the quantile, each based on independent replications — in the quasi–Monte Carlo context, this can be done by creating m copies of a randomized point set and then using the first point of each of those m randomized point sets to construct the first estimator, the second point of each randomized point set to construct the second estimator, and so on — but with a dependence across those n estimators. The idea is that each of these n estimators then has the same expectation as the Monte Carlo estimator, but we expect that the correlation across the different estimators $\hat{q}_{p,i}$, $i = 1, \ldots, n$ will contribute to providing an overall estimator (6.15) with smaller variance. One problem with this approach is that quantile estimators can be quite inaccurate when they are based on too small a sample size, especially when p is near 0 or 1. In the numerical experiments reported in [23], the estimator that uses the n points of a Latin hypercube sample to compute $\hat{q}_p$ performs much better than those of the form (6.15).

Examples

We end this section with a detailed description of how to use a randomly shifted Korobov point set on a variant of the simple bank example of Chap. 1 and then on a finance problem.

Example 6.12. For the bank example discussed in Chap. 1, we can use a randomly shifted Korobov point set to run the simulations as follows. Each shifted point provides the uniform numbers required to generate the interarrival and service times for one simulation of the bank. To simplify things for now, we assume that instead of fixing the simulation horizon, the goal is to estimate the number of clients among the first 300 that will wait more than 5 minutes. By doing so, the dimension s of the problem becomes 599 (one interarrival time and one service time per client, except for the last one, who only needs an interarrival time).

The code to run these simulations is given in Figs. 6.3 and 6.4. Results are given in Table 6.4. What we see there is that the randomized quasi–Monte Carlo estimator reduces the half-width of the 95% confidence interval by a factor greater than 2, while the computation time is also smaller. In general, randomly shifted Korobov lattices and digital nets in base 2 require less computation time than Monte Carlo.

```
RunAllSim(a, n, m, 599)
    InitKorobov(a, n, 599, z)
    for k ← 1 to m
        for j = 1 to 599
            v_j ← Rand01()
        u ← 0
        result[1] ← OneSimBank(v)
        for i = 2 to n do
            NextKorobov(n, z, u)
            w ← (u + v) mod 1
            result[i] ←OneSimBank(w)
        x[k] ← ave(result)
    print("average is",ave(x))
    hw ← 1.96 × sqrt(var(x)/m)
    print ("95% CI half-width is", hw)
```

Fig. 6.3 Running simulations of the bank example based on a shifted Korobov point set with m randomizations.

```
OneSimBank(u_1, u_2, ..., u_599)
    time ← 0
    NbWait5 ← 0
    w ← 0
    a ← GenExpon(u_1,1)
    t ← 1 // number of clients simulated so far
    for t = 2 to 300 do
        s ← GenExpon(u_{2(t-1)},0.75)
        a ← GenExpon(u_{2t-1},1)
        w ← max(0, w + s - a)
        if (w > 5) then
            NbWait5 ← NbWait5 + 1
    return NbWait5
```

Fig. 6.4 Pseudocode for the function OneSimBank.

Table 6.4 Comparison of Monte Carlo and randomly shifted Korobov point set with $n = 1024$, $a = 139$, and $m = 25$ for the bank example. Shown are the estimates $\hat{\mu}$ for the number of clients (among the first 300) who will wait more than 5 minutes, the corresponding 95% confidence half-width (HW), and the time in CPU required for the computation.

	$\hat{\mu}$	HW	CPU(sec.)
MC	39.15	0.422	7.42
rKOR	38.99	0.189	4.66

Example 6.13. We use the setup from Prob. 1.12, where the value of one share of IBM stock at time t, denoted $S(t)$, is assumed to follow a lognormal distribution (i.e., $\ln S(t)$ has a normal distribution with mean $\ln(S(0)) + (r - \sigma^2/2)t$ and variance $\sigma^2 t$ under our pricing measure), where r is the risk-free interest rate and σ is the volatility of the stock price. Now consider an *Asian call option* on this stock, which is a financial contract whose payoff at expiration time T is given by

$$C(T) = \max\left(0, \frac{1}{s}\sum_{j=1}^{s} S(t_j) - K\right),$$

where K is the *strike price* and $0 \le t_1 < \ldots < t_s = T$ are observation dates where the price of the stock is recorded. It can be shown — and we will explain this in more detail in Chap. 7 — that the value $C(0)$ of this option at time 0 is given by

$$C(0) = \mathrm{E}\left(e^{-rT} C(T)\right).$$

There is no known analytical formula for this price, but Monte Carlo and (randomized) quasi–Monte Carlo can be used to estimate $C(0)$. An early reference on the use of Monte Carlo for this problem is [215].

In Figs. 6.5 and 6.6, we give code for estimating $C(0)$ based on a randomly shifted Korobov point set. We assume there that the observation dates are of the form $t_j = jT/s$ for $j = 1, \ldots, s$.

```
RunAllSim(a, n, m, s)
    InitKorobov(a, n, s, z)
    for k ← 1 to m
        for j = 1 to s
            v_j ← Rand01()
        u ← 0
        result[1] ← AsianCall(v)
        for i = 2 to n do
            NextKorobov(n, z, u)
            w ← (u + v) mod 1
            result[i] ←AsianCall(w)
        x[k] ← ave(result)
    print("average is",ave(x)
    hw ← 1.96 ×√(var(x)/m)
    print ("95% CI half-width is", hw)
```

Fig. 6.5 Running simulations of the Asian call option example based on a shifted Korobov point set with m randomizations.

Table 6.5 gives results for the case where the expiration time is $T = 1$ year, $s = 32$, $r = 0.05$, $\sigma = 0.3$, $S(0) = 50$, and $K = 50$. In addition to using randomly shifted Korobov point sets (with the same generator a as in Table 6.4), we also test a randomly digitally shifted Sobol' point set and generalized Halton point sets based on the multiplicative factors suggested in [115], which can also be found at [498]. Note that we have made no effort to try to improve the computation time of the generalized Halton sequence in these experiments. A more careful implementation could certainly reduce this computation time.

6.5 Suggestions for practitioners

To conclude this chapter, we wish to offer a few tips that can be useful to practitioners when applying quasi–Monte Carlo. Generally, one of the most

```
AsianCall(r, σ, S(0), s, T, K, u_1, u_2, ..., u_s)
  a ← (r − σ^2/2) × (T/s)
  b ← σ × √(T/s)
  S[0] ← S(0)
  sum ← 0
  for t = 1 to s do
    z ← Norm01(u_t)
    S[t] ← S[t − 1] × exp^(a+bz)
    sum ← sum +S[t]
  sum ← sum/s
  C ← exp(−rT)× (sum −K)
  if C > 0 then
    return C
  else
    return 0
```

Fig. 6.6 Code for evaluating the discounted payoff of the Asian call option.

Table 6.5 Comparison of Monte Carlo (MC), randomly shifted lattice point set (rKOR), randomly digitally shifted Sobol' sequence (rSOB), and randomly digitally shifted generalized Halton sequence (rGHal) with $n = 1024$ and $m = 25$ for the Asian call option example. Shown are the option price estimates $\hat{\mu}$, the corresponding 95% confidence interval half-width (HW), and the CPU time required.

	$\hat{\mu}$	HW	CPU(sec.)
MC	7.029	0.102	0.575
rKOR	7.067	0.014	0.456
rSOB	7.063	0.012	0.466
rGHal	7.070	0.015	1.93

important decisions to make is the choice of construction. The first thing to do in this selection process is to decide whether we should use a sequence or a point set of fixed size. A sequence should clearly be used if the user wants to be able to increase the size of n if desired, to improve the accuracy of the approximation. A sequence might also be preferable if the user does not want to be restricted in terms of specific values of n to use. That is, point sets of a fixed size (e.g., lattices, whether they are standard or polynomial) are sometimes "offered" only for certain values of n, typically given by prime powers. If this is too restrictive and the user wants to be able to take n =10,000 or n =20,000, for example, then a sequence should be used. This is because taking the first 10,000 points of a point set with, say, n =16,384, provides no guarantee on the quality of the subset chosen.

A second consideration is to look at what kind of function is to be integrated. For instance, if the function can be shown to belong to a certain class of integrands (like the weighted classes discussed in the appendix to this chapter), then constructions specifically built for these classes should be used, such as the rank-1 lattices given in [409, 410]. If not much is known on the integrand, then a more "general-purpose" construction like the Sobol' sequence might be more appropriate. A third consideration is the dimension of f: Is it known or is it unbounded? If it is unbounded, then recurrence-based point sets are a good choice.

In terms of implementation, there are a few libraries that contain quasi–Monte Carlo routines [490, 491, 500, 496], and more links can be found on the Web site [490]. In addition, simple constructions such as rank-1 lattices (including Korobov) and generalized Halton sequences such as those discussed in [115] can be implemented from scratch rather easily.

In practical settings, users generally like to be able to estimate the error of their approximations, which means a randomization should be applied to the point set. For lattices, a random shift is typically used. For digital nets, our point of view is that if the underlying point set is believed to be of good quality, then using a random digital shift is a reasonable choice. Otherwise, a random scrambling is probably more appropriate.

Problems

6.1. Consider the bank example from Chap. 1 also discussed in Example 6.12, where we fixed the number of clients at 300. Suppose now that, as in Chap. 1, the bank is simulated for 5 hours, so that the number of clients is random. (a) Estimate the expected number of clients that will wait more than 5 minutes in a given day at the bank using 25 repetitions of a randomly shifted Korobov point set based on $n = 1021$ and $a = 76$. (b) Determine by how much the (estimated) variance of this estimator reduces the variance of the corresponding Monte Carlo estimator based on 1021×25 independent simulations.

6.2. Write a program that, given two s-dimensional points $\mathbf{u}$ and $\mathbf{v}$ in $[0,1)^s$, computes the point obtained by performing a b-ary digital addition of $\mathbf{u}$ and $\mathbf{v}$ (as in the random digital shift method in base b), where $b > 2$. (b) Repeat (a) with $b = 2$.

6.3. Show that the random digital shift in base b does not satisfy Property 2(b) on p. 214 but that the random linear scrambling does.

6.4. Consider the bank example as implemented in Prob. 6.1. Compare the following randomized quasi–Monte Carlo methods based on $m = 25$ randomizations through their empirical variance: (i) randomly shifted Korobov lattice with $n = 1024$ and $a = 139$; (ii) first 1024 points of a randomly digitally $(b_1, \ldots, b_s)$-shifted Halton sequence; and (iii) same as (ii) but for a generalized Halton sequence implemented in Prob. 5.5.

6.5. Consider the Asian option problem studied in Example 6.13. Compare the empirical variance of the estimator based on $m = 25$ randomizations of a Korobov point set with parameters $n = 1021$ and $a = 76$ when (i) no periodization is applied; (ii) the periodization proposed by Sidi and mentioned on p. 196 is applied; (iii) the baker transformation is applied; and (iv) no periodization but the use of the Brownian bridge technique.

6.6. Show that $\hat{\sigma}^2_{\{j\}}$ as given in (6.9) is an unbiased estimator of $\sigma^2_{\{j\}}$.

6.7. Consider the function

$$f(\mathbf{u}) = \prod_{i=1}^{s}(1 + c(u_j - 1/2)).$$

(a) Determine the ANOVA components of this function. (b) Give an expression for σ_I^2 for all $I \subseteq \{1,\ldots,s\}$. (c) Give an expression for the average dimension in the superposition sense. (d) Give an expression for the effective dimension in the truncation sense and the superposition sense.

6.8. Using the estimator $\hat{\gamma}_I$ given in (6.11), estimate $\gamma_{\{1,\ldots,d\}}$ for $d = 1,\ldots,31$ for the Asian option problem from Example 6.13, and use your results to estimate the effective dimension of the underlying function (in proportion 0.99). (b) Repeat but with paths generated using the Brownian bridge technique, as in Prob. 6.5.

6.9. Prove that the equality (6.12) holds and then that (6.11) is an unbiased estimator for γ_I.

6.10. Consider the component-by-component rank-1 lattice point set described in [409]. Compare the empirical variance of the randomly shifted version of this point set based on $m = 25$ randomizations obtained for (a) the Asian option problem and (b) the function

$$f(\mathbf{u}) = \prod_{j=1}^{20} \frac{|4u_j - 2| + j}{1 + j}$$

with that of (i) the Monte Carlo estimator (based on the same total number of function evaluations) and (ii) the first 2003 points of the extensible Korobov lattice sequence in base 2 using the generator $a = 14471$.

6.11. Consider the bank example described in Example 6.12. Compare the performance (using the empirical variance) of (i) array-RQMC based on the two-dimensional Korobov point set with $n = 1021$ and $a = 76$, 25 random shifts, and using as in [263] the underlying Markov chain defined by

$$X_1 = W_1 = 0; X_2 = W_1 + S_1; X_3 = W_2; X_4 = W_2 + S_2, \text{ etc.},$$

(ii) the Latin supercube sampling method also based on using $d_j = 2$ and a randomly shifted two-dimensional Korobov point set based on $n = 1021$ and $a = 76$, and (iii) a randomly shifted Korobov point set with $n = 1021$, $a = 76$, and $s = 599$.

Appendix: Tractability, weighted spaces, and component-by-component constructions

Another way to study low-discrepancy sequences is through the concept of *computational complexity* for multivariate integration (see [189, 190, 470, 479], for example, and the survey [406]). The goal here is to determine, for a certain class of functions, the minimum number n of function evaluations required to build an approximation whose worst case error for that class is ϵ times smaller than the trivial approximation by zero and to look at how this number n behaves as a function of the dimension s.

More precisely, since here we are interested in asymptotics not only with respect to n but also with respect to s, we will rewrite the point set P_n as $P_{n,s}$ and consider families of the form $\{P_{n,s}\}$, where $n, s \geq 1$. That is, we are considering a construction that can be extended both in the number of points and the dimension. Second, we consider the worst-case error $e(P_{n,s})$ of a point set $P_{n,s} = \{\mathbf{u}_1, \ldots, \mathbf{u}_n\}$ over some class of functions $\mathcal{F}_s$ equipped with a norm $\|\cdot\|_s$. This worst-case error is defined as

$$e(P_{n,s}) = \sup_{f \in \mathcal{F}_s, \|f\|_s \leq 1} |I(f) - Q_n|$$

for $n, s \geq 1$, where, as usual,

$$Q_n = \frac{1}{n} \sum_{i=1}^{n} f(\mathbf{u}_i).$$

We also define

$$e_{0,s} = \sup_{f \in \mathcal{F}_s, \|f\|_s \leq 1} |I(f)|.$$

We then define $n = n_{\min}(\epsilon, s, P_{n,s})$ as the smallest n for which the worst-case error satisfies $e(P_{n,s}) \leq e_{0,s}\epsilon$. So, for example, if $\epsilon = 0.001$, then $n_{\min}(\epsilon, s, P_{n,s})$ is the smallest value of n such that the first n points of our construction can be used to build an estimator Q_n whose corresponding error will be at least 1000 times smaller than the trivial approximation $Q_0 := 0$ for all $f \in \mathcal{F}_s$.

Definition 6.12. A family $\{P_{n,s}\}$ is *tractable* if and only if there exist non-negative constants C, q, and p such that

$$n_{\min}(\epsilon, s, P_{n,s}) \leq C s^q \epsilon^{-p} \tag{6.16}$$

for all $s \geq 1$ and for all ϵ in $(0,1)$. If (6.16) holds with $q = 0$, then we say $\{P_{n,s}\}$ is *strongly tractable.*

Furthermore, integration in $\mathcal{F}_s$ is said to be *QMC-tractable* if there exists a family $\{P_{n,s}\}$ that is tractable and similarly for strong tractability. If no such family exists, then integration over that space is said to be *QMC-intractable.*

Since $\epsilon < 1$, this means we want p to be as small as possible in (6.16). The smallest (infimum) power p in the bound (6.16) is called the *ϵ-exponent* and the *strong ϵ-exponent,* for tractability and strong tractability, respectively. Also, the smallest (infimum) power q for which the desired bound holds is called the *s-exponent.* A possible variation is to replace the worst-case error by an average error using some measure on the space of functions under study [470, 469, 479, 480], but we will not discuss this further here.

An important property to point out is that it can be shown that

$$e(P_{n,s}) = \mathcal{D}_s(P_{n,s}),$$

where $\mathcal{D}_s(\cdot)$ is the discrepancy measure that corresponds to the space of functions $\mathcal{F}_s$ under study and its accompanying norm $\|\cdot\|_s$. That is, one can explicitly construct a function f such that $\|f\|_s = 1$ and for which the upper bound

$$|Q_n - I(f)| \leq \mathcal{D}_s(P_{n,s}) \times \|f\|_s = \mathcal{D}_s(P_{n,s})$$

is in fact an equality. Based on this, if we go back to the Koksma-Hlawka inequality, it is clear that strong tractability cannot be achieved in this setting because by definition all low-discrepancy point sets are such that

$$D^*(P_{n,s}) \in O(n^{-1}(\log n)^s),$$

and therefore $e(P_{n,s})$ clearly depends on s in this case.

To remove the dependence on the dimension s that arises in bounds such as the Koksma-Hlawka inequality, weights γ_j can be used to assess the importance of each variable u_j. This is in contrast with the class of functions that are of bounded variation in the sense of Hardy and Krause, for which there is an implicit assumption that all the variables u_j are equally important. The use of weights is consistent with our discussion in Sect. 6.3.4, where we argued that, in practice, problems can sometimes be formulated (or engineered) so that the variables $u_1, u_2, \ldots, u_s$ are of decreasing importance.

In what follows, we will be using sequences $\gamma = \{\gamma_j\}$ of weights such that

$$\gamma_1 \geq \gamma_2 \geq \ldots \geq \gamma_j \geq \ldots \geq 0.$$

With such weights, we can obtain an error bound similar to the Koksma-Hlawka inequality but with a weighted version of the discrepancy $D_2(P_n)$ introduced in (5.23). That is, the *weighted L_2-discrepancy*

$$D_{2,\gamma}(P_n) = \left[\sum_I \gamma_I \int_{[0,1)^d} |\alpha(P_n(I), \mathbf{v}_I) - v_{i_1} \dots v_{i_d}|^2 \, d\mathbf{v}_I \right]^{1/2}$$

is used, where $I = \{i_1, \dots, i_d\}$ and

$$\gamma_I = \prod_{j \in I} \gamma_j.$$

(Weighted discrepancy measures are also considered in [180, 181, 182] with the assumption $\gamma_j = \gamma$ for all j in [180, 181].) The corresponding norm $\| \cdot \|_s$ is defined as

$$V_{2,\gamma}(f) = \left[\sum_I \gamma_I^{-1} \int_{[0,1)^d} \left(\left. \frac{\partial^d f}{\partial \mathbf{u}_I} \right|_{\mathbf{u}_{-I}=(1,\dots,1)} \right)^2 d\mathbf{u}_I \right]^{1/2}, \qquad (6.17)$$

and one can get the error bound

$$|Q_n - I(f)| \leq D_{2,\gamma}(P_n) V_{2,\gamma}(f).$$

The class of functions $\mathcal{F}_{s,\gamma}$ for which $V_{2,\gamma}(f)$ is finite is in the *Sobolev space* $W_2^{(1,1,\dots,1)}([0,1)^s)$, which consists of all functions defined over $[0,1)^s$ for which each mixed partial derivative of the form

$$\frac{\partial^d f}{\partial u_{i_1} \dots \partial u_{i_d}}, \qquad 1 \leq i_1 < \dots < i_d \leq s,$$

is square-integrable. In fact, this space can be defined as a *weighted Sobolev space* and could be made more general by introducing more parameters in the definition of its associated norm. Our brief overview of tractability results does not require this added generality, and for this reason it will not be discussed further here. Another widely used type of space in tractability results are the weighted Korobov spaces, which are used to study periodic functions and are thus relevant when studying deterministic lattice point sets.

Note that even if the choice of the weights γ_j does not influence whether or not a function belongs to $\mathcal{F}_{s,\gamma}$ — that is, if $f \in \mathcal{F}_{s,\gamma}$, then $f \in \mathcal{F}_{s,\gamma'}$ as long as γ and γ' both represent a sequence of positive weights — the norm $V_{2,\gamma}(f)$ increases as the weights γ_j decrease through the terms γ_I^{-1} included in the definition (6.17). In turn, since $e(P_{n,s})$ is defined as the worst case error for functions with $V_{2,\gamma}(f) \leq 1$, it means that as the weights γ_j decrease, this worst-case is taken over a smaller set of functions, which is why the choice of γ_j affects what kind of tractability result we get. Equivalently, the choice of γ_j influences the value of the corresponding discrepancy $D_{2,\gamma}(P_n)$, which can be more easily seen by looking at the following formula, given in [413]:

$$D_{2,\gamma}(P_n) = \left[\prod_{j=1}^{s}\left(1+\frac{\gamma_j}{3}\right)^2 - \frac{2}{n}\sum_{i=0}^{n-1}\prod_{j=1}^{s}\left(1+\frac{\gamma_j}{2}(1-u_{i,j}^2)\right)\right.$$
$$\left. + \frac{1}{n^2}\sum_{i,i'=1}^{n}\prod_{j=1}^{s}(1+\gamma_j \min(1-u_{i,j}, 1-u_{i',j}))\right]^{1/2}.$$

Hence, for a given function f in $\mathcal{F}_{s,\gamma}$, as we decrease the weights γ_j, the discrepancy $D_{2,\gamma}(P_n)$ decreases at the expense of an increase in the norm $V_{2,\gamma}(f)$.

We can now state the following important result.

Theorem 6.13. *[413] (i) Multivariate integration in $\mathcal{F}_{s,\gamma}$ is strongly QMC-tractable if and only if*

$$\sum_{j=1}^{\infty}\gamma_j < \infty,$$

and in that case the ϵ-exponent is in $[1,2]$. *(ii) Multivariate integration in $\mathcal{F}_{s,\gamma}$ is QMC-tractable if and only if*

$$a := \limsup_{s\to\infty}\frac{\sum_{j=1}^{s}\gamma_j}{\ln s} < \infty.$$

If a is finite, then the d-exponent belongs to $[a/12, a/6]$ *and the ϵ-exponent belongs to* $[1,2]$. *(iii) Let $n_{\gamma,\min}(\epsilon, s, P_{n,s})$ be the minimal number of sample points needed to reduce the initial error by a factor of ϵ by a quasi–Monte Carlo algorithm. Then*

$$n_{\gamma,\min}(\epsilon, s, P_{n,s}) \le \left\lfloor \frac{\exp(\frac{1}{6}\sum_{j=1}^{s}\gamma_j)-1}{\epsilon^2}\right\rfloor$$

and

$$n_{\gamma,\min}(\epsilon, s, P_{n,s}) \ge (1-\epsilon^2)1.055^{\sum_{j=1}^{s}\gamma_j}.$$

Hence, this result shows that the weights γ_j must decrease fast enough that $\sum_{j=1}^{\infty}\gamma_j$ is finite in order for integration to be strongly QMC-tractable in the corresponding space. For instance, weights of the form $\gamma_j = \gamma^j$ for some $0<\gamma<1$ satisfy this condition.

Typically, results in this area are not constructive. That is, the existence of a construction that can get the error below ϵ with a certain number of points is demonstrated, but the specific construction achieving this is not given. However, such results are useful to understand better which types of functions are difficult or easy for multivariate integration. Also, results in this area have been used to a large extent in several papers on *component-by-component* constructions. Here the idea is to start with a class of functions known to be tractable or QMC-tractable, identify the type of quasi–Monte

Carlo construction that can achieve these tractability results, and then use the corresponding discrepancy measure to guide a search where the parameters defining the construction for a given n are found one dimension at a time by minimizing this discrepancy measure. We will illustrate this approach with an example coming from [409]. (Our setting is not as general as in [409], as we do not exploit the full generality of weighted Sobolev spaces, but is sufficiently broad to include the specific numerical example given in that paper.)

Component-by-component construction of a rank-1 lattice

We consider as before the class $\mathcal{F}_{s,\gamma}$, where we assume that $\sum_{j=1}^{\infty} \gamma_j < \infty$, so that integration is known to be strongly QMC-tractable for that space. The first step toward finding a construction that achieves the corresponding bound on its discrepancy for a given n (recall that the discrepancy is equal to the worst-case error $e(P_{n,s})$ for which a bound $\epsilon \times e_{0,s}$ independent of s has been established via the strong tractability result) is to narrow the choice of possible constructions. This can be done using a result in [414], which shows that if n is sufficiently large, then a *shifted lattice point set* can achieve the bound. From there, one possibility is to try to find that shifted lattice, and this is the approach used in [410]. Alternatively, one can try to find an unshifted lattice and study the error of its randomly shifted version. This can be achieved by looking at the *mean-square discrepancy* of the lattice, given by [414]

$$E_{\mathbf{v}}(D^2_{2,\gamma}(P_{n,s} + \mathbf{v})),$$

where $P_{n,s}$ denotes the unshifted lattice and the expectation $E_{\mathbf{v}}$ is taken over the random shift $\mathbf{v}$. In [409], it is shown that

$$E(D^2_{2,\gamma}(P_{n,s} + \mathbf{v})) \leq \frac{1}{n} \left(\prod_{j=1}^{s} (1 + \gamma_j/2) - \prod_{j=1}^{s} (1 + \gamma_j/3) \right), \tag{6.18}$$

and because it can be shown that for this class of functions we have [413]

$$e_{0,s} = \prod_{j=1}^{s} \left(1 + \frac{\gamma_j}{3}\right),$$

then it can be proved that

$$\frac{E_{\mathbf{v}}(D^2_{2,\gamma}(P_{n,s} + \mathbf{v}))}{e_{0,s}}$$

is bounded independently of s. Therefore, strong tractability can be achieved in a probabilistic sense by a randomly shifted lattice.

Based on this, a component-by-component construction algorithm for finding such lattices (for a given n) is given in [409]. The algorithm rests on the fact that if we have a generating vector $(z_1, \ldots, z_j)$ for which (6.18) is satisfied for $s = j$, then a successive component z_{j+1} can be found so that same bound will hold with $s = j + 1$. It suffices to choose this next component z_{j+1} by simply searching the one that minimizes $E(D^2_{2,\gamma}(P_{n,j+1} + \mathbf{v}))$, which is given by

$$E(D^2_{2,\gamma}(P_{n,s} + \mathbf{v})) = \frac{1}{n} \sum_{i=1}^{n} \prod_{j=1}^{s} \left[1 + \gamma_j \left(B_2 \left(\frac{iz_j}{n} \bmod 1 \right) + \frac{1}{3} \right) \right] - \prod_{j=1}^{s} \left(1 + \frac{\gamma_j}{3} \right),$$

where $B_2(\cdot)$ is the Bernoulli polynomial of degree 2.

Examples of parameters with γ_j of the form 0.9^j for $n = 2003$ are given in [409, Table 5.1] up to dimension $s = 100$.

Chapter 7
Financial Applications

Financial problems such as option pricing form a rich class of applications for simulation, variance reduction techniques, and quasi–Monte Carlo sampling. They provide a unique opportunity to present these topics in an applied setting and therefore represent a valuable learning tool that we believe will be useful to the reader. Readers interested in a more extensive treatment of Monte Carlo simulation in finance are referred to [145, 202, 314].

The problems studied in this chapter all fit in the following framework. We start with a market model where we have q *underlying assets* and denote by $S_j(t)$ the value of the jth asset at time t for $j = 1, \ldots, q$. We also have a *bank account*, which pays interest at a rate $r_t \geq 0$ at time t. Most of the time, we assume that $r_t = r$ is constant, and the corresponding value of r is called the *risk-free rate*. We think of an *option* in a loose sense as a security that entitles its holder to a certain payoff whose value depends on one or more of the q underlying assets. We are interested in determining different quantities related to the option, the most important one being its value at a given time, for a given model of the underlying assets.

We start in Sect. 7.1 by considering the special case of *European option pricing* under the lognormal model and then explain in Sect. 7.2 how to handle more complex models. In Sect. 7.3, we discuss the use of quasi–Monte Carlo methods and then describe in Sect. 7.4 how variance reduction techniques can be used in finance. We conclude with two sections on more complex estimation problems, starting with *American option pricing* in Sect. 7.5 and then sensitivity and percentile estimates — including *value-at-risk* — in Sect. 7.6.

7.1 European option pricing under the lognormal model

In this section, we assume that the type of contract we are interested in is a *European option.* This type of contract has a specified expiration time T

C. Lemieux, *Monte Carlo and Quasi–Monte Carlo Sampling*,
Springer Series in Statistics 692, DOI: 10.1007/978-0-387-78165-5_7,

and gives its owner the right — but not the obligation — to perform certain actions at time T involving the option's underlying asset(s), which produces a certain payoff $H(T, \mathbf{S})$, where $\mathbf{S} = \{(S_1(t), \ldots, S_q(t)), t \geq 0\}$. For example, a *European call* on a single asset gives its owner the right to buy the asset at time T for a predetermined price K called the *strike price.* In other words, the *payoff* at time T of an option like this is given by

$$H(T, S) = \max(0, S(T) - K).$$

A European put option is a similar contract, but where the owner is instead given the right to *sell* the asset at the strike price value K.

The two options mentioned above are *path-independent* options, which means their associated payoff only depends on the price of the underlying asset at the expiration time T. Later, we will see examples of *path-dependent* options, whose payoff at expiration depends not only on the final value of the underlying asset(s) but also on earlier values at $t < T$.

In this section, we assume that each underlying asset has a price $S_j(t)$ at time t that is lognormally distributed and that the bank account pays a fixed rate r. This corresponds to the model used by Black and Scholes in their seminal work [33]. Formally, this means our vector $\mathbf{S}$ of assets is a *multivariate geometric Brownian motion.* More precisely, suppose $(\Omega, \mathcal{F}, P)$ is a complete probability space. Let $\mathbf{B}$ be a vector of q independent standard Brownian motions on $(\Omega, \mathcal{F}, P)$. Also, we let $\{\mathcal{F}_t, t \geq 0\}$ denote the (P-augmented) *natural filtration* generated by $\mathbf{B}$. (We will not define this notion in detail here. $\mathcal{F}_t$ can be thought of as the information gathered by observing $\{\mathbf{B}(s), 0 \leq s \leq t\}$. See [212, 350] for more information.)

The behavior of $\mathbf{S}$ is then described by the stochastic differential equation (SDE)

$$d\mathbf{S}(t) = \boldsymbol{\mu}\mathbf{S}(t)dt + M\mathbf{S}(t)d\mathbf{B}(t), \tag{7.1}$$

where $\boldsymbol{\mu}^{\mathrm{T}} = (\mu_1, \ldots, \mu_q)$ is the vector of return rates for $\mathbf{S}$ and M is a $q \times q$ matrix such that $C = MM^{\mathrm{T}}$ is the covariance matrix of $\mathbf{S}$. For instance, if the q underlying assets are independent, then $C_{ij} = \sigma_i^2$ if $i = j$ and is zero otherwise, where σ_i is the *volatility* of the ith asset. More generally, if asset i and asset j have a correlation ρ_{ij} and a volatility σ_i and σ_j, respectively, then $C_{ij} = \rho_{ij}\sigma_i\sigma_j$.

For readers that are not too familiar with SDEs, (7.1) probably does not give much intuition about the behavior of $\mathbf{S}(\cdot)$. To help give some insight, we will focus for a moment on the one-dimensional case, where $\mathbf{S} = S$ is a single asset. Equation (7.1) then becomes

$$dS(t) = \mu S(t)dt + \sigma S(t)dB(t), \tag{7.2}$$

whose solution can be proved to be

$$S(t) = S(0)e^{(\mu - \sigma^2/2)t + \sigma B(t)}$$

using Ito's lemma [350]. Hence $S(t)$ has a lognormal distribution because $\ln S(t) = \ln S(0) + (\mu - \sigma^2/2)t + \sigma B(t)$ and $B(t) \sim N(0,t)$. In particular, $\mathrm{E}(S(t)) = S(0)e^{\mu t}$. A stochastic process that satisfies an SDE of the form (7.2) is called a *geometric Brownian motion.*

Going back to the multivariate case described by (7.1), this model implies that the vector $(\ln S_1(t), \ldots, \ln S_q(t))$ has a multinormal distribution with parameters that can be inferred from (7.1). Equivalently, a description that turns out to be useful when manipulating this model is to say that $S_j(t) = S_j(0)e^{X_j(t)}$ for $j = 1, \ldots, q$, where the vector $(X_1(t), \ldots, X_q(t))$ has a multinormal distribution with marginal means $\mathrm{E}(X_j(t)) = (\mu_j - \sigma_j^2/2)t$ and covariance terms $\mathrm{Cov}(X_i(t), X_j(t)) = \rho_{ij}\sigma_i\sigma_j t$, where, as before, μ_j is the rate of return for asset j, σ_j is its volatility, and ρ_{ij} is the correlation term between asset i and asset j. Hence this model is completely specified by the parameters μ_j, σ_j, for $j = 1, \ldots, q$, and ρ_{ij} for $1 \leq i < j \leq q$.

Now that we have a model for $\mathbf{S}(\cdot)$, the goal is to find the value V_0 at time 0 of a given European option with payoff $H(T, \mathbf{S})$. In order to do that, we use the theory of option pricing. Here, we only give a very brief overview of this theory and refer the reader to [92, 150, 370] for more details.

To derive a formula for V_0, we first assume that the model is specified in a way that prevents the existence of *arbitrage opportunities*, which are strategies involving the construction of a portfolio with an initial value less than or equal to 0 and with a future payoff that is nonnegative and takes a positive value with nonzero probability. In turn, the no-arbitrage assumption implies the existence of a *risk-neutral probability measure* Q — also sometimes called the *equivalent martingale measure* — under which for each asset the *discounted value process* $\{Z_j(t) := e^{-rt}S_j(t), t \geq 0\}$ is a martingale. In particular, the martingale property means that we must have $\mathrm{E}_Q(Z_j(t)|\mathcal{F}_s) = Z_j(s)$ for $t > s$.

In addition, we assume the parameters in our model have been chosen so that the market is *complete*, which means that any payoff $H(T, \mathbf{S})$ at time T can be *replicated* by constructing an appropriate portfolio over the underlying assets. The fundamental theorem of asset pricing [370] states that this assumption is equivalent to the existence of a *unique* risk-neutral probability measure.

Under these assumptions, we have that

$$V_0 = \mathrm{E}_Q(e^{-rT}H(\mathbf{S}(T))). \tag{7.3}$$

That is, the value at time 0 of the option is given by the expected value — under the measure Q — of its discounted payoff. From now on, we will drop the Q in the notation E_Q because all expectations are computed under this measure unless otherwise stated.

Now, in order to use (7.3), we need to know the behavior of $\mathbf{S}$ under the new measure Q. It turns out that for the lognormal model (7.1), $\mathbf{S}$ under Q still obeys an equation of the form (7.1), but where the vector $\boldsymbol{\mu}$ of rates of return is replaced by $\mathbf{r} = (r, \ldots, r)^{\mathrm{T}}$. In other words, under Q, we simply assume that the return on each asset $S_j(\cdot)$ is r rather than μ_j.

The following example illustrates how to use (7.3) in the case of a European call option on a single asset. We then look at two more complex examples.

Example 7.1. For a call option under the lognormal model, we want to compute

$$C_0 = \mathrm{E}(e^{-rT} \max(0, S(T) - K)),$$

where S satisfies

$$dS(t) = rS(t)dt + \sigma S(t)dB(t).$$

Hence $S(t) = S(0)e^{(r-\sigma^2/2)t+\sigma B(t)}$, and it can be proved that

$$C_0 = S(0)\Phi(d_1) - Ke^{-rT}\Phi(d_2), \tag{7.4}$$

where

$$d_1 = \frac{\ln(S(0)/K) + (r + \sigma^2/2)T}{\sigma\sqrt{T}}$$

and $d_2 = d_1 - \sigma\sqrt{T}$ (Prob. 7.2 asks you to verify this). This is the formula derived by Black and Scholes in [33] but using a different approach. It is usually referred to as the *Black-Scholes-Merton formula* to underline the important contribution of Merton [316], who expanded and enhanced the work of Black and Scholes shortly after the publication of their work in 1973.

Example 7.2. Another common type of option is an *Asian option.* An Asian call option has a payoff defined by

$$H(T, S) = \max\left(0, \frac{1}{d}\sum_{j=1}^{d} S(t_j) - K\right), \tag{7.5}$$

where, as before, K is the strike price and the variables t_j are observation times where the value of the asset is recorded and satisfy $0 \le t_1 < \ldots < t_d = T$. Hence, for the Asian option, we compare an *average value* of the underlying asset with the strike price rather than only looking at the value at expiration. This type of option is thus path-dependent. Here the theoretical value of the option at time 0 is given by

$$C_{\mathrm{as},0} = \mathrm{E}\left[e^{-rT} \max\left(0, \frac{1}{d}\sum_{j=1}^{d} S(t_j) - K\right)\right],$$

which has no closed-form expression.

Example 7.3. If in the previous example we use a geometric average instead of the arithmetic average $\sum_{j=1}^{d} S(t_j)/d$, then a closed-form expression for the value

$$C_{g,\mathrm{as},0} = \mathrm{E}\left[e^{-rT} \max\left(0, \left(\prod_{j=1}^{d} S(t_j)\right)^{1/d} - K\right)\right]$$

can be found. Informally, the geometric average makes things easier because a product of lognormal random variables is itself lognormal, whereas a sum of lognormal random variables does not have a known distribution. Hence, for an Asian call option on the geometric average, the value at time 0 has a Black-Scholes-Merton–like formula given by

$$C_{g,\mathrm{as},0} = e^{-rT}(e^{a+0.5b}\Phi(d_1) - K\Phi(d_2)), \tag{7.6}$$

where

$$\begin{aligned}
a &= \ln(S(0)) + (r - 0.5\sigma^2) \times T(d+1)/2d,\\
b &= \sigma^2(T/d)(d+1)(2d+1)/6d,\\
d_1 &= (-\ln(K) + a + b)/\sqrt{b},\\
d_2 &= d_1 - \sqrt{b},
\end{aligned}$$

and, for simplicity, we assume that $t_j = jT/d$ for $j = 1, \ldots, d$. Values of t_j that are not equally spaced can be handled similarly.

In the three examples discussed so far, in two cases we were able to analytically solve the expression

$$V_0 = \mathrm{E}(e^{-rT} H(T, \mathbf{S})).$$

For cases like the Asian call option on the arithmetic average, where we cannot obtain a closed-form expression for the time-0 value V_0, the Monte Carlo method can be used to provide an estimate of the expectation above. This idea was first proposed by Boyle in his seminal paper [37]. Before describing this approach in general, let us look at how it can be applied to estimate C_0 for the plain call option and then the time-0 value of the Asian call option $C_{\mathrm{as},0}$. Even if we have an analytical expression for C_0 for the plain call option, it is helpful to use this as a first example describing how to use the Monte Carlo method.

So, for the plain call option, we can estimate C_0 by

$$\frac{1}{n}\sum_{i=1}^{n} e^{-rT} \max(0, S^i(T) - K), \tag{7.7}$$

where $\{S^i(T), i = 1, \ldots, n\}$ is an i.i.d. sample from the lognormal distribution with parameters $(\ln S(0) + (r - \sigma^2/2)T, \sigma^2 T)$. More precisely, this sample can be obtained from an i.i.d. sample $\{Z_1, \ldots, Z_n\}$ of $N(0,1)$ random variables as follows:

$$S^i(T) = S(0)e^{(r-\sigma^2/2)T+\sigma\sqrt{T}Z_i}, \qquad i = 1, \ldots, n.$$

In turn, as was seen in Chap. 2, the variables Z_i can be generated by inverting the $N(0,1)$ CDF. That is, we let $Z_i = \Phi^{-1}(u_i)$, where $u_i \sim U(0,1)$.

For the Asian call option, we need to generate not only the final value of the underlying asset but also all the values that enter the average in (7.5). For that purpose, we can use the recursive relation

$$S(t_j) = S(t_{j-1})e^{(r-\sigma^2/2)\Delta_j+\sigma\sqrt{\Delta_j}Z_j}, \qquad j = 1, \ldots, s, \tag{7.8}$$

where the Z_j's are i.i.d. $N(0,1)$ and $\Delta_j = t_j - t_{j-1}$, $j = 1, \ldots, s$. The pseudocode given in Fig. 7.1 explains how to construct the Monte Carlo estimator for $C_{\text{as},0}$ based on a random point set P_n and where we assume $t_j = jT/s$ just to simplify things.

```
AsianCall(P_n, r, σ, T, d, S(0))
a ← (r − σ²/2)T/d
b ← σ√(T/d)
sum2 ← 0
prevS ← S(0)
for i = 1 to n do
    sum ← 0
    for j = 1 to d do
        Z ← Norm01(u_{i,j})
        S ← prevS ×e^{a+bz}
        sum ← sum + S
        prevS ← S
    x ← sum/d − K
    if x > 0 then
        sum2 ← sum2 +xe^{−rT}
return sum2/n
```

Fig. 7.1 Pseudocode for estimating $C_{\text{as},0}$ with Monte Carlo.

Hence the pseudocode given in Fig. 7.1 returns the estimator

$$\hat{C}_{\text{as},0} = \frac{1}{n}\sum_{i=1}^{n} e^{-rT} \max\left(0, \frac{1}{d}\sum_{j=1}^{d} S^i(t_j) - K\right), \tag{7.9}$$

where the $\{S^i(t_1), \ldots, S^i(t_d)\}$, for $i = 1, \ldots, n$, represent n i.i.d. realization paths for the underlying asset.

So far, we have only seen options on one asset. An example of an option on several assets is a call option on the maximum of q assets, which is sometimes called a *rainbow option*. Its payoff is defined by

$$H(T, \mathbf{S}) = \max\left(0, \max_{1 \le j \le q} S_j(T) - K\right).$$

In other words, at expiration, the holder of the option has the right to buy at a price K any of the q underlying assets and rationally chooses to buy the most expensive one. The payoff is thus given by the difference between the highest-valued asset and the strike price K. To use the Monte Carlo method for estimating the value at time 0 of this option, denoted $C_{m,0}$, we need to generate observations of correlated lognormal random variables based on (7.1). This can be done as follows. We let

$$S_j(T) = S_j(0)e^{(r-\sigma_j^2/2)T+\sqrt{T}W_j},$$

where

$$W_j = \sum_{l=1}^{q} M_{j,l} Z_l$$

and the variables Z_l are i.i.d. $N(0,1)$. Note that multiasset models are usually specified by giving the covariance matrix C — that is, the volatilities σ_j and correlation terms ρ_{ij} are given — rather than the matrix M such that $MM^{\mathrm{T}} = C$. One can then get M by performing a Cholesky decomposition of C, thus finding a lower-triangular matrix M such that $MM^{\mathrm{T}} = C$.

Putting it all together, $C_{m,0}$ can be estimated using the pseudocode given in Fig. 7.2.

```
RainbowCall(P_n, r, σ, M, S(0), σ, q)
for j = 1 to q do
    a[j] ← (r − σ_j^2/2)T
sum ← 0
for i = 1 to n do
    for j = 1 to q do
        Z[j] ← Norm01(u_{i,j})
    max ← 0
    for j = 1 to q do
        w ← 0
        for l = 1 to q do
            w ← w + M[j][l]Z[l]
        S ← S_j(0) × e^{a+w√T}
        if S > max then
            max ← S
    x ← max − K
    if x > 0 then
        sum ← sum +xe^{−rT}
return sum/n
```

Fig. 7.2 Pseudocode for estimating $C_{m,0}$ with point set P_n.

In general, the Monte Carlo method for European pricing in the lognormal model can be applied as follows. Assume we have a payoff function $H(T, \mathbf{S})$ that depends on $S_j(t_1), \ldots, S_j(t_d)$, for $j = 1, \ldots, q$.

(1) For $i = 1, \ldots, n$:

a. Generate observations under the risk-neutral measure Q for

$$S_j^i(t_1), \ldots, S_j^i(t_d)$$

for each security $j = 1, \ldots, q$. (The pseudocode given in Figs. 7.1 and 7.2 can be combined to do that. More details are given below.)

b. Compute the payoff

$$H_i = H(T, S_1^i(t_1), \ldots, S_1^i(t_d), \ldots, S_q^i(t_1), \ldots, S_q^i(t_d)).$$

(2) Return the estimate

$$\frac{1}{n} \sum_{i=1}^{n} e^{-rT} H_i.$$

We now wish to establish the correspondence between this "simulation" formulation and the underlying integration problem over $[0, 1)^s$ that is solved when we estimate

$$V_0 = \mathrm{E}(e^{-rT} H(T, \mathbf{S}))$$

with Monte Carlo. Using the same notation as in Fig. 1.6 of Chap. 1, we can write

$$V_0 = \int_{[0,1)^s} f(\mathbf{u}) d\mathbf{u}$$

and use two intermediate functions $g(\cdot)$ and $h(\cdot)$ such that $f(\mathbf{u}) = h(g(\mathbf{u}))$ and a random vector $\mathbf{X}$ corresponding to the random variables that need to be simulated. In our case, we can choose $\mathbf{X}$ to be

$$\mathbf{X} = (S_1(t_1), \ldots, S_1(t_d), \ldots, S_q(t_1), \ldots, S_q(t_d)).$$

In that case, we define $h(\mathbf{X}) = e^{-rT} H(T, \mathbf{X})$. Also, if we use the standard path generation method described in Fig. 7.1 combined with the approach used in Fig. 7.2 for generating correlated asset prices, then we can obtain $\mathbf{X}$ by applying the following function g to a vector $\mathbf{u}$ of $s = dq$ uniform random numbers in $[0, 1)^s$. Here we assume $t_j - t_{j-1} = T/d$ for $j = 1, \ldots, d$ to simplify the notation. Also, since each component of $\mathbf{X}$ depends on several uniform numbers, we use intermediate functions $g_{j,l} : [0, 1)^s \rightarrow \mathbb{R}$ and $w_j : [0, 1)^q \rightarrow \mathbb{R}$ to describe g, where j indexes assets and l indexes time. More precisely, we write

$$\mathbf{X} = g(\mathbf{u}) = (g_{1,1}(\mathbf{u}), \ldots, g_{1,d}(\mathbf{u}), \ldots, g_{q,1}(\mathbf{u}), \ldots, g_{q,d}(\mathbf{u})), \quad (7.10)$$

where

$$g_{j,l}(\mathbf{u}) = S_j(0)e^{la_j+b_j(w_j(u_1,\ldots,u_q)+\ldots+w_j(u_{(l-1)q+1},\ldots,u_{lq}))}$$

for $j = 1, \ldots, q$ and $l = 1, \ldots, d$, with

$$a_j = (r - \sigma_j^2/2)T/d,$$
$$b_j = \sqrt{T/d},$$
$$w_j(u_{(k-1)q+1}, \ldots, u_{kq}) = \sum_{p=1}^{q} M_{j,p}\Phi^{-1}(u_{(k-1)q+p}),$$

where M is such that $MM^{\mathrm{T}} = C$. That is, here we chose to use the first set $u_1, \ldots, u_q$ of q random numbers to generate a vector of q i.i.d. standard normal random variables, transform them into correlated normals $w_1(u_1, \ldots, u_q), \ldots, w_q(u_1, \ldots, u_q)$, and use them to generate observations for the q prices at time $t_1 = T/d$. Then the next q random numbers are used for the prices at time $t_2 = 2T/d$ and so on.

As mentioned above, the dimension s is equal to the number q of processes that need to be simulated multiplied by the number of observations d per path that are required. This quantity d stems either from the payoff definition (e.g., the number of prices that enter the average for an Asian option) or the size of the time steps chosen when discretizing the process when it is not possible to generate observations directly from the price dynamics. The need for discretization typically arises with more complex models such as those discussed in Sect. 7.2, for example when the volatility itself is a stochastic process.

Hence, high-dimensional problems in finance can come either from payoff functions that are based on several observations, a large number of securities, or a fine discretization possibly combined with a large maturity T. For instance, *mortgage-backed security* problems tend to have a large associated dimension since the maturity is typically between 20 and 30 years, and monthly cash flows need to be simulated (e.g., s is between 240 and 360). An example will be given in Sect. 7.3. As for the number of steps in the discretization, a rule of thumb is to take $d \in O(\sqrt{n})$, so that the $O(1/d)$ error from the discretization process is about the same as the error produced from Monte Carlo sampling, which is $O(1/\sqrt{n})$ [93].

Going back to the notation above for g and f, to help understand it better we will reuse the three option examples that we have seen so far and in each case describe explicitly what the function f is.

Example 7.4. For a plain call option, we have that

$$f(\mathbf{u}) = f(u_1) = e^{-rT}\max(0, S(0)e^{(r-\sigma^2/2)T+\sigma\sqrt{T}\Phi^{-1}(u_1)} - K).$$

For the Asian call option, we have

$$
\begin{aligned}
f(\mathbf{u}) &= f(u_1, \ldots, u_d) \\
&= e^{-rT} \max\left(0, \frac{1}{d}\sum_{j=1}^{d} S(0)e^{(r-\sigma^2/2)T/d+\sigma\sqrt{T/d}\Phi^{-1}(u_j)} - K\right).
\end{aligned}
$$

For the rainbow call option on the maximum of q assets, we have

$$
\begin{aligned}
f(\mathbf{u}) &= f(u_1, \ldots, u_q) \\
&= e^{-rT} \max\left(0, \max_{1\le j\le q}\left(S_j(0)e^{(r-\sigma_j^2/2)T+\sum_{l=1}^{q} M_{j,l}\Phi^{-1}(u_l)}\right) - K\right).
\end{aligned}
$$

Once a model for $\mathbf{S}$ and a payoff function have been chosen, the main factor that affects the definition of the function f is the choice of what we could call the *path generation* method. In Example 7.4 above, when successive observations $S(t_1), \ldots, S(t_d)$ need to be generated for pricing the Asian call option, we use the standard method where the prices $S(t_j)$ are generated in chronological order using the recursive formula (7.4). As was seen in Sect. 6.3, alternative methods are the generalized Brownian bridge techniques such as those used in [2, 4, 200, 327]. The method chosen for path generation can greatly affect the effective dimension of f and therefore the performance of quasi–Monte Carlo methods for pricing the corresponding option. Choosing a generation method can also be formulated in terms of the choice of the matrix M satisfying $MM^{\mathrm{T}} = C$, as was discussed in Sect. 6.3 of Chap. 6. Another factor that affects the definition of the function f is the method chosen for generating normal variates. Above, we chose inversion for reasons discussed in Chap. 2 having to do mainly with the fact that it is best suited for quasi–Monte Carlo.

7.2 More complex models

Although the lognormal model used by Black and Scholes [33] and Merton [316] is still quite popular due to its simplicity, several more realistic models have been proposed over the years as alternatives to this model. Often, these more complicated models include added sources of randomness — for instance, the volatility is assumed to be stochastic instead of being constant — which make the market *incomplete* [92], and thus there is more than one risk-neutral probability measure to choose from for pricing options. In this text, we do not discuss how to choose the risk-neutral probability measure in those cases and assume that for such models a specific measure has been chosen. See, for instance, [20, 103, 138, 314] and the references therein for methods of choosing an appropriate martingale measure.

In what follows, we will illustrate with three models how paths can be generated under models more complex than the lognormal one. For these models, we often need to discretize the associated SDE. To do so, there are several methods available (see, for instance, [219]), but here we use a Euler scheme to keep things simple.

7.2.1 Heston's process

This model replaces the constant volatility in the lognormal model by a stochastic volatility [49, 179],

$$dS(t) = rS(t)dt + \sigma(t)S(t)\left[\rho dB_1(t) + \sqrt{1-\rho^2}dB_2(t)\right],$$
$$d\sigma^2(t) = \kappa\left[\theta - \sigma^2(t)\right]dt + \sigma_v\sigma(t)dB_1(t),$$

where $B_1(\cdot)$ and $B_2(\cdot)$ are two independent standard Brownian motions, κ is the *speed of mean reversion*, $\theta > 0$ is the *long-run mean variance*, $\sigma_v > 0$ is the *volatility of the volatility process*, and ρ is the correlation between the Brownian motions driving $S(\cdot)$ and $\sigma^2(\cdot)$.

A closed-form expression can be derived for the price of a plain call option under that model [179], but for more complicated options we might need to use Monte Carlo and a discretization of the process. Suppose we use a Euler scheme with d steps to discretize both $S(\cdot)$ and $\sigma(\cdot)$. Let $\Delta = T/d$. Then, using a uniform random point $\mathbf{u} = (u_1, u_2, \ldots, u_{2d})$, paths can be generated as in Fig. 7.3 [49, 476].

```
HestonPaths(σ(0), κ, θ, ρ, σ_v, T, d, S(0), u)
σ[0] ← σ(0)
S[0] ← S(0)
t_0 ← 0
for l = 1 to d
    S ← S[l − 1]
    σ ← σ[l − 1]
    Z_1 ← Norm01(u_{2l−1})
    Z_2 ← Norm01(u_{2l})
    Z ← ρZ_1 + √(1 − ρ²) Z_2
    S[l] ← S(1 + rΔ + σ√Δ Z)
    if S[l] < 0 then
        S[l] ← 0
    σ²[l] ← σ² + κ(θ − σ²)Δ + σ_v σ√Δ Z_1
    if σ²[l] < 0 then
        σ²[l] ← 0
```

Fig. 7.3 Pseudocode for generating discretized paths under Heston's process, where $S[l]$ and $\sigma[l]$ represent $S(t_l)$ and $\sigma(t_l)$, respectively.

Note how we chose to assign the uniform variates u_j in the chronological order in which they are required. That is, u_1 and u_2 are used to generate $S(t_1)$ and $\sigma(t_1)$, then u_3 and u_4 are used to generate $S(t_2)$ and $\sigma(t_2)$, and so on. This is of course arbitrary, and another "natural" choice would have been to assign $u_1, \ldots, u_d$ to generate $Z_1(\cdot)$ and then $u_{d+1}, \ldots, u_{2d}$ to generate $Z_2(\cdot)$. The assignment is irrelevant in the Monte Carlo context, but it can make a difference in the quasi–Monte Carlo context, as we will see in Sect. 7.3. This issue was also briefly investigated in [29].

7.2.2 Regime switching model

The underlying idea here is to assume that the parameters describing the behavior of the market are themselves random and change according to an unobservable (hidden) Markov process. For example, in [103], the model consists of a risky underlying asset driven by a Markov-modulated geometric Brownian motion. That is, there is a Markov chain $\mathbf{X}(t)$ whose state space is the set of N unit vectors $e_1, \ldots, e_N$ (i.e., e_i is an N-dimensional vector of zeros with a one in the ith position), and then we have

$$dS(t) = \mu(t)S(t)dt + \sigma(t)S(t)dB(t),$$

where $\mu(t) = \mathbf{X}^{\mathrm{T}}(t) \cdot \boldsymbol{\mu}$, $\sigma(t) = \mathbf{X}^{\mathrm{T}}(t) \cdot \boldsymbol{\sigma}$, and then $\boldsymbol{\mu} = (\mu_1, \ldots, \mu_N)^{\mathrm{T}}$ and $\boldsymbol{\sigma} = (\sigma_1, \ldots, \sigma_N)^{\mathrm{T}}$ are the N possible values for the return and volatility parameters of the asset, respectively.

We can thus view the N possible states of the Markov process $\mathbf{X}(t)$ as N different business cycles, where μ_i and σ_i are the return and volatility of the asset associated with the ith business cycle. Similarly, the risk-free rate is assumed to take a value in $\mathbf{r}^{\mathrm{T}} = (r_1, \ldots, r_N)$ depending on the business cycle. The $N \times N$ infinitesimal generator matrix for $\mathbf{X}(\cdot)$ is denoted by A. That is, if $\mathbf{X}(t) = e_i$, then for $j \neq i$, a transition to state e_j occurs according to a Poisson process with rate $A_{ij} \geq 0$ and $A_{ii} = \sum_{j \neq i} A_{ij}$. Equivalently, we can say that, while in state e_i, the time until the next transition follows an exponential distribution with mean $-1/A_{ii}$ and will be in state e_j with probability $-A_{ij}/A_{ii}$ for $j \neq i$.

In [103], option pricing formulas are derived under a risk-neutral probability measure for which the underlying asset obeys the SDE

$$dS(t) = r(t)S(t)dt + \sigma(t)S(t)dB(t).$$

Under that measure, we have that

$$\ln S(T)/S(t) \sim N\left(\int_t^T (r(s) - \sigma^2(s)/2)ds, \int_t^T \sigma^2(s)ds\right).$$

Equivalently, if we define $J_i(t,T)$ to be the occupation time of $\mathbf{X}(t)$ in state e_i over the time interval $[t,T]$, then we can write

$$P_{t,T} := \int_t^T r(s)ds = \sum_{i=1}^N r_i J_i(t,T),$$

$$U_{t,T} := \int_t^T \sigma^2(s)ds = \sum_{i=1}^N \sigma_i^2 J_i(t,T),$$

and we have that $\ln S(T)/S(t) \sim N(P_{t,T} - U_{t,T}/2, U_{t,T})$.

The pseudocode given in Fig. 7.4 outlines an approach for generating a terminal price $S(T)$ for this model using a uniform random point $\mathbf{u} = (u_1, u_2, \ldots)$.

```
RegimeSwitchPath(X[0], A, r, σ, T, u)
I ← X[0] // state of the chain
t ← 0
j ← 2
for i = 1 to N
    J[i] ← 0 // occupation time
while t < T do
    τ ← ln(1 − u_j)/A_{I,I} // time until next transition
    p ← 0
    m ← 1
    while u_{j+1} > p do // generate next state
        if m ≠ I then
            p ← p − A_{I,m}/A_{I,I}
        m ← m + 1
    if t < T then
        J[I] ← J[I] + τ
    else
        J[I] ← J[I] + T − t
    t ← t + τ
    I ← m − 1 // update state
    j ← j + 2
P ← 0, U ← 0
// reached time T
for i = 1 to N
    P ← P + r_i J[i]
    U ← U + σ_i^2 J[i]
S(T) ← S(0)e^{(P−U/2)+√U Φ^{−1}(u_1)}
```

Fig. 7.4 Pseudocode for generating a terminal price under regime switching, given an input point $\mathbf{u}$.

Note how we chose to assign the first uniform number u_1 to generate the price at time T, and used the subsequent uniform variates to simulate the underlying Markov chain. The dimension of $\mathbf{u}$ is unbounded because in the simulation approach described in Fig. 7.4 we do not know a priori how many times the chain will change its state before we reach the expiration time T. Note that if the regime changes were instead modeled using a discrete-time Markov chain where, say, we assume there is a transition every month, then the dimension would be bounded.

7.2.3 Variance gamma model

Financial models for which randomness is input through the Brownian motion are such that the price paths are almost surely continuous. Sometimes prices move in an abrupt way that cannot be captured by a continuous model. It is therefore of interest to study models that allow jumps to occur in the price paths. An example of this is the *jump-diffusion* model that was proposed by Merton in 1976 [317], in which a jump process is added to the components of the geometric Brownian motion model.

In what follows, we describe instead the *variance gamma model* proposed by Madan et al. [299]. It works as follows [150]. We first write $S(t) = S(0)e^{X(t)}$ and then model $X(t)$ as $X(t) = B(G(t))$, where $B(\cdot)$ is a Brownian motion with drift μ and diffusion coefficient σ and $G(\cdot)$ is a gamma process with parameters a and b. That is, for fixed times $s < t$, we have that the increments $G(t) - G(s) \sim \text{Gamma}(a(t-s), b)$, where $a, b > 0$, and these increments are independent. The model proposed by Madan et al. is defined so that $a = 1/b$, which means that the expectation of the increment $G(t) - G(s)$ is $t - s$. We can then simulate a discretized path of $S(\cdot)$ as shown in Fig. 7.5 [150]. Since we work under the risk-neutral probability measure Q there, we take $\mu = r$.

7.3 Randomized quasi–Monte Carlo methods in finance

We start with a short discussion recalling the difference between quasi–Monte Carlo and Monte Carlo simulation adapted to the problem of pricing European options.

With Monte Carlo, we use a set of n independent points in $[0, 1)^s$ to generate n independent paths $\mathbf{S}^1, \ldots, \mathbf{S}^n$ of the vector of underlying assets, compute the payoff $H(T, \mathbf{S}^i)$ obtained on each path, discount it back to time 0, and then take the average over all paths. Here the dimension s is influenced by the number q of underlying assets, the number of prices per path that need to be simulated, and the model used for the underlying assets. For instance,

```
VarGammaPath(b, r, σ, u)
t_0 ← 0
X[0] ← 0
for l = 1 to d
    X ← X[l]
    Y ← Gamma(u_{2l-1}, (t_l − t_{l-1})/b, b)
    Z ← Norm01(u_{2l})
    X[l] ← X[l − 1] + μY + σ√Y Z
    S[l] ← exp(X[l])
```

Fig. 7.5 Pseudocode for generating discretized paths of a variance gamma process. The function Gamma(u, a, b) uses inversion to generate from $u \sim U(0,1)$ a gamma random variable with parameters (a, b).

with one asset and a stochastic volatility or variance gamma model, we saw in Sect. 7.2 that $s = 2d$, where d is the number of discretization steps used.

With quasi–Monte Carlo sampling, we use n points in $[0,1)^s$ that come from a low-discrepancy point set P_n instead. If P_n has been randomized according to the description made in Sect. 6.2, then each point $\mathbf{u}_i$ used to generate a path of $\mathbf{S}$ in the quasi–Monte Carlo setting is uniformly distributed in $[0,1)^s$, and thus the path generated has the same distribution properties as with Monte Carlo. The difference is that now the n paths are correlated. Numerical examples illustrating the use of this approach will be given throughout the rest of this chapter.

We now discuss a few important topics related to the use of quasi–Monte Carlo in finance.

Choice of path generation method, assignment of coordinates, and dimension reduction

We already discussed in Sect. 6.3 how techniques like the Brownian bridge, principal components, and the approach from [200, 201] could be used as path generation methods aimed at reducing the effective dimension. These approaches can be useful and should be investigated when quasi–Monte Carlo is used. But they should not be applied blindly either, as the study from [373] shows us.

Also, unless we are working with one asset driven by only one random process — for instance, the lognormal model — we usually have to deal with simulations where several underlying random processes need to be simulated, and in that case we must decide how to assign the uniform numbers u_j in $\mathbf{u}$ to these various processes.

The first possibility is an *interleaved* (or sequential) assignment, where the numbers u_j are assigned as needed. For example, for a stochastic volatility model, we assign u_1, u_2 to the generation of $S(t_1)$ and $\sigma(t_1)$, u_3, u_4 to the generation of $S(t_2)$ and $\sigma(t_2)$, and so on, as shown in Fig. 7.3. The second possibility is a *block assignment*, where we break down $\mathbf{u} = (u_1, \ldots, u_s)$ into successive blocks, which are then assigned to the various processes. Again using the example of a stochastic volatility model, this means we assign $u_1, \ldots, u_d$ to the generation of the first Brownian motion and $u_{d+1}, \ldots, u_{2d}$ to the second Brownian motion. Hence, in that case u_1, u_{d+1} are used to generate $S(t_1)$ and $\sigma(t_1)$, u_2, u_{d+2} are used to generate $S(t_2)$ and $\sigma(t_2)$, and so on.

Although there is no clear answer to how this assignment should be done on a given problem, here are a few things to take into account. If the different processes that need to be simulated are used to define variables that do not interact with each other too much, then an assignment by block might be better suited. An example of a problem like this is the pricing of a European option on the maximum of several assets that each follow an independent jump process.

Also, if a generalized Brownian bridge approach is applied to only some of the processes, then it makes sense to use a separate block of uniform numbers u_j for each of the processes simulated with that approach and another one for the other processes [29].

If the different processes give rise to random variables that interact more strongly, then interleaving might be better. An example of this is when two Brownian motions need to be simulated for an underlying asset that follows a stochastic volatility model. In addition, an assignment by block might not be a good choice if the underlying point set is such that the quality of its projections $P_n(I)$ deteriorates as the indices in I increase. This is because the processes will be assigned blocks of different quality and, for instance, the last block might be of poor quality. Hence, for block assignments, it might be better to use a dimension-stationary point set since by definition we then have that $P_n(\{1, \ldots, d\})$,$P_n(\{d+1, \ldots, 2d\})$, and any projections of the form $P_n(\{ld+1, \ldots, (l+1)d\})$ are the same, and thus they all have the same quality.

If a more rigorous approach for choosing the assignment is desired, then one possibility is to first perform a study of the ANOVA components to determine which of the situations above prevails. Techniques such as those discussed in Sect. 6.3.3 can be used for that purpose.

Table 7.1 gives results for the problem of pricing an Asian option under Heston's process, where the number of steps d in the discretization is assumed to correspond to the number of prices entering the average. We use the parameters $S(0) = K = 100$, $r = 0$, $\kappa = 2$, $\sigma(0) = 0.1$, $\theta = 0.01$, $\sigma_v = 0.1$, $\rho = 0.5$, and $T = 0.5$ year [476]. We compare the Monte Carlo method with the Sobol' sequence and a polynomial Korobov lattice, both of which have been randomly digitally shifted using $m = 25$ repetitions. The polynomial Korobov lattice is based on a combined Tausworthe generator with two components, defined respectively by ($\nu_1 = 1, P_1(z) = z^3 + z + 1$) and ($\nu_2 = 4, P_2(z) = z^7 + z^3 + 1$).

Table 7.1 Asian call option under Heston's process: price estimate $\hat{\mu}$ and 95% confidence interval half-width (HW) with $n = 1024$, $m = 25$, and different numbers of time steps d in the discretization.

	$d = 32$		$d = 64$		$d = 128$	
	$\hat{\mu}$	HW	$\hat{\mu}$	HW	$\hat{\mu}$	HW
MC	1.674	2.85e−2	1.642	3.36e−2	1.640	3.60e−2
Sob leave	1.659	1.52e−2	1.627	1.36e−2	1.631	1.80e−2
Sob block	1.655	1.36e−2	1.632	2.07e−2	1.637	1.96e−2
pKor leave	1.659	8.02e−3	1.638	8.62e−3	1.628	8.15e−3
pKor block	1.659	9.51e−3	1.638	6.18e−3	1.631	7.04e−3

A few things should be mentioned about these results. First, both randomized quasi–Monte Carlo methods perform better than Monte Carlo, reducing the width of the 95% confidence intervals by factors ranging between about 2 and 4. The polynomial Korobov lattice generally performs better than the Sobol' sequence. Using an interleaved assignment or one by block does not seem to make a consistent difference for this particular problem.

As a final note, recall that the generalized Brownian bridge technique is meant to reduce the effective dimension of the integrand. A related idea is to use methods that result in the need for a point set with smaller dimension. At least two methods fall in that category: array-RQMC, and conditional Monte Carlo. With array-RQMC, an example of Asian call option pricing under the lognormal model is discussed in [263], where a two-dimensional point set is required instead of an s-dimensional one, where s in the number of prices entering the average. Using array-RQMC instead of Monte Carlo in that case results in variance reduction factors between 1500 and 40,000. With conditional Monte Carlo, as discussed in Sect. 7.4.4, a stochastic volatility model with d time steps in the discretization can be handled by a d-dimensional point set instead of a $2d$-dimensional one. This fact combined with the added smoothness of the integrand that is obtained when applying conditional Monte Carlo can result in important gains when using randomized quasi–Monte Carlo instead of Monte Carlo [476]. Numerical results illustrating this will be given in Sect. 7.4.4.

Problems of unbounded dimension

We already discussed in Sect. 7.1 some reasons that can cause the dimension s of the function f associated with a given problem to be large. In what follows, we mention a few cases where the dimension is unbounded.

For European pricing, typically the dimension is finite because the simulation horizon T is fixed. Cases where we could have an unbounded problem are if the simulation model requires an unbounded number of random variates. An example of this is the regime switching model discussed in Sect. 7.2, assuming we use the straightforward simulation approach described there. This can also happen with models that include jumps if instead of using the discretization approach mentioned in Sect. 7.2 we decide to explicitly simulate the jumps themselves. Since the number of jumps is random and at least one uniform variate is needed for each, this causes the dimension to be unbounded.

Outside the framework of European option pricing, problems of unbounded dimension arise when we need to run financial simulations until a certain event takes place and this event occurs at a random time. For instance, problems in risk theory involving the computation of the probability of *ruin* of an insurance company can give rise to simulations having this property. More precisely, one of the approaches used to estimate ruin probabilities is based on *regenerative simulation*, which is such that the end of the simulation is determined by a random stopping time, much like the bank simulation discussed in Chap. 1. The following example discusses this specific application.

Example 7.5. An insurance company receives claims at random times and of random size. These claims define an *aggregate claim process*

$$L(t) = \sum_{k=1}^{N(t)} Y_k, \qquad t > 0,$$

where $Y_k > 0$ is the size of the kth claim received and $N(t)$ is the number of claims received during the time interval $(0, t]$. In addition, we let t_k be the time at which the kth claim is received. We assume here that the aggregate claim process is a *compound Poisson process.* That is, the times $T_k = t_k - t_{k-1}, k \geq 1$, between two successive claims are i.i.d. exponential random variables and the claim sizes Y_k are i.i.d. random variables.

In exchange for the payment of the claims, the company charges a premium at a rate $c(\cdot)$. We let $U(t)$ be the company's surplus at time t. The rate function $c(\cdot)$ is allowed to depend on the surplus value, so that we have

$$dU(t) = c(U(t))dt - dL(t).$$

That is, $U(t)$ grows at the rate specified by $c(\cdot)$ until a claim comes and makes $U(t)$ drop by the value of the claim.

The goal is to estimate the *probability of ruin* of the insurance company, given by

$$\psi_u = P(U(t) \leq 0 \text{ for some } t \geq 0 | U(0) = u).$$

A naive approach for estimating ψ_u is to simulate the surplus process for a large number L of claims and verify for each claim if it causes $U(t)$ to become smaller than or equal to 0. We can then estimate ψ_u by the proportion of simulation runs in which ruin occurred. Since ψ_u is typically very small, this naive approach can be highly inefficient, and importance sampling should be used to improve the accuracy of this approach. Asmussen wrote several papers describing how this can be done using exponential twisting — which was discussed in Sect. 4.5 — and other well-known tools in risk theory [13, 15, 16].

Another approach is to use the duality between the surplus process and its associated *storage process* $\{X(t), t \geq 0\}$, described by

$$dX(t) = \begin{cases} -c(X(t)) + dU(t) & \text{if } X(t) > 0, \\ dU(t) & \text{if } X(t) = 0, \end{cases} \tag{7.11}$$

and an initial value $X(0) = u$ [13, 14]. That is, $X(\cdot)$ starts at the same point as $U(\cdot)$, but then its change is the negative of the change observed for $U(\cdot)$, except that when $X(\cdot)$ reaches 0, it stays there until the next jump. Figure 7.6 illustrates the difference between the two processes.

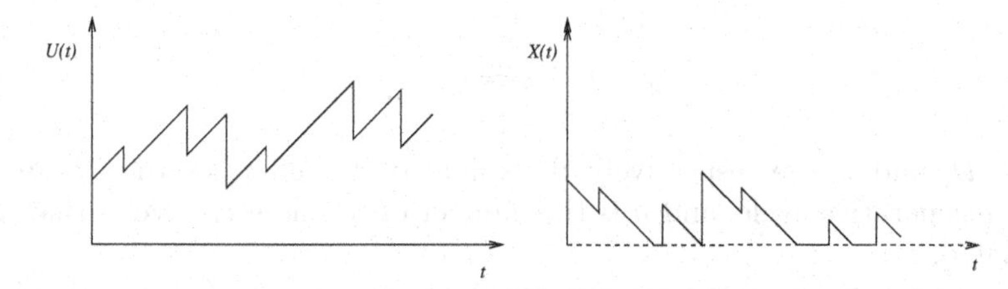

Fig. 7.6 The surplus process $U(\cdot)$ versus the storage process $X(\cdot)$.

It can be shown that the probability of ruin ψ_u satisfies

$$\psi_u = 1 - \lim_{d\to\infty} \frac{D_d(u)}{\sum_{k=1}^{d} T_k}, \tag{7.12}$$

where $D_d(u)$ is the total time that $X(\cdot)$ spent below u before the dth claim arrived. One can then estimate the ratio on the right-hand side of (7.12) by running simulations with a very large number of claims [322]. Alternatively, as in [458], we can use the fact that the process $X(\cdot)$ is a *regenerative process*, where the regenerative epochs are the times where $X(\cdot)$ hits (going down) the level u. The regenerative epochs for a given realization of $X(\cdot)$ are shown in Fig. 7.7.

The fact that $X(\cdot)$ is a regenerative process implies that we have

$$\lim_{d\to\infty} \frac{D_d(u)}{\sum_{k=1}^{d} T_k} = \frac{\mathrm{E}(D)}{\mathrm{E}(\tau)},$$

where

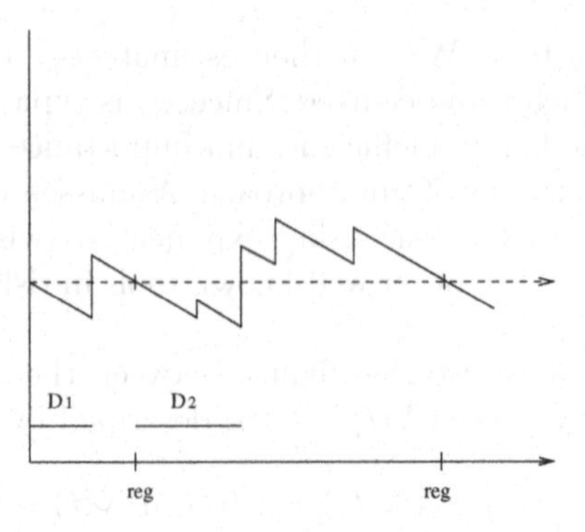

Fig. 7.7 The regenerative epochs for the process $X(\cdot)$, along with $D_1(u)$ and $D_2(u)$.

D = amount of time spent by $X(\cdot)$ below u during one regenerative cycle,
τ = length of a regenerative cycle.

This means we can estimate ψ_u as

$$\hat{\psi}_u = 1 - \frac{\sum_{i=1}^n D_i/n}{\sum_{i=1}^n \tau_i/n}, \tag{7.13}$$

where D_i and τ_i are respectively the values of D and τ for the ith simulated regenerative cycle, and n is the number of regenerative cycles that are simulated.

When we simulate a regenerative cycle, we start with $X(0) = u$, simulate claim sizes and arrival times, and update $X(t)$ according to (7.11) — determining also at what time D the process $X(\cdot)$ goes above u — until the time τ where $X(\tau) = u$. Hence the number N of claims that need to be simulated per cycle is a random variable with, in our case, a Poisson distribution. Since the dimension of this problem is $2N$ — for each of the N claims, we need one uniform number to generate the interarrival time between two claims and one uniform number to generate the claim size — it means the dimension is unbounded for this type of problem.

Note that the estimator (7.13) is biased because the expectation of the ratio of two random variables is not generally equal to the ratio of their expectations. However, this estimator is strongly consistent [243]. Also, to construct a confidence interval for $\bar{\psi}_u := \mathrm{E}(D)/\mathrm{E}(\tau)$, we can use the following approach [243, pp. 532–533]. Define

$$Z_i = D_i - \bar{\psi}_u \tau_i, \qquad i = 1, \dots, n.$$

Then the variables Z_i are i.i.d. with mean zero and variance

$$\sigma_Z^2 = \sigma_D^2 + \bar{\psi}_u^2 \sigma_\tau^2 - 2\bar{\psi}_u \sigma_{D,\tau},$$

where $\sigma_D^2 = \mathrm{Var}(D)$, $\sigma_\tau^2 = \mathrm{Var}(\tau)$, and $\sigma_{D,\tau} = \mathrm{Cov}(D,\tau)$. Hence, by the central limit theorem,

$$\frac{\sum_{i=1}^n Z_i/n}{\sigma_Z/n} \Rightarrow N(0,1) \text{ as } n \to \infty.$$

It can then be shown that

$$\frac{\bar{D}}{\bar{\tau}} - \bar{\psi}_u \Rightarrow N\left(0, \frac{\hat{\sigma}_Z^2}{n\bar{\tau}}\right),$$

where

$$\bar{D} = \frac{1}{n}\sum_{i=1}^n D_i,$$
$$\bar{\tau} = \frac{1}{n}\sum_{i=1}^n \tau_i,$$
$$\hat{\sigma}_Z^2 = \hat{\sigma}_D^2 + \left(\frac{\bar{D}}{\bar{\tau}}\right)^2 \hat{\sigma}_\tau^2 - 2\frac{\bar{D}}{\bar{\tau}}\hat{\sigma}_{D,\tau}.$$

Figure 7.8 gives pseudocode to simulate one regenerative cycle, assuming that the premium rate function is of the form $c(u) = cu + \delta$. This is equivalent to a fixed premium rate c and the assumption that the surplus earns interest at a rate δ. Under this assumption, we have that, for $t < T_1$ [322],

$$X(t) = X(0)e^{-\delta t} - \frac{c}{\delta}(1 - e^{-\delta t}).$$

Furthermore, we assume the interarrival times between claims have a mean of $1/\lambda$ and that the claim sizes Y_k are exponential with mean β. The value θ computed in this code represents the time that elapses between the moment where $X(\cdot)$ reaches u (going down) and the last claim that occurred before that and can thus be found by solving

$$u = X(\theta) = Xe^{-\theta\delta} - \frac{c}{\delta}\left(1 - e^{-\theta\delta}\right),$$

thus obtaining

$$\theta = \frac{1}{\delta}\ln\left(\frac{\delta X + c}{\delta U + c}\right).$$

Note that for a simulation like this, where the dimension is unbounded and uniform numbers are used for two different purposes — claim size and claim time — it is more convenient to use the interleaving approach described in the pseudocode above to assign the uniform numbers to the variables to be simulated since we do not know beforehand the size of the blocks required for each purpose.

```
RuinRegen(δ,c,λ,β,u,u)
X ← u
j ← 1
D ← 0
sumT ← 0
done ← 0
while (done = 0)
    T ← GenExpon(u_j, 1/λ)
    x ← Xe^(−δ×T) − c × (1 − e^(−δ∗T))/δ
    // x is the new value of X(·) just before claim Y arrives
    Y ← GenExpon(u_{j+1}, β)
    if x > 0 then
        X̃ ← x + Y
    else
        X̃ ← Y
    // X̃ is the tentative new value for X(·)
    if X ≤ u then
        D ← D + T
    if X > u and x < u then
        done ← 1
        θ ← (1/δ) ln((δX + c)/(δU + c))
        sumT ← sumT+θ
    else
        sumT ← sumT +T
    j ← j + 2
    X ← X̃
return(D,sumT)
```

Fig. 7.8 Pseudocode describing how to estimate the ruin probability ψ_u based on the regenerative approach.

In Table 7.2, we give results comparing the performance of Monte Carlo ($n = 1024$), randomly shifted Korobov lattices ($n = 1021, a = 76$), and randomly digitally shifted polynomial Korobov lattices — with $n = 1024$, the same construction as the one used in Table 7.1 — for the estimator based on regenerative simulation, as done in [458]. Similar results and additional ones, including the importance sampling approach of [13] mentioned at the beginning of the example, are compared for both Monte Carlo and quasi–Monte Carlo in [281].

What we see in Table 7.2 is that both randomized quasi–Monte Carlo methods perform better than Monte Carlo, the half-width of the 95% confidence interval being reduced by factors between 1.6 and 2.4 when using a polynomial Korobov lattice instead of Monte Carlo.

Table 7.2 Results for the ruin probability estimation problem using the regenerative method based on $n \approx 1021$ and $m = 25$. Shown are the probability estimate $\hat{\mu}$ and the corresponding 95% confidence interval half-width (HW).

	$u = 0$		$u = 10$	
	$\hat{\mu}$	HW	$\hat{\mu}$	HW
MC	0.8419	4.70e−3	0.0377	1.11e−3
Kor	0.8399	3.04e−3	0.0390	9.34e−4
pKor	0.8410	1.92e−3	0.0382	6.98e−4

Valuation of mortgage-backed securities

This problem has been used by several researchers to test the performance of quasi–Monte Carlo methods for high-dimensional finance problems [4, 51, 56, 278, 348, 375]. For this reason, we believe that our discussion of the use of quasi–Monte Carlo methods in finance would be incomplete without a description of this problem.

Here the goal is to evaluate the time-0 value of the cash flow received by the holder of a *mortgage-backed security*, which is a product that banks sell to investors and in which the cash flows come from the payments made by mortgage holders.*

This type of product can become quite complex depending on how the bank pools the mortgages and distributes the cash flows. The problem description used in [51, 348] keeps things simple and assumes that the only source of uncertainty when valuing such cash flows is the interest rate. More precisely and following [51], the problem here is to estimate an expectation of the form

$$M_0 = \mathrm{E}\left(\sum_{l=1}^{M} v_l c_l\right),$$

which represents the time-0 value of this contract. Here v_l is the discount factor for month l and c_l is the cash flow for month l. Both of these quantities depend on the interest rate process in the following way. Let i_l be the interest rate for month l. As in [51], we use the interest rate model

$$i_l = K_0 e^{\xi_l} i_{l-1}, \qquad l \geq 1,$$

where $\xi_l \sim N(0, \sigma^2)$. Then

* These products have received a lot of attention recently because of their role in the mortgage crisis in the United States.

$$v_l = \prod_{k=0}^{l-1} (1 + i_k)^{-1}$$

and

$$c_l = cr_l((1 - w_l) + w_l f_l),$$

where

$$\begin{aligned} c &= \text{monthly mortgage payment}, \\ w_l &= \text{fraction of remaining mortgages prepaying in month } l \\ &= K_1 + K_2 \arctan(K_3 \times i_l + K_4), \\ r_l &= \text{fraction of remaining mortgages at month } l, \\ &= \prod_{k=1}^{l-1} (1 - w_k), \\ f_l &= (\text{remaining annuity at month } l)/c \\ &= \sum_{k=0}^{M-l} (1 + i_0)^{-k}. \end{aligned}$$

Therefore the problem is completely specified by the parameters (i_0, K_0, σ^2) for the interest rate model and (K_1, K_2, K_3, K_4) for the prepayment model. As in [51], we choose $K_0 = \exp(-\sigma^2/2)$ so that $\mathrm{E}(i_k) = i_0$. Hence, overall, we need to specify $(K_1, K_2, K_3, K_4, \sigma, i_0)$.

In [51], two sets of parameters are chosen. The first one is given by

$$(K_1, K_2, K_3, K_4, \sigma, i_0) = (0.01, -0.005, 10, 0.5, 0.02, 0.007)$$

and is such that the 360-dimensional function $f(\cdot)$ satisfying

$$M_0 = \int_{[0,1)^{360}} f(\mathbf{u}) d\mathbf{u}$$

is almost linear in its 360 inputs $u_1, \ldots, u_s$. (This is based on the assumption that the normal random variables ξ_l are generated by inversion.) The second choice,

$$(K_1, K_2, K_3, K_4, \sigma, i_0) = (0.04, 0.0222, -1500, 7, 0.02, 0.007),$$

does not have such a strong linear component. Following [51], they are referred to as “nearly linear” and “nonlinear”, respectively, in what follows.

Figures 7.9 and 7.10 provide results of experiments made on these two sets of parameters to compare the performance of different digital sequences on this problem. Here we chose to study the difference between the original constructions of Faure and Halton and improved versions of these sequences, as discussed in Sect. 5.4.4. We also compare the Sobol’ sequence with well-chosen initial direction numbers — in our case, they are coming from [279] — and the naive choice of initializing all of them to one, as is sometimes done in comparative studies [348]. The generalized Halton sequence used in these experiments comes from [115], while the generalized Faure sequence is based on current work with H. Faure that is not yet published. All estimators based on digital sequences are randomly digitally shifted.

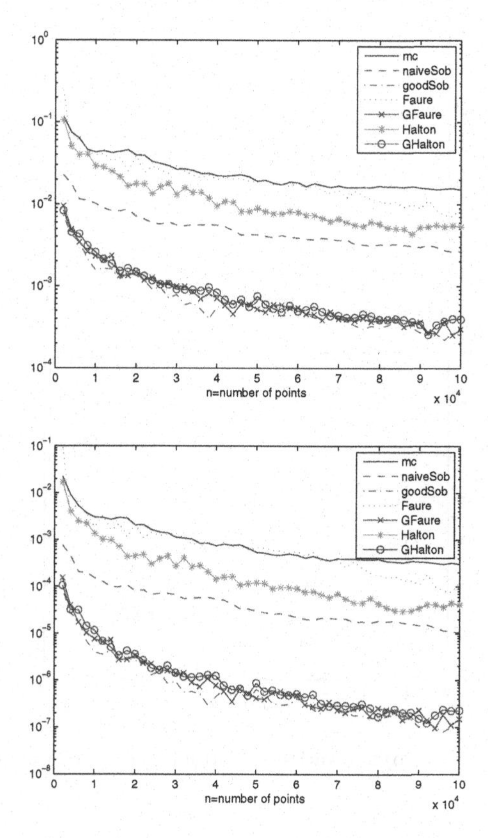

Fig. 7.9 Absolute error (top) and variance (bottom) based on 25 randomizations for the "nearly linear" case.

In each figure, the top graph shows the absolute error of the approximation based on $m = 25$ copies of an estimator based on n simulations, where n is plotted on the x-axis. To compute the absolute error, we use the approximations for the real value M_0 given in [51], which are 131.78706 and 130.712365, for the nearly linear and nonlinear sets of parameters, respectively. As is typically done when studying this type of problem [51, 348], we chose to show the absolute error and estimated variance as functions of the number of points n, shown at every multiple of 2000 between 0 and 100,000.

This is especially convenient when comparing digital sequences that have different bases b. (With only one base b, we could have chosen to restrict ourselves to values of n that are powers of b.)

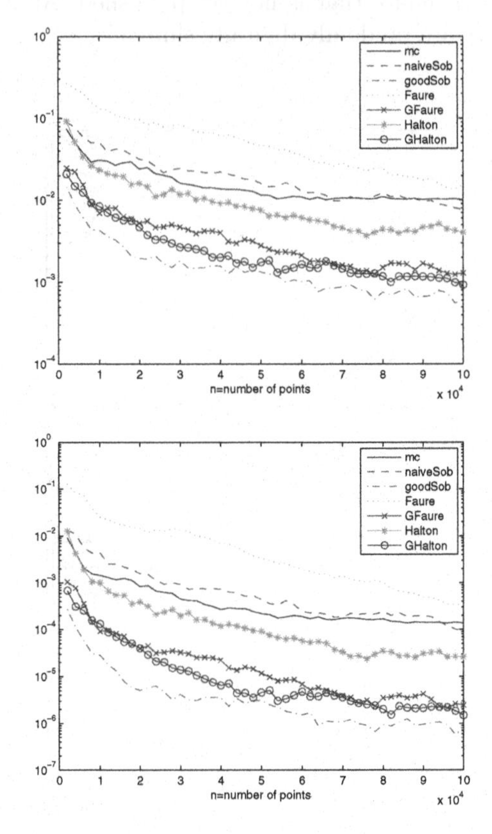

Fig. 7.10 Absolute error (top) and variance (bottom) based on 25 randomizations for the "nonlinear" case.

Looking at Figs. 7.9 and 7.10, we see that there is a clear difference between the original Faure and Halton sequence and their generalized version, the latter being much better. The same can be said for the Sobol' sequence with or without appropriate direction numbers. In fact, for the nonlinear problem, the original Faure sequence and the Sobol' sequence with direction numbers initialized to one are worse than Monte Carlo.

7.4 Commonly used variance reduction techniques

Option pricing has been and continues to be an excellent source of applications for variance reduction techniques [38, 150, 202, 314]. In this section, we illustrate with different examples how the variance reduction techniques seen in Chap. 4 can be used to provide more accurate estimators. We also discuss a few additional techniques that turned out to be useful in the context of finance.

7.4.1 Antithetic variates

In the context of option pricing, using antithetic variates amounts to replacing each realization path of a given underlying asset by a *pair* of antithetic paths. The discounted payoff for each path is then computed, and the average of the two values thus obtained is returned. The idea is that, in an antithetic pair, if we have one path where the asset's value has an upward trend, then the other one will have a downward trend so that the two paths average out to a behavior closer to what is expected.

In Fig. 7.11, we give pseudocode where antithetic variates are used to price Asian options. There, we use the fact that $\Phi^{-1}(1-u) = -\Phi^{-1}(u)$, which comes from the symmetry of the normal density function, as discussed in Chap. 4.

7.4.2 Control variates

This technique has been used quite successfully in finance and was in fact already discussed in the seminal paper by Boyle [37]. Translated into the financial context, using control variates means finding a variable related to the payoff of the option that needs to be priced and for which the expectation can be computed. For instance, in [215], an Asian option based on a geometric average is used as a control variate to price the corresponding option on the arithmetic average. More precisely, let

$$\hat{C}_{g,\mathrm{as},0} = \frac{1}{n}\sum_{i=1}^{n} e^{-rT} \max\left[0, \left(\prod_{j=1}^{d} S^i(t_j)\right)^{1/d} - K\right].$$

Then the control variate estimator for an Asian call option on the arithmetic average is given by

$$\hat{C}_{\mathrm{as},0} + \hat{\beta}\left(C_{g,\mathrm{as},0} - \hat{C}_{g,\mathrm{as},0}\right),$$

```
AntitAsianCall(P_n, r, σ, S(0), T, d)
    sum ← 0
    S[0] ← S(0)
    S̄[0] ← S(0)
    for i = 1 to n
        for j = 1 to d
            x ← Norm01(u_{i,j})
            S[j] ← S[j − 1]e^{(r−σ²/2)(T/d)+σx√(T/d)}
            S̄[j] ← S̄[j − 1]e^{(r−σ²/2)T/d−σx√(T/d)}
        temp ← (S[1] + ... + S[d])/d
        if temp > K then
            Z ← e^{−rT}(temp − K)
        else
            Z ← 0
        temp2 ← (S̄[1] + ... + S̄[d])/d
        if temp2 > K then
            Z̄ ← e^{−rT}(temp2 − K)
        else Z̄ ← 0
        sum ← sum + 0.5(Z + Z̄)
    return sum/n
```

Fig. 7.11 Using antithetic variates for an Asian call option based on a point set P_n.

where $\hat{\beta}$ is computed as in (4.7), with Y_i given by the summand in $\hat{C}_{\text{as},0}$, as defined in (7.9), C_i given by the summand in $\hat{C}_{g,\text{as},0}$, $\hat{\mu}_{\text{mc}}$ by $\hat{C}_{\text{as},0}$, and $\hat{\mu}_c$ by $\hat{C}_{g,\text{as},0}$.

Several other examples where control variates are used in finance can be found in, for instance, [29, 48] and the references given in [38, 150]. Recent applications with American options can be found in [34, 101], and novel approaches for option pricing that allow one to use control variates for which the expectation is not known are studied in [139, 426].

In Table 7.3, we give numerical results illustrating the performance of antithetic variates and control variates for the Asian option problem using the same parameters as in Table 6.5 from Chap. 6. That is, the expiration time is $T = 1$ year, $s = 32$, $r = 0.05$, $\sigma = 0.3$, and $S(0) = 50$. The label "QMC" refers to a randomly digitally shifted Sobol' sequence.

What we see in Table 7.3 is that for this problem the control variate works very well, reducing the half-width of the 95% confidence interval by a factor of about 20 or 25 for the Monte Carlo estimator. The reduction factor brought by the control variate is not as important for the randomized quasi–Monte Carlo estimator, but whether we use antithetic variates and/or control variates, the randomized quasi–Monte Carlo estimator always has a smaller error than that for the corresponding Monte Carlo estimator.

Table 7.3 Comparison of naive simulation (MC/QMC) for the Asian call option example and then for both MC and QMC, the combination with antithetic variates (AV), control variate (CV), and the pair AV+CV, with $n = 1024$ and $m = 25$ repetitions. Given are the price estimates $\hat{\mu}$ and the corresponding 95% confidence interval half-width (HW).

	$K = 45$		$K = 55$	
	$\hat{\mu}$	HW	$\hat{\mu}$	HW
MC	7.0854	0.048	2.1403	0.0304
AV	7.0835	0.018	2.1431	0.0188
CV	7.0640	0.0019	2.1229	0.0014
AV+CV	7.0647	0.0014	2.1235	0.0010
QMC	7.0657	0.0063	2.1166	0.0085
AV	7.0645	0.0046	2.1238	0.0047
CV	7.0643	0.0013	2.1238	0.0010
AV+CV	7.0646	0.0008	2.1236	0.0006

7.4.3 Importance sampling

A typical setup under which importance sampling is useful in finance is for pricing out-of-the-money options. This is because in such cases realization paths simulated under naive Monte Carlo will most of the time yield zero-valued payoffs, thus causing the resulting estimator to have a large relative error. In such cases, importance sampling can be used to change the measure and generate paths where the final payoff is positive more often. Deciding how to change the measure is usually nontrivial, but in some cases it is possible to establish a theoretical justification for a given measure change. An example of this is given in [146] and works as follows. We rewrite the payoff function in terms of a vector $\mathbf{Z}$ of independent standard normal random variables. That is, we suppose the goal is to estimate

$$\mathrm{E}(e^{-rT}\phi(\mathbf{Z})).$$

For instance, for an Asian call option, we have

$$\phi(\mathbf{Z}) = e^{-rT} \max\left[0, \frac{1}{d}\left(\sum_{j=1}^{d} S(0)e^{(r-\sigma^2/2)t_j+\sigma\sqrt{\Delta t}(Z_1+\ldots+Z_j)} - K\right)\right].$$

Importance sampling is then applied by changing the mean of $\mathbf{Z}$ from $\mathbf{0}$ to some vector $\boldsymbol{\theta} = (\theta_1, \ldots, \theta_d)$. The likelihood ratio thus has the form

$$L(\mathbf{Z}) = \frac{\exp(-\sum_{j=1}^{d} z_j^2/2)}{\exp(-\sum_{j=1}^{d}(z_j - \theta_j)^2/2)}$$

$$= \exp\left(-\sum_{j=1}^{d} z_j\theta_j + d\boldsymbol{\theta}^{\mathrm{T}}\boldsymbol{\theta}/2\right).$$

The vector $\boldsymbol{\theta}$ is determined by maximizing the function $G(\boldsymbol{\theta})\exp(-\boldsymbol{\theta}^{\mathrm{T}}\boldsymbol{\theta}/2)$, where $G(\boldsymbol{\theta})$ is such that the payoff can be written as $\phi(\mathbf{Z}) = G(\mathbf{Z})\mathbf{1}_{G(\mathbf{Z})\geq 0}$. For instance, for an Asian call option, we have

$$G(\mathbf{Z}) = \frac{1}{d}\left(\sum_{j=1}^{d} S(0)e^{(r-\sigma^2/2)t_j+\sigma\sqrt{\Delta t}(Z_1+\ldots+Z_j)} - K\right).$$

In Table 7.4, we give some results obtained with the change of measure above when the parameters are $T = 1$ year, $r = 0.05$, $S(0) = 50$, $K = 55$, $\sigma = 0.1$, $t_j = j/16$, $j = 1,\ldots,16$. For each method, we give the estimate for $C_{\mathrm{as},0}$ and the half-width of a 95% confidence interval obtained with 100 repetitions based on 4096 runs each [280].

Table 7.4 Comparison of naive simulation (MC/QMC–Sobol') and importance sampling (IS) for the Asian call option based on $n = 4096$ and $m = 100$. Shown are the option estimates $\hat{\mu}$ and half-width (HW) of the corresponding 95% confidence interval.

	MC		QMC	
	$\hat{\mu}$	HW	$\hat{\mu}$	HW
plain	0.202	1.11e−3	0.202	3.13e−4
IS	0.202	2.76e−4	0.203	1.54e−4

What we see in Table 7.4 is similar to what was observed for the control variate studied in Table 7.3 in that importance sampling brings a larger error reduction for Monte Carlo (about 4) than for randomized quasi–Monte Carlo (about 2), but with or without importance sampling, the latter always dominates the former.

Another approach for choosing a good importance sampling measure is to formulate the problem as an optimization one and then use techniques such as *infinitesimal perturbation analysis* and *stochastic approximation* to solve it [430]. More precisely, the goal is to find an importance sampling measure P equivalent to the risk-neutral probability measure Q but for which the corresponding importance sampling estimator has a smaller variance. That is, consider the importance sampling estimator for a European option price with payoff $H(T) = H(T,\mathbf{S})$ given by

$$\hat{\mu}_{\mathrm{is}} = \frac{1}{n}\sum_{i=1}^{n}\left(\frac{dQ}{dP}\right)_i e^{-rT}\tilde{H}^i(T),$$

where the payoffs $\tilde{H}^1(T), \ldots, \tilde{H}^n(T)$ are simulated under P and $(dQ/dP)_i$ is the likelihood ratio (or Radon-Nikodym derivative) observed in the ith simulation. This estimator has a variance

$$\mathrm{Var}(\hat{\mu}_{\mathrm{is}}) = \frac{1}{n}\left(\mathrm{E}_P\left(\left(\frac{dQ}{dP}\right)e^{-rT}\tilde{H}^i(T)\right)^2 - V_0^2\right).$$

Assume P is parameterized by some variable θ. If we let

$$\nu(\theta) = \mathrm{E}_P\left(\left(\frac{dQ}{dP}\right)^2 e^{-2rT}(\tilde{H}^i(T))^2\right), \tag{7.14}$$

then our goal is to determine the value of θ that solves the optimization problem

$$\min_{\theta\in\Theta} \nu(\theta), \tag{7.15}$$

where $\Theta = \{\theta : P_\theta \text{ is equivalent to } Q\}$. As was shown in [430], this problem can be simplified by reformulating the expectation in $\nu(\theta)$ as an expectation under Q rather than P. That is, we use the fact that

$$\begin{aligned}\nu(\theta) &= \int_\Omega e^{-2rT}(H(T))^2\left(\frac{dQ}{dP}\right)^2 dP\\ &= \int_\Omega e^{-2rT}(H(T))^2\left(\frac{dQ}{dP}\right) dQ\\ &= \mathrm{E}_Q\left(e^{-2rT}(H(T))^2\frac{dQ}{dP}\right).\end{aligned} \tag{7.16}$$

The advantage of (7.16) over (7.14) is that under Q the payoff $H(T)$ no longer depends on θ, and it is therefore easier to apply techniques such as IPA to solve (7.15) as these techniques require derivation of the term inside the expectation with respect to θ.

The optimization problem (7.15) can then be solved using *stochastic approximation* [133, 234, 385] as follows:

1. Initialize θ_0.
2. Iteratively compute

$$\theta_{n+1} = \Pi_\Theta(\theta_n - a_n\hat{h}_n)$$

until some stopping criterion is met, where $\{a_n, n \geq 1\}$ is a sequence of positive numbers that goes to 0, $\hat{h}_n$ is an estimate of the gradient $\nabla\nu(\theta)$ at θ_n, and Π_Θ is a projection operator on Θ.

Typically, the sequence $\{a_n, n \geq 1\}$ is chosen so that $\sum_n a_n = \infty$ and $\sum_n a_n^2 < \infty$, as these conditions are required (along with other conditions) to guarantee the convergence of the stochastic approximation algorithm. The stopping criterion is met when either $n > N_1$ or $a_n \hat{h}_n < \epsilon$ for some prespecified threshold values $N_1, \epsilon > 0$. For instance, in the numerical experiments reported in [430], the values $N_1 = 100$ and $\epsilon = 0.0005$ are used.

To get the estimate $\hat{h}_n$, we can use the *IPA estimator* [133, 144, 150, 192], which is based on the identity

$$\frac{\partial}{\partial \theta}\mathrm{E}(Y(\theta)) = \mathrm{E}\left(\frac{\partial}{\partial \theta}Y(\theta)\right), \tag{7.17}$$

which holds under certain continuity conditions and the existence of finite moments. In our case,

$$Y(\theta) = e^{-2rT}(H(T))^2 \frac{dQ}{dP_\theta},$$

and since the expectation $\nu(\theta) = \mathrm{E}_Q(Y(\theta))$ is computed under Q, only the Radon-Nikodym derivative dQ/dP_θ in $Y(\theta)$ depends on θ, as mentioned before. Furthermore, the conditions allowing (7.17) to hold are typically met for most payoffs [430], and thus in this case we get that

$$\frac{\partial}{\partial \theta}\mathrm{E}(Y(\theta)) = e^{-2rT}(H(T))^2 \frac{\partial}{\partial \theta}(dQ/dP_\theta).$$

Example 7.6 illustrates this approach on the Asian call option example as done in [430].

Example 7.6. Suppose we want to price an Asian call option that is out of the money under the lognormal model. In that case, we can define P_θ to be the measure under which

$$dS(t) = (r + \theta)S(t)dt + \sigma S(t)dW(t),$$

where $W(\cdot)$ is a standard Brownian motion under P_θ defined as

$$W(t) = B(t) - \theta t$$

and where $B(\cdot)$ is a standard Brownian motion under Q. From Girsanov's theorem [350], we know that

$$\frac{dQ}{dP_\theta} = \exp^{-\theta W(T) - \theta^2 T/2}.$$

But we wish to use (7.16) rather than (7.14), so it is appropriate to work with the Q-Brownian motion $B(\cdot)$ instead of $W(\cdot)$, and thus we obtain

$$\frac{dQ}{dP_\theta} = \exp^{-\theta(B(T) - \theta T) - \theta^2 T/2} = \exp(-\theta B(T) + \theta^2 T/2).$$

Now, we also need to compute

$$\frac{\partial(dQ/dP_\theta)}{\partial\theta} = (-B(T) + \theta T)\exp^{-\theta B(T)+\theta^2 T/2}.$$

Hence, an IPA estimator for $\hat{h}_n$ can be constructed as

$$\hat{h}_n = \frac{1}{N_2}\sum_{i=1}^{N_2} e^{-2rT} H^i(T)(-B^i(T) + \theta_n T)\exp(-\theta_n B^i(T) + \theta_n^2 T/2), \quad (7.18)$$

where $H^i(T)$ and $B^i(T)$ are respectively the payoff under Q and the terminal value of the Brownian motion for the ith simulation.

As was discussed in [430], the fact that $H^i(T)$ is simulated under Q implies that the estimator $\hat{h}_n$ given in (7.18) will suffer from the same kind of problems as the original option price estimator. To avoid this drawback, we can simply perform another change of measure bringing us back to P and use instead

$$\hat{h}_n = \frac{1}{N_2}\sum_{i=1}^{N_2} e^{-2rT}(\tilde{H}^i(T))^2(-W^i(T) - \theta_n T)\exp(-2\theta_n W^i(T) - \theta_n^2 T),$$

where we wrote everything in terms of the P-Brownian motion $W(\cdot)$. The factor of 2 in the second exponential comes from the fact that we must multiply again by dQ/dP_θ when going from Q to P_θ. Typically, the number of simulation runs used to construct $\hat{h}_n$ is relatively small. For instance, in [430], the value $N_2 = 100$ is used.

7.4.4 Conditional Monte Carlo

When the underlying assets follow a multivariate geometric Brownian motion, the distribution of $S(t)$ given $S(0)$ is known. For more complicated models such as those discussed in Sect. 7.2, the distribution of $S(t)$ may not be known, for instance because the volatility is modeled by a stochastic process. Conditional Monte Carlo can be useful in that context because even if the distribution of $S(t)$ given $S(0)$ is not known, in some cases the conditional distribution of $S(t)$ given $S(0)$ *and* $(\sigma(u), 0 \le u \le t)$ is known [199, 476].

More precisely, for models of the form

$$dS(t) = rS(t)dt + \sigma(t)S(t)\left(\rho dB_1(t) + \sqrt{1-\rho^2}dB_2(t)\right),$$
$$d\sigma^2(t) = \gamma(t)dt + \eta(t)dB_1(t),$$

we have that

$$\ln S(T)|(\sigma(u), u \le t, S(0)) \sim N(a, b)$$

with

$$a = \ln S(0)\xi + rT - \frac{1-\rho^2}{2}\int_0^T \sigma^2(t)dt,$$
$$b = (1-\rho^2)\int_0^T \sigma^2(t)dt,$$

and where

$$\xi = \exp\left(-\frac{\rho^2}{2}\int_0^T \sigma^2(t)dt + \rho\int_0^T \sigma(t)dB_1(t)\right).$$

Based on this, we have that the price of a plain call option conditional on the path $\{B_1(t), 0 \leq t \leq T\}$ has a closed-form expression similar to the Black-Scholes-Merton option price (7.4) and is given by

$$S(0)\xi\Phi(\tilde{d}_1) - Ke^{-rT}\Phi(\tilde{d}_2), \tag{7.19}$$

where

$$\tilde{d}_1 = \frac{\ln S(0)\xi/K + (r+\tilde{\sigma}^2/2)T}{\tilde{\sigma}\sqrt{T}},$$
$$\tilde{d}_2 = \tilde{d}_1 - \tilde{\sigma}\sqrt{T},$$
$$\text{and } \tilde{\sigma}^2 = \frac{1}{T}\int_0^T \sigma^2(t)dt.$$

Hence we can apply conditional Monte Carlo by conditioning on $\{\sigma(t), 0 \leq t \leq T\}$. However, in practice, we cannot simulate the whole path $\{\sigma(t), 0 \leq t \leq T\}$ but only a discretized version of it. Hence our conditioning vector is

$$\mathbf{Z} = (\sigma(t_1), \ldots, \sigma(t_d)),$$

which means we use the approximations

$$\tilde{\sigma}^2 \approx \frac{1}{T}\sum_{j=1}^{d} \sigma^2(t_{j-1})\Delta_j,$$
$$\xi \approx \exp\left(-\frac{\rho^2}{2}\sum_{j=1}^{d}\sigma^2(t_{j-1})\Delta_j + \rho\sum_{j=1}^{d}\sigma(t_{j-1})(B_1(t_j) - B_1(t_{j-1}))\right)$$

in the Black-Scholes-Merton–like formula (7.19), where $\Delta_j = t_j - t_{j-1}$.

Summing up, we can use the (approximate) conditional Monte Carlo estimator

$$\hat{\mu}_{\text{cmc}} = \frac{1}{n}\sum_{i=1}^{n} S(0)\xi^i N(\tilde{d}_1^i) - Ke^{-rT}N(\tilde{d}_2^i),$$

where ξ^i, $\tilde{d}_1^i$, and $\tilde{d}_2^i$ are calculated based on the ith path $\{\sigma^i(t_1),\ldots,\sigma^i(t_d)\}$ for $i=1,\ldots,n$. Hence, with conditional Monte Carlo, we only need to generate $\{\sigma(t_j), j=1,\ldots,d\}$ and not $\{S(t_j), i=1,\ldots,d\}$.

Table 7.5 gives results comparing Monte Carlo, a randomized Sobol' point set, and a randomized polynomial Korobov lattice for the problem of pricing a plain call option under Heston's process with or without conditional Monte Carlo. The parameters for Heston's process are the same as in Table 7.1. That is, $r=0$, $\kappa=2$, $\sigma(0)=0.1$, $\theta=0.01$, $\sigma_v=0.1$, and $T=0.5$ year. In Table 7.5, we experiment with two different values of $S(0)$ and ρ. We see that in contrast with the control variate and importance sampling applications seen previously, here conditional Monte Carlo brings a *larger* reduction of the 95% confidence interval's half-width for randomized quasi–Monte Carlo (reduction factors of 10 and more) than Monte Carlo (reduction factors of about 2).

Table 7.5 Using conditional Monte Carlo to price a simple call option under Heston's process, with $n=1024$ and $m=25$. Shown are the option price estimates $\hat{\mu}$ and half-width of the corresponding 95% confidence interval (HW).

	$\rho=-0.5$				$\rho=0.5$			
	$S(0)=90$		$S(0)=110$		$S(0)=90$		$S(0)=110$	
	without CMC							
	$\hat{\mu}$	HW	$\hat{\mu}$	HW	$\hat{\mu}$	HW	$\hat{\mu}$	HW
MC	0.122	8.39e−3	10.42	1.07e−1	0.288	2.00e−2	10.21	1.08e−1
Sobol'	0.120	8.33e−3	10.41	1.86e−2	0.287	1.69e−2	10.22	3.64e−2
pKor	0.125	4.70e−3	10.40	1.42e−2	0.291	9.69e−3	10.21	1.08e−2
	with CMC							
	$\hat{\mu}$	HW	$\hat{\mu}$	HW	$\hat{\mu}$	HW	$\hat{\mu}$	HW
MC	0.123	1.26e−3	10.40	4.33e−2	0.291	8.30e−3	10.21	5.10e−2
Sobol'	0.123	1.59e−4	10.40	4.65e−3	0.287	3.61e−3	10.21	8.06e−3
pKor	0.123	1.13e−4	10.40	2.63e−3	0.289	3.57e−3	10.21	3.98e−3

7.4.5 Common random numbers

A natural setting for this method in finance is for estimating "greeks", which are partial derivatives of the option's value with respect to a parameter. For instance, for an option whose value at time 0 is denoted by $V_0 := V_0(S(0))$, its *delta* is given by

$$\mu_d = \frac{\partial V_0(S(0))}{\partial S(0)},$$

and its *gamma* is given by the second derivative

$$\mu_g = \frac{\partial^2 V_0(S(0))}{\partial S(0)^2}.$$

Section 7.6 discusses some of the reasons for studying these quantities.

In some cases, closed-form expressions can be found for these greeks. But in more complex settings, they must be estimated. In [47], common random numbers are used within the *finite difference* method to estimate various greeks. Using this approach, we can, for example, estimate the delta by

$$\hat{\mu}_d = \frac{[V_0(S(0)+h) - V_0(S(0)]}{h}$$

where h is a small quantity; for instance, $h = 0.0001$. The way common random numbers are applied here is that the same random numbers are used to generate paths starting at $S(0)$ and $S(0)+h$. The use of randomized quasi–Monte Carlo for such problems is discussed in [283].

7.4.6 Moment-matching methods

These methods have been used in finance, in particular within a method called *empirical martingale simulation* [91, 90]. They are based on the idea of adjusting a given set of underlying variables, after the simulation is done, so that their empirical mean equals their theoretical expectation. For instance, suppose we want to estimate an option on one asset for which the prices $S(t_1), \ldots, S(t_d)$ need to be simulated. Using the fact that $\mathrm{E}(e^{-rt_j}S(t_j)) = S(0)$ for each t_j under the risk-neutral measure, we can define a modified version $\{\tilde{S}^1(t_j), \ldots, \tilde{S}^n(t_j)\}$ of the sample $\{S^1(t_j), \ldots, S^n(t_j)\}$ for each t_j as follows. First, let

$$Z^i(t_1) = S^i(t_1), \qquad i = 1, \ldots, n,$$
$$Z_0(t_1) = \frac{1}{n} e^{-rt_1} \sum_{i=1}^{n} Z^i(t_1),$$

$$\tilde{S}^i(t_1) = S(0)\frac{Z^i(t_1)}{Z_0(t_1)}, \qquad i = 1, \ldots, n.$$

Then, recursively define, for $j = 2, \ldots, d$,

$$Z^i(t_j) = \tilde{S}^i(t_{j-1})\frac{S^i(t_j)}{S^i(t_{j-1})},$$

$$Z_0(t_j) = \frac{1}{n}e^{-rt_j}\sum_{i=1}^{n} Z^i(t_j),$$

$$\tilde{S}^i(t_j) = S(0)\frac{Z_i(t_j)}{Z_0(t_j)}.$$

7.5 American option pricing

As mentioned before, European options can only be exercised at expiration time. While this feature simplifies the task of pricing these contracts, in practice options can usually be exercised before expiration. Such options are called *American options.*

For a while, it was thought that American options could not be priced using the Monte Carlo method, but since then several techniques based on Monte Carlo have been proposed, for instance in [53, 134, 298, 48, 169, 387, 481]. A recent survey can be found in [150, Chap. 8].

Formally, the problem is to estimate

$$V_0(\mathbf{S}(0)) = \sup_{0\le\tau\le T} \mathrm{E}\left(e^{-r\tau}H(\tau,\mathbf{S})\right),$$

where τ is a stopping time that represents the moment when the option is exercised, thereby resulting in a payoff $H(\tau,\mathbf{S})$. Hence this problem qualifies as an *optimal stopping problem.*

In what follows, we make the assumption that there is a finite set of equally spaced exercise times $t_1,\ldots,t_b = T$ where the option can be exercised. Such options are usually called *Bermudan options.* As before, we use the notation $\Delta_j = t_j - t_{j-1}$ for $j = 1,\ldots,b$.

Several methods for pricing American options — including those based on Monte Carlo — that use this type of discretization first formulate the pricing problem using dynamic programming. This is a natural idea since for a given realization path of the underlying process, the first information that can be extracted is the terminal value of the option, given by $H(T,\mathbf{S})$. Using this, we can then work our way backward, determining the value at time t_j of the option, given by

$$V_j(\mathbf{S}(t_j)) = \max(H(t_j,\mathbf{S}), C_j(\mathbf{S}(t_j))) \tag{7.20}$$

for $j = b-1,\ldots,0$, where $H(t_j,\mathbf{S})$ is the *exercise value* of the option and

$$C_j(\mathbf{S}(t_j)) := e^{-r\Delta_j}\mathrm{E}(V_{j+1}(\mathbf{S}(t_{j+1}))|\mathcal{F}_{t_j})$$

is the *continuation value* of the option, given by the discounted expected option value at the next exercise time, conditioned on the prices observed so far. It is also implicit in this definition of the continuation value that we are conditioning on the event that the option has not been exercised before time t_j. The time-0 value of the option is then given by $V_0(\mathbf{S}(0))$.

Hence, to solve the problem based on dynamic programming, we need to have an estimate for the continuation value. To achieve this with Monte Carlo, we can first generate n i.i.d. paths $\{\mathbf{S}^i(t_1), \ldots, \mathbf{S}^i(t_b)\}$, $i = 1, \ldots, n$. But then, at time t_j, to estimate the continuation value $C_j(\mathbf{S}^i(t_j))$ for path i, we only have one path where $\mathbf{S}(t_{j+1})$ is simulated conditioned on $\mathbf{S}(t_j)$. If we want to use all paths in our estimate, then paths other than path i must be weighted accordingly, much like in the construction of an importance sampling estimator. This *stochastic mesh* approach is studied in [48].

Another possibility is to use (nonlinear) regression to estimate the continuation value $C_j(\mathbf{S}(t_j))$ on each path [53, 298, 452]. The idea here is to think of the current prices $\mathbf{S}^1(t_j), \ldots, \mathbf{S}^n(t_j)$ as the *independent variables* and the option values at the next time step, given by $V_{j+1}(\mathbf{S}^1(t_{j+1})), \ldots, V_{j+1}(\mathbf{S}^n(t_{j+1}))$, as the *dependent variables.* By choosing a basis of M multivariate functions $\psi_l(\mathbf{x}_j), l = 0, \ldots, M-1$, where $\mathbf{x}_j$ is a vector of variables each of which is a function of the prices observed so far, we can approximate the continuation value $C_j(\mathbf{S}(t_j))$ as

$$C_j(\mathbf{S}(t_j)) \approx \sum_{l=0}^{M-1} \hat{\beta}_{l,j} \psi_l(\mathbf{x}_j),$$

where $\hat{\beta}_{l,j}$ is an approximation for the regression coefficient $\beta_{l,j}$. That is,

$$(\hat{\beta}_{0,j}, \ldots, \hat{\beta}_{M-1,j})^{\mathrm{T}} = (\Psi_j^{\mathrm{T}} \Psi_j)^{-1} \Psi_j^{\mathrm{T}} (y_1, \ldots, y_n)^{\mathrm{T}}, \qquad j = 1, \ldots, b,$$

where $y_i = e^{-r\Delta_j} V_{j+1}(\mathbf{S}^i(t_{j+1}))$, and the element on the ith row and lth column of Ψ_j is given by $\psi_l(\mathbf{x}_j^i)$ for $i = 1, \ldots, n, l = 0, \ldots, M-1$.

When applying this type of method, one needs to decide (i) which polynomials to use; (ii) how to define $\mathbf{x}$; (iii) whether to include all paths or not when estimating the regression coefficients $\beta_{l,j}$; and (iv) whether the regression coefficients should be estimated beforehand or if everything should be done using the same set of n paths. In [298], it is suggested for (iii) to keep only the paths that are *in the money* at each time t_j. That is, we only keep the paths for which the payoff $H(t_j, \mathbf{S}^i)$ is positive. However, other authors have found that this approach was sometimes less accurate than the one where all paths are used [150]. As for (iv), an advantage of precomputing the regression coefficients $\hat{\beta}_{l,j}$ for $l = 0, \ldots, M-1$ and $j = 1, \ldots, b$ is that when a second set of simulation runs is performed, the resulting estimator is based on an average of i.i.d. discounted payoffs when using Monte Carlo simulation. This is not the case if everything is done using the same paths because then the $\hat{\beta}_{j,l}$ introduce dependence across the discounted payoffs that form the estimator. As for (i) and (ii), for a plain put option, one can sometimes simply take $\mathbf{x}_j = x_j = S(t_j)$ and use the first few powers $1, x_j, \ldots, x_j^{M-1}$ of x_j as basis functions. For more complicated payoffs, there are several possibilities. For instance, in [298], the authors suggest taking

$$\mathbf{x} = (S(t_j), (S(t_1) + \ldots + S(t_j))/j)$$

for an American-Asian option and then the eight basis functions

$$\nu_0(1), \nu_1(x_1), \nu_2(x_1), \nu_1(x_2), \nu_2(x_2), \nu_1(x_1)\nu_1(x_2), \nu_1(x_1)\nu_2(x_2), \nu_2(x_1)\nu_1(x_2),$$

where $\nu_l(x)$ is a (weighted) Laguerre polynomial of degree l satisfying

$$\nu_0(x) = 1, \qquad \nu_1(x) = e^{-x/2}, \qquad \nu_2(x) = e^{-x/2}(1-x).$$

Once we have a way of estimating the continuation value, an estimate of the American option's price at time 0 can be obtained by simulating n paths and estimating in a backward recursive way the optimal exercise time on each path. This is the *least-squares Monte Carlo* approach of Longstaff and Schwartz [298]. The code in Fig. 7.12 describes this approach in detail. There, we assume that the regression coefficients have been precomputed.

```
LeastSquaresMC(P_n, β)
for i ← 1 to n
    t*(i) ← T
    generate S^i(t_1), ..., S^i(t_b) based on u_i
for j ← b − 1 downto 1
    for i ← 1 to n
        C^i_j ← Σ_{l=0}^{M−1} β̂_{l,j} ψ_l(x^i_j)
        if H(t_j, S^i) > C^i_j then
            t*(i) ← t_j
return (1/n) Σ_{i=1}^{n} e^{−r t*(i)} H(t*(i), S^i)
```

Fig. 7.12 Pseudocode describing least-squares Monte Carlo with paths generated by a point set P_n.

The estimator returned by this type of method is *low-biased* — that is, the bias is negative — since it uses for each path an *estimate* of the optimal exercise time, hence resulting in a possibly suboptimal exercise policy that causes the option to be undervalued. It can be shown that as the number of simulation paths n and the number of basis functions M go to infinity, the bias of the estimator goes to 0 [59]. But for finite n and M, there is a bias and so typically it is of interest to use another approximation that is high-biased, so that we can have a lower bound *and* an upper bound on the true price.

The stochastic mesh approach of Broadie and Glasserman [48] mentioned above produces a high-biased estimator. It can be coupled with variance reduction techniques and quasi–Monte Carlo methods to produce accurate bounds [39, 48].

Another family of methods that produce high-biased estimators are the *dual pricing methods* introduced independently in [387] and [169]. Our presentation here follows [387] and [150]. Dual approaches rely on an important result that states that the American option pricing problem can be written as

$$V_0(\mathbf{S}(0)) = \inf_{M(\cdot)\in\mathcal{H}_0^1} \mathrm{E}\left(\sup_{0\le t\le T}(e^{-rt}H(t,\mathbf{S}) - M(t))\right), \tag{7.21}$$

where $\mathcal{H}_0^1$ is the set of all martingales $M(\cdot)$ for which $\mathrm{E}(\sup_{0\le t\le T}|M_t|) < \infty$ and such that $M(0) = 0$. Based on (7.21), for a given martingale $M(\cdot) \in \mathcal{H}_0^1$, the quantity

$$\mathrm{E}\left(\sup_{0\le t\le T}(e^{-rt}H(t,\mathbf{S}) - M(t))\right) \tag{7.22}$$

gives us an upper bound on the American option's price $V_0(\mathbf{S}(0))$. If we let M^* be the martingale that achieves the infimum in (7.21), then the goal is to find a good approximation for M^* so as to obtain an upper bound that is close to the true value $V_0(\mathbf{S}(0))$.

The dual formulation (7.21) has an interesting interpretation, which is to view the process $M(\cdot)$ as the discounted value of a trading strategy for the option. The quantity inside the expectation then represents the largest possible loss that could be produced by this strategy. That is, if an investor sells the American option and enters the trading strategy described by $M(\cdot)$, then in the worst case he or she loses $\sup_{0\le t\le T} e^{-rt}H(t,\mathbf{S}) - M(t)$ (value at time 0). The price of the option must then be given by the smallest possible value that this discounted worst-case loss can take, where the optimization is done over all trading strategies.

Going back to the problem of identifying a martingale M that can be used to construct the upper bound (7.22), here we discuss one possible approach, following the presentation in [150, pp. 474–475]. The main idea is to define the discrete-time martingale $M = \{M_j, j = 0,\ldots,b\}$, where

$$M_j = D_1 + \ldots + D_j$$

and D_j is the difference

$$D_j = e^{-r\Delta_j}V_j(\mathbf{S}(t_j)) - C_{j-1}(\mathbf{S}(t_{j-1})). \tag{7.23}$$

Since by definition

$$C_{j-1}(\mathbf{S}(t_{j-1})) = \mathrm{E}(e^{-r\Delta_j}V_j(\mathbf{S}(t_j))|\mathbf{S}(t_{j-1})),$$

we have that $\mathrm{E}(D_j) = 0$ and M is indeed a martingale. In fact, it can be shown that this martingale is the discretized version of the martingale that solves the optimization problem (7.21). However, the exact continuation value usually is not known in practice. Furthermore, if we replace $V_j(\mathbf{S}(t_j))$ and $C_{j-1}(\mathbf{S}(t_{j-1}))$ by estimates in (7.23), then the difference may not have a zero expectation, and therefore this simulated version of M may not be a martingale.

An alternative approach is to define

$$\hat{D}_j = e^{-r\Delta_j}(\hat{V}_j(\mathbf{S}(t_j)) - \mathrm{E}(\hat{V}_j(\mathbf{S}(t_j))|\mathcal{F}_{t_{j-1}})), \tag{7.24}$$

where

$$\hat{V}_j(\mathbf{S}(t_j)) = \max(H(t_j, \mathbf{S}), \hat{C}_j(\mathbf{S}(t_j)))$$

and the estimated continuation value $\hat{C}_j(\mathbf{S}(t_j))$ can be obtained by regression, as in the least-squares Monte Carlo approach of Longstaff and Schwartz. The second term $\mathrm{E}(\hat{V}_j(\mathbf{S}(t_j))|\mathcal{F}_{t_{j-1}})$ can be estimated using an inner set of simulations of the prices $\mathbf{S}(t_j)$ given $\mathbf{S}(t_{j-1})$. Figure 7.13 gives pseudocode for computing the dual estimator in this fashion. Note that, depending on the choice of basis functions and variables $\mathbf{x}$, it might be possible to compute exactly the expectation in (7.24) [150].

```
DualApproach(P_n, β)
for i ← 1 to n
    M[i,0] ← 0 // ith simulated martingale
    generate S^i(t_1),...,S^i(t_b) based on u_i
    V̂^i_b ← H^i_b
    for p ← 1 to N
        generate S̃^p(t_b) given S^i(t_{b-1})
        Ṽ^p_b ← H̃^p_b
    D̂[i,b] ← e^{-rΔ_b}(V̂^i_b − (1/N) Σ_{p=1}^N Ṽ^p_b)
    for j = b − 1 downto 1
        C^i_j ← Σ_{l=0}^M β̂_{l,j} ψ_l(x^i_j)
        V̂^i_j ← max(H^i_j, C^i_j)
        for p = 1 to N
            generate S̃^p(t_j) given S^i(t_{j-1})
            C̃^p_j ← Σ_{l=0}^{M-1} β̂_{l,j} ψ_l(x̃^p_j)
            Ṽ^p_j ← max(H̃^p_j, C̃^p_j)
        D̂[i,j] ← e^{-rΔ_j}(V̂^i_j − (1/N) Σ_{p=1}^N Ṽ^p_j)
    for j = 1 to b
        M[i,j] ← M[i,j−1] + D̂[i,j]
        if (e^{-rt_j} H^i_j − M[i,j]) > max^i then
            max^i ← e^{-rt_j} H^i_j − M[i,j]
return (1/n) Σ_{i=1}^n max^i
```

Fig. 7.13 Pseudocode describing dual approach based on martingales and approximate value functions. We assume the regression coefficients have been precomputed, and use the notation $H^i_j = H(t_j, \mathbf{S}^i(t_j))$, $\tilde{H}^p_j = H(t_j, \tilde{\mathbf{S}}^p(t_j))$.

Table 7.6 gives results using least-squares Monte Carlo on a Bermudan-Asian option problem that was studied in [298], where $T = 2$ years and $\Delta_j = 1/100$ for $j = 1, \ldots, 200$. The option cannot be exercised during the first three months of the contract, but the prices observed during that period enter the average that determines the payoff. In fact, taking 0 to be the valuation time, the average used to determine the payoff at time $0 \leq t \leq T$ is taken from time -0.25 years until time t. Thus, in addition to the strike price K and the initial stock price $S(0)$, we also need to know the average stock price A from time -0.25 until time 0. In Table 7.6, we fix $A = 90$, $K = 100$, and the stock price is assumed to follow a lognormal model with volatility $\sigma = 0.2$. The risk-free rate is $r = 0.06$. The regression coefficients are precomputed using 5000 runs and Monte Carlo.

Table 7.6 Comparison of Monte Carlo and randomized Sobol' methods for pricing a Bermudan-Asian option using the (low-biased) least-squares approach with $n = 1024$ points and $m = 25$ repetitions. Shown are the option price estimates $\hat{\mu}$ and half-width of the corresponding 95% confidence interval (HW).

	$S(0) = 90$		$S(0) = 100$		$S(0) = 110$	
	$\hat{\mu}$	HW	$\hat{\mu}$	HW	$\hat{\mu}$	HW
MC	3.346	0.101	7.943	0.162	14.532	0.217
QMC	3.341	0.060	7.927	0.054	14.527	0.053

As seen in Table 7.6, in all cases considered, the Sobol' estimator provides a more precise estimator, with a reduction of the 95% confidence interval's half-width by factors ranging between 1.7 and 4.

7.6 Estimating sensitivities and percentiles

We saw in Sect. 7.4 one way of estimating sensitivities of option prices — called the *greeks* — using finite differences and common random variates. Determining the value of the greeks is an important task in mathematical finance because, for example, these quantities are required for constructing portfolios that *hedge* a given instrument. We illustrate this with a simple example.

Example 7.7. Suppose an investor sells a call option on a stock and wants to establish a trading strategy where positions a_t in the stock and b_t in the riskless investment at time t are continuously monitored and that replicates the option's value at any time $0 \leq t \leq T$. For a European option, the trading strategy should be *self-financing*. That is, no money should be added or withdrawn from the replicating portfolio between time 0 and the expiration time T. Formulated alternatively, this means that the (continuous) rebalancing of the portfolio should be done at a zero net cost. In this way, the no-arbitrage argument implies that the time-0 value of the trading strategy must be equal to the option's price at time 0. Furthermore, it can be shown that the number a_t of shares of stock that should be held at time t is given by the *delta* of the option [198]. That is, we should have

$$a_t = \frac{\partial V(S(t))}{\partial S(t)},$$

where $V(S(t))$ is the value of the option at time t. Hence the option and the replicating portfolio not only have the same value at any time t but also the same delta.

In simple settings, this quantity can be calculated exactly. For example, for a plain call option under the lognormal model, we have that

$$\frac{\partial V(S(t))}{\partial S(t)} = \Phi(d_1),$$

where, as in the Black-Scholes-Merton option pricing formula (7.4),

$$d_1 = \frac{\ln(S(t)/K) + (r + \sigma^2/2)(T - t)}{\sigma\sqrt{T - t}}.$$

Now, even when the delta can be computed exactly, in practice it is not possible to continuously update the portfolio's composition. If the portfolio is updated only a discrete number of times, then a discretization error between the option and the replicating portfolio's value is introduced, which causes the rebalancing cost to be larger than zero. Although frequent rebalancing transactions can allow the replicating portfolio's value to stay relatively close to the option's value, the high transaction costs associated with such strategies are an incentive to choose less frequent rebalancing transactions.

An alternative way of reducing the discretization error caused by a non-continuous rebalancing is to add more securities in the replicating portfolio and try to match not only the value and delta of the option but other derivatives as well. A development in multivariate Taylor series can be used to show how this can help reduce the discretization error [495]. For example, one might want to add a call option on the same stock but with a different strike price, so that the *gamma* of the option, given by the second derivative

$$\frac{\partial^2}{\partial S^2(t)} V(S(t)),$$

and the portfolio's gamma match. To establish such strategies, greeks other than the delta must thus be computed, which is one reason why it is important to know how to provide good estimators for these quantities when exact formulas are not available.

As we saw in the example above, under the lognormal model and for a simple call (or put) option, the greeks can be computed exactly. But for more complex models and/or option payoffs, this might not be the case. For such cases, finite differences can be used, as discussed on p. 282. Alternative approaches are *infinitesimal perturbation analysis* (IPA — also called *pathwise differentiation*) [133, 144, 150, 192] and the *likelihood ratio method* (LR — also called the *score function method*) [133, 152]. We already explained IPA in the section on importance sampling, but we discuss it again here in the context of greeks estimation.

Estimating sensitivities using IPA

As we mentioned on p. 278, the idea of IPA is to estimate a quantity of the form

$$\frac{\partial}{\partial\theta}\mathrm{E}(Y(\theta))$$

using the estimator

$$\frac{1}{n}\sum_{i=1}^{n}\frac{\partial}{\partial\theta}Y^i(\theta).$$

For this method to work, we need the relation

$$\mathrm{E}\left[\frac{\partial}{\partial\theta}Y(\theta)\right] = \frac{\partial}{\partial\theta}\mathrm{E}(Y(\theta)) \tag{7.25}$$

to hold, and we need the derivative

$$\frac{\partial}{\partial\theta}Y(\theta)$$

to exist almost everywhere. Note that one of the conditions required for (7.25) to hold is that $Y(\theta)$ must be a continuous function of θ.

In the context of option price sensitivities, $\mathrm{E}(Y(\theta))$ represents the option's price based on the risk-neutral pricing formula (7.3), and thus $Y(\theta)$ is the discounted payoff of the option. Hence, for IPA to be applicable, we need the payoff to be differentiable (with probability 1) with respect to θ.

We illustrate this approach with the example of estimating delta for an Asian call option under the lognormal model [150, pp. 389–390].

Example 7.8. Here the goal is to estimate

$$\frac{\partial}{\partial S_0}\mathrm{E}(v(S_0)),$$

where

$$v(S_0) = e^{-rT} \max\left(0, \frac{1}{d}\sum_{j=1}^{d} S_0 e^{(r-\sigma^2/2)t_j+\sigma B(t_j)} - K\right).$$

The function $v(S_0)$ is continuous and differentiable almost everywhere — except in $S_0 = K$ — and therefore we can write

$$\frac{\partial}{\partial S_0}\mathrm{E}(v(S_0)) = \mathrm{E}\left(\frac{\partial}{\partial S_0}v(S_0)\right),$$

where

$$\frac{\partial}{\partial S_0}v(S_0) = \begin{cases} e^{-rT}\frac{1}{d}\sum_{j=1}^{d} e^{(r-\sigma^2/2)t_j+\sigma B(t_j)} & \text{if } v(S_0) > 0 \\ 0 & \text{otherwise.} \end{cases} \tag{7.26}$$

Note that $v(S_0) > 0$ is equivalent to having

$$S_0 > K\left(\frac{1}{d}\sum_{j=1}^{d} e^{(r-\sigma^2/2)t_j+\sigma B(t_j)}\right)^{-1}.$$

Since (7.26) still involves a sum of lognormal random variables, no closed-form expression can be found for $\mathrm{E}(\partial v(S_0)/\partial S_0)$, and it must therefore be estimated by simulation. Table 7.7 gives numerical results comparing the performance of Monte Carlo and randomized quasi–Monte Carlo methods for this problem.

Estimating sensitivities using LR

The likelihood ratio method can be applied to estimate

$$\frac{\partial}{\partial\theta}\mathrm{E}(Y(\theta))$$

when θ is a parameter of the pdf of Y; that is, when $Y(\theta) = h(\mathbf{X})$ and $\mathrm{E}(Y(\theta))$ can be written as

$$\frac{\partial}{\partial\theta}\mathrm{E}(Y(\theta)) = \frac{\partial}{\partial\theta}\int_{\Omega} h(\mathbf{X})\varphi_\theta(\mathbf{x})d\mathbf{x},$$

where $\varphi_\theta(\mathbf{x})$ is the pdf of $\mathbf{X}$. Assuming that the order of the derivative and integral can be interchanged, we can then write

$$\begin{aligned}\frac{\partial}{\partial\theta}\mathrm{E}(Y(\theta)) &= \int_{\Omega} h(\mathbf{X})\frac{\partial}{\partial\theta}\varphi_\theta(\mathbf{x})d\mathbf{x}\\ &= \int_{\Omega} h(\mathbf{X})\frac{\partial\varphi_\theta(\mathbf{x})/\partial\theta}{\varphi_\theta(\mathbf{x})}\varphi_\theta(\mathbf{x})d\mathbf{x}\\ &= \mathrm{E}\left(h(\mathbf{X})\frac{\partial\varphi_\theta(\mathbf{x})/\partial\theta}{\varphi_\theta(\mathbf{x})}\right).\end{aligned}$$

The LR estimator is then defined as

$$\frac{1}{n}\sum_{i=1}^{n} h(\mathbf{x}_i)\frac{\partial\varphi_\theta(\mathbf{x}_i)/\partial\theta}{\varphi_\theta(\mathbf{x}_i)}.$$

Hence the LR estimator has the same form (up to a constant) as an importance sampling estimator, if we think of φ_θ as the "new measure" and $\partial\varphi_\theta/\partial\theta$ as the "original measure".

One advantage that the LR estimator has over the IPA estimator is that it can handle discontinuous payoff functions. However, in settings where both methods can be applied, the IPA estimator tends to provide estimators with much smaller variances [133, 150]. This can be seen in Table 7.7.

We illustrate the use of LR for the Asian call option discussed in Example 7.8.

Example 7.9. Recall that the goal here is to estimate

$$\frac{\partial}{\partial S_0}\mathrm{E}(v(S_0)),$$

where

$$v(S_0) = e^{-rT}\max\left(0, \frac{1}{d}\sum_{j=1}^{d} S_0 e^{(r-\sigma^2/2)t_j+\sigma B(t_j)} - K\right).$$

To apply LR, we need to rewrite the function $v(S_0)$ so that S_0 becomes a parameter of a pdf. This can be done as

$$v(S_0) = e^{-rT}\max\left(0, \frac{1}{d}\sum_{j=1}^{d} e^{X_1(S_0)+X_2+\ldots+X_j} - K\right),$$

where

$$\begin{aligned}X_1(S_0) &\sim N(\ln S_0 + (r-\sigma^2/2)t_1, \sigma^2\Delta_1),\\ X_j &\sim N((r-\sigma^2/2)\Delta_j, \sigma^2\Delta_j), j = 2,\ldots,d,\end{aligned}$$

$\Delta_j = t_j - t_{j-1}$, and thus the variables X_j are independent. Hence, we have

$$\varphi_{S_0}(\mathbf{x}) = \frac{\exp(-(x_1 - (\ln S_0 + r - \sigma^2/2)t_1)^2/2\sigma^2 t_1)}{\sqrt{2\pi\sigma^2 t_1}} \times \prod_{j=2}^{d} \frac{\exp(-(x_j - (r - \sigma^2/2)\Delta_j)^2/2\sigma^2\Delta_j)}{\sqrt{2\pi\sigma^2\Delta_j}},$$

and therefore

$$\frac{\partial}{\partial S(0)}\varphi_{S(0)}(\mathbf{x}) = \frac{(x_1 - (\ln S(0) + (r - \sigma^2/2)\Delta_1))}{S(0)\sigma^2\Delta_1} \times \varphi_{S(0)}(\mathbf{x}).$$

Hence the LR estimator has the form

$$\frac{e^{-rT}}{n}\sum_{i=1}^{n} \max\left(0, \frac{1}{d}\sum_{j=1}^{d} e^{x_{i,1}+\ldots+x_{i,j}} - K\right) \times \frac{(x_{i,1} - (\ln S_0 + (r - \sigma^2/2)\Delta_1))}{S(0)\sigma^2\Delta_1}.$$

Table 7.7 gives numerical results comparing the performance of Monte Carlo and a randomized Sobol' point set for this problem. We use the same parameters as in Table 7.3: $r = 0.05$, $\sigma = 0.3$, $T = 1$ year, $S(0) = 50$.

Table 7.7 Performance of Monte Carlo and quasi–Monte Carlo for IPA and LR estimators. Shown are the estimates $\hat{\mu}$ for delta and the half-width of the corresponding confidence interval (HW) based on $n = 1024$ and $m = 25$.

	$s = 32$				$s = 64$			
	$K = 45$		$K = 55$		$K = 45$		$K = 55$	
	$\hat{\mu}$	HW	$\hat{\mu}$	HW	$\hat{\mu}$	HW	$\hat{\mu}$	HW
IPA								
MC	0.771	5.88e−3	0.366	7.55e−3	0.773	4.96e−3	0.364	6.76e−3
Sobol'	0.775	2.83e−3	0.371	2.86e−3	0.775	2.95e−3	0.365	3.18e−3
LR								
MC	0.732	4.64e−2	0.358	2.45e−2	0.742	6.98e−2	0.361	3.82e−2
Sobol'	0.773	1.09e−2	0.366	1.14e−2	0.769	1.69e−2	0.366	1.73e−2

As expected, the IPA estimators generally have smaller variances than the LR estimators. The Sobol' estimators have a confidence interval with an half-width that is smaller than for Monte Carlo by factors ranging between about 2 and 4.

Estimating percentiles, including value-at-risk

In addition to security pricing and sensitivity estimates, another type of problem in finance is to study the tail behavior of large portfolios. More precisely, let $\Pi_t(\mathbf{S}(t))$ be the value at time t of a large portfolio containing different instruments with values depending on $\mathbf{S}(t)$, and let Δt be a certain period of time. We are interested in studying the loss random variable

$$L(\Delta t) := \Pi_t(\mathbf{S}(t)) - \Pi_{t+\Delta t}(\mathbf{S}(t+\Delta t))$$

of the portfolio over the interval of time $[t, t+\Delta t)$. Based on this loss variable, we give three possible *risk measures* for the portfolio. Note that these risk measures are typically computed under the actual probability measure rather than the risk-neutral probability measure.

(1) Fix a level $\alpha \in (0,1)$ and find the smallest value L_α such that

$$P(L(\Delta t) > L_\alpha) > \alpha.$$

The value L_α is the *value-at-risk* (VaR) of the portfolio at the level α.

(2) Fix a loss value L, and calculate the probability $p_L = P(L(\Delta t) > L)$. This measures the probability of losing more than L over an interval of length Δt.

(3) Fix a loss value L, and compute the *conditional tail expectation* (CTE) $\mathrm{E}(L(\Delta t)|L(\Delta t) > L)$.

The value-at-risk is a widely used risk measure. Its importance stems from the fact that government regulations in several countries require banks to estimate their value-at-risk on a daily basis [315, Chap. 1]. However, the value-at-risk has been criticized by certain authors, in particular because it fails to be a *coherent risk measure*, as defined in [10]. The conditional tail expectation — also called *TailVaR* — is an alternative to value-at-risk that fulfills the conditions for being a coherent risk measure [10, 478]. This risk measure has been studied by several researchers in actuarial science (see [46] and the references therein). In what follows, we focus on estimating the value-at-risk and the corresponding loss probability p_L. Clearly, the techniques used to perform these estimations could also be used for the TailVaR and other related risk measures.

Generally speaking, quantile estimation is technically more difficult than estimating a probability. In addition, if we can estimate p_L for several large values of L, then an estimate of the value-at-risk can be obtained. Hence we will first discuss the problem of estimating p_L and then talk about how the value-at-risk can be derived.

Estimating the loss probability p_L by simulation is hard not only because it deals with rare events but also because the loss random variable $L(\Delta t)$ is typically associated with portfolios that include a large number of options and derivatives that must all be simulated. It is thus crucial for such problems to find ways of improving the accuracy of the plain Monte Carlo estimator of the form

$$\hat{p}_L = \frac{1}{n}\sum_{i=1}^{n} \mathbf{1}_{L^i > L},$$

where the variables L^i form an i.i.d. sample of loss observations obtained as

$$L^i := \Pi_t(\mathbf{S}(t)) - \Pi_{t+\Delta t}(\mathbf{S}^i(t+\Delta t)), \qquad i = 1, \ldots, n,$$

where the values $\mathbf{S}^i(t+\Delta t)$ are distributed conditionally on $\mathbf{S}(t)$. Since we are interested in the tail of the loss distribution, it is natural to use importance sampling in order to improve the efficiency of the estimator for p_L. Although this idea has been studied by other authors, here we restrict our attention to the approach proposed by Glasserman, Heidelberger, and Shahabuddin [147, 148], which is also discussed in [150].

First, to identify the change of measure to be used with importance sampling, a *delta-gamma approximation* based on the assumption of a market consisting of normal factors can be used. That is, we make the assumption that the vector $\Delta\mathbf{S}(t) = \mathbf{S}(t + \Delta t) - \mathbf{S}(t)$ representing the change in the underlying assets over $[t, t + \Delta t)$ is normally distributed. The delta-gamma approximation then consists in writing

$$L(\Delta t) \approx D := -\frac{\partial \Pi(\mathbf{S}(t), t)}{\partial t} \times \Delta t - \Delta_\Pi^{\mathrm{T}} \times \Delta\mathbf{S}(t) - \frac{1}{2}(\Delta\mathbf{S}(t))^{\mathrm{T}} \Gamma \Delta\mathbf{S}(t), \quad (7.27)$$

where

$$\Delta_\Pi^{\mathrm{T}} = \left(\frac{\partial}{\partial S_1(t)}\Pi_t(\mathbf{S}(t)), \ldots, \frac{\partial}{\partial S_q(t)}\Pi_t(\mathbf{S}(t))\right)$$

is the vector of deltas and Γ is a matrix whose element in position (j, l) is given by the mixed partial derivative

$$\Gamma_{j,l} = \frac{\partial^2 \Pi(t)}{\partial S_j(t) \partial S_j(t)}, \qquad j, l = 1, \ldots, q.$$

By using an appropriate change of variables, the delta-gamma approximation D can be rewritten as a function of a vector $\mathbf{Z}$ of i.i.d. standard normals [150, p. 486]. The mechanism used to do that is quite similar to the one described in Sect. 2.6 for generating a vector of multinormal random variables. More precisely, assume $\Delta\mathbf{S}(t) \sim N(\mathbf{0}, \Sigma)$. Here, the mean of $\Delta\mathbf{S}(t)$ is taken to be zero because typically Δt is small (e.g., one week). Now let C be such that $CC^{\mathrm{T}} = \Sigma$. For example, C can be obtained by performing a Cholesky decomposition of the covariance matrix Σ. Hence, we can write $\Delta\mathbf{S}(t) = C\mathbf{Z}$, where $\mathbf{Z} \sim N(\mathbf{0}, I_q)$, and the delta-gamma approximation then becomes

$$D = -\frac{\partial \Pi(\mathbf{S}(t), t)}{\partial t} \times \Delta t - \Delta_\Pi^{\mathrm{T}} \times C\mathbf{Z} - \frac{1}{2}\mathbf{Z}^{\mathrm{T}} C^{\mathrm{T}} \Gamma C \mathbf{Z}.$$

By choosing C appropriately, we can make the matrix $(1/2)C^{\mathrm{T}}\Gamma C$ diagonal. For example, we can take $C = \tilde{C}U$, where $\tilde{C}$ is obtained by Cholesky decomposition of Σ and U is the matrix whose columns are formed by the eigenvectors of $(1/2)\tilde{C}^{\mathrm{T}}\Gamma\tilde{C}$. We then get

$$(1/2)C^{\mathrm{T}}\Gamma C = (1/2)U^{\mathrm{T}}\tilde{C}^{\mathrm{T}}\Gamma\tilde{C}U = (1/2)\Lambda,$$

where Λ is the diagonal matrix containing the eigenvalues $\lambda_1, \ldots, \lambda_q$ of $(1/2)\tilde{C}^{\mathrm{T}}\Gamma\tilde{C}$. With this choice of matrix C, we get the delta-gamma approximation

$$D = -\frac{\partial \Pi(\mathbf{S}(t), t)}{\partial t} \times \Delta t - \Delta_{\Pi}^{\mathrm{T}} \times C\mathbf{Z} - \frac{1}{2}\mathbf{Z}^{\mathrm{T}}\Lambda\mathbf{Z}. \tag{7.28}$$

This rewriting is helpful to derive a closed-form expression for the cumulant generating function $G(\theta)$ of D [150, p. 487]. More precisely, using the formulation (7.28), it can be shown that as long as $\max_j \theta\lambda_j < 1/2$, we have

$$G(\theta) = -\theta \times \frac{\partial}{\partial t}\Pi(\mathbf{S}(t), t) + \frac{1}{2}\sum_{j=1}^{q}\left(\frac{\theta^2 b_j^2}{1 - 2\theta\lambda_j} - \log(1 - 2\theta\lambda_j)\right), \tag{7.29}$$

where the vector $(b_1, \ldots, b_q)$ is given by $\Delta_{\Pi}^{\mathrm{T}} \times C$.

In turn, this can be used to choose an importance sampling measure by using the following key point: Applying importance sampling by performing an exponential twisting with parameter θ on the distribution of the delta-gamma approximation D is equivalent to modifying the mean and covariance matrix of the underlying multinormal vector $\mathbf{Z}$ in (7.28) according to θ [148]. More precisely, it corresponds to using $\mathbf{Z} \sim N(\boldsymbol{\mu}, \tilde{\Sigma})$, where

$$\mu_j = \frac{\theta b_j}{1 - 2\lambda_j\theta}, \qquad j = 1, \ldots, q,$$

and $\tilde{\Sigma}$ is a diagonal matrix whose jth element is given by $(1 - 2\lambda_j\theta)^{-1}$. Furthermore, the choice of the "twisting parameter" θ can be determined using the general technique outlined in Chap. 4 on p. 114 since, as we just saw, under the normality assumption on $\Delta\mathbf{S}(t)$, we have the closed-form expression (7.29) for the cumulant generating function of D. Interestingly, the parameter θ_L^* chosen in this way — that is, θ_L^* is the solution to $G'(\theta_L^*) = L$, where L is the value for which we want p_L — is such that, under the new twisted measure, we have

$$\mathrm{E}_{\theta_L^*}(L(\Delta t)) = L.$$

Hence, rather than being the $1 - p_L$ quantile of the loss distribution, L is now its expectation.

With this approach, the probability of loss is estimated by an estimator of the form

$$\hat{p}_{L,\mathrm{is}} = \frac{1}{n}\sum_{i=1}^{n} e^{-\theta_L^* \tilde{D}^i + G(\theta_L^*)} \mathbf{1}_{\tilde{L}^i > L}, \tag{7.30}$$

where the loss $\tilde{L}^i$ and its corresponding delta-gamma approximation $\tilde{D}^i$ are simulated under the new measure, for which $\mathbf{Z} \sim N(\boldsymbol{\mu}, \tilde{\Sigma})$. Figure 7.14 gives pseudocode for constructing the estimator (7.30), and Table 7.8 gives results comparing the performance of Monte Carlo and quasi–Monte Carlo for estimating p_L with or without importance sampling. The portfolio used is taken from [147] and consists of a short position in ten call options on ten different stocks, each with initial value 100, rate of return of 5%, volatility of 0.3, and strike price of 100. The correlation between each pair of stocks is 0.2. The options' expiration time is 0.5, and the time period Δt used is equal to 10 trading days, where it is assumed that there are 250 trading days per year.

As we can see, for the naive estimators (without importance sampling), the randomized quasi–Monte Carlo estimator based on the Sobol' sequence performs better than the Monte Carlo one, reducing the confidence interval's half-width by factors between 1.6 and 1.8. When importance sampling is applied, the improvement is marginal. One reason might be that the change of measure used is (approximately) optimal for the Monte Carlo estimator but not necessarily for the quasi–Monte Carlo one. The improvement obtained by using importance sampling is interesting, with reduction factors ranging between about 3 and 7. It works especially well for the case where the probability p_L of loss is smaller.

```
ProbLossIS(L, P_n)
obtain C̃ by Cholesky decomposition of Σ
find eigenvalues λ_1, ..., λ_q of (1/2)C̃^T Γ C̃
let U be formed with eigenvectors of (1/2)C̃^T Γ C̃
C ← C̃U
compute the vector of deltas Δ_Π
(b_1, ..., b_q) ← Δ_Π^T C
find the solution θ_L^* to G'(θ) = L
compute μ and Σ̃
for i ← 1 to n
    for j ← 1 to q
        Z_j^i ← μ_j + Σ̃_jj Φ^{-1}(u_ij)
    compute D̃^i based on Z^i as in (7.28)
    L̃^i ← Π_t(S(t)) − Π_{t+Δt}(S(t) + CZ^i)
return (1/n) Σ_{i=1}^n exp(−θ_L^* D̃_i + G(θ_L^*)) 1_{L̃^i > L}
```

Fig. 7.14 Pseudocode for estimating p_L with IS.

Similarly, the value-at-risk at the level α can be estimated by

Table 7.8 Probability of loss estimates based on $n = 1024$ and $m = 25$. Shown are the estimates $\hat{p}_L$ and the corresponding 95% confidence interval half-width.

	no IS				IS			
	$L = 30.9$		$L = 41.9$		$L = 30.9$		$L = 41.9$	
	$\hat{p}_L$	HW	$\hat{p}_L$	HW	$\hat{p}_L$	HW	$\hat{p}_L$	HW
MC	0.0496	1.40e−3	0.010	5.82e−4	0.0496	4.91e−4	0.0099	8.68e−5
Sobol'	0.0502	7.79e−4	0.050	3.58e−4	0.0500	3.58e−4	0.0100	8.18e−5

$$\hat{L}_{\alpha,\text{is}} = \hat{F}_{n,\text{is}}^{-1}(\alpha) = \inf\{L : \hat{F}_{n,\text{is}}(L) \geq \alpha\}, \tag{7.31}$$

where

$$\hat{F}_{n,\text{is}}(L) = 1 - \frac{1}{n}\sum_{i=1}^{n} e^{-\theta_L \tilde{D}^i + G(\theta_L^*)} \mathbf{1}_{\tilde{L}^i > L} \tag{7.32}$$

is an approximation for the CDF of L based on importance sampling. Note that in order to compute the value-at-risk estimate $\hat{L}_{\alpha,\text{is}}$, we need to determine $\hat{F}_{n,\text{is}}(L)$ for a range of values of L. As discussed in [150], the same change of measure can be used for these different values of L, so that the same samples $\{\tilde{L}^i, i = 1, \ldots, n\}$ and $\{\tilde{D}_i, i = 1, \ldots, n\}$ can be used for all values of L when constructing the empirical CDF (7.32).

Glynn [154] discusses other ways of constructing an empirical CDF based on importance sampling which in turn yield alternative ways of estimating quantiles through (7.31).

Problems

7.1. Show that, as stated on p. 249, if $S(t)$ follows the lognormal model, then $\mathrm{E}(S(t)) = S(0)e^{\mu t}$.

7.2. Prove the validity of the formula for C_0 given in Example 7.1.

7.3. Prove the validity of the Black-Scholes-Merton–like formula (7.6) for an Asian call option on the geometric average.

7.4. Consider a rainbow call option on the maximum of two assets that both have initial value $S(0) = 100$, $\sigma = 0.2$. Assume that the correlation between the two assets is 0.5, $r = 0.05$, $K = 90$, and $T = 1$. (a) What is the covariance matrix C in this case? (b) Find the Cholesky decomposition of C. (c) Construct a 95% confidence interval for $C_{m,0}$ based on $n = 1000$ runs.

7.5. Give an expression for $\mathrm{E}(S(t))$ for the variance gamma model discussed in Sect. 7.2.

7.6. Consider the mortgage-backed security problem discussed in Sect. 7.3. Implement the Brownian bridge technique for this problem. Compare the variance of the (randomly digitally shifted) Sobol' sequence with or without the Brownian bridge, using $n = 8192$ points and $m = 25$ repetitions, for both the "nearly linear" and "nonlinear" parameter sets.

7.7. Verify if the monotonicity conditions that are sufficient for antithetic variates to reduce the variance are satisfied for the Asian call option problem.

7.8. Apply antithetic variates to the Monte Carlo method for the "nearly linear" mortgage-backed security problem using n =10,000. Comment on the variance reduction obtained. Would you expect to get the same kind of reduction on the "nonlinear" problem?

7.9. Write a program that can compute the estimator $\hat{C}_{\mathrm{as},0}$ for the Asian call option. (a) Compute a 95% confidence interval for $C_{\mathrm{as},0}$ based on $n = 1000$ runs for $S(0) = 50$, $K = 45$, $r = 0.05$, $T = 1$, $s = 32$, and $\sigma = 0.3$. (b) Implement the control variate based on the call option on the geometric average. Compare the empirical variance (again based on $n = 1000$ runs) of the control variate estimator with that of the naive estimator using the same parameters as in (a). (c) In addition to the control variate described in (b), use also the terminal price $S(T)$ as a control variate, and compare the empirical variance (based on $n = 1000$ runs) with the estimators from (a) and (b).

7.10. Implement the moment-matching method described in Sect. 7.4.6 to estimate the Asian call option with the same parameters as in the previous problem.

7.11. Write a program that can compute the estimator for the plain put option. (a) Using the parameters $S(0) = 50$, $K = 55$, $r = 0.05$, $T = 1$, and $\sigma = 0.2$, construct a 95% confidence interval for the time-0 value P_0 of the put option. (b) Construct an importance sampling estimator by changing r to $r = 0.06$. (i) What is the likelihood ratio for your estimator. (ii) Construct a 95% confidence interval for P_0 using the importance sampling estimator, again with $n = 1000$. (c) Construct a 95% confidence interval for P_0 by instead estimating C_0 and using the *put-call parity*, which says that $C_0 + Ke^{-rT} = S(0) + P_0$. Compare the half-width with that of the naive and IS estimators.

7.12. Using the same example as in Table 7.7, estimate *delta* for an Asian option using finite differences and common random numbers with (i) $h = 0.01$ and (ii) $h = 0.0001$.

7.13. Derive expressions for both the IPA and LR estimators in the case of an Asian put option.

7.14. Apply the importance sampling approach described in Sect. 7.6 to estimate the conditional tail expectation using the same setup as in Table 7.8.

7.15. Determine the value-at-risk for $p = 0.01$ using the empirical CDF based on (i) naive Monte Carlo and (ii) the importance sampling estimator using the same setup as in Table 7.8.

Chapter 8
Beyond Numerical Integration

In this chapter, we discuss areas of application for quasi–Monte Carlo that go beyond numerical integration. Taking a step back, we recall that the general task discussed in this book is that of *sampling*. As mentioned before, we can think of numerical integration as using the produced sample average to approximate the true mean of the distribution of interest. But sampling can be used for many other tasks. For example, we briefly discussed percentile/quantile estimation in Chaps. 1 and 7.

Here we want to focus on a few important statistical approaches that rely on random sampling and see how to replace this by quasi-random sampling. The topics we discuss in this chapter are Markov chain Monte Carlo (MCMC), sequential Monte Carlo, and computer experiments. A common feature that the first two topics share is that there is some sort of dynamic updating process done on the simulated processes in order to produce a sample from a complicated distribution, in contrast with the fixed models we have assumed so far. On the other hand, computer experiments deal with problems where a very complicated function needs to be studied in order to better understand a given system. Typically, this function can be evaluated using a computer program, but the valuation is expensive and therefore needs to be done at well-chosen sample points. The task of choosing these points falls under the umbrella of *experimental design*. As a consequence, the idea of using sampling methods that are more uniform than random sampling has been studied extensively in this area. For instance, the method of Latin hypercube sampling, which we briefly described in Chap. 6, was introduced in the context of computer experiments in [313]. This *sampling* aspect of computer experiments offers a first connection with the quasi-random methods discussed in this text. More generally, the task of evaluating a function's integral, determining its most important variables, or constructing a good approximation for it are of interest in both fields, which is why we thought a brief discussion of computer experiments would fit well in this last chapter.

In order to better relate the topics discussed in the present chapter with the stochastic simulation setup used so far, we provide in Table 8.1 a simpli-

C. Lemieux, *Monte Carlo and Quasi–Monte Carlo Sampling*,
Springer Series in Statistics 692, DOI: 10.1007/978-0-387-78165-5_8,

fied description that outlines the similarities and differences between these different topics.

Table 8.1 Overview of the tasks discussed in the current chapter and how they relate to the ongoing topic of simulation.

	Goal: estimate properties of $h(\mathbf{X})$ by sampling		
Model:	Can draw from the distribution of $\mathbf{X}$, but not from $Y = h(\mathbf{X})$ directly.	Cannot draw directly from $\mathbf{X}$; h might be simple.	$\mathbf{X}$ not necessarily stochastic, but h is very complicated; need well-chosen points $\mathbf{X}$ where h will be evaluated to better understand $h(\mathbf{X})$.
Method:	stochastic simulation	MCMC and seq. MC	computer experiments

In our treatment of MCMC, we review two quasi–Monte Carlo versions of Metropolis-Hastings type algorithms that have been proposed recently. In the first case, successive draws from a quasi-random sample are used at each time step, while in the second one, quasi-random sampling is used at each time step to search the state-space for a good "proposal". We also discuss the exact (or perfect) sampling algorithm proposed by Propp and Wilson and its quasi-random versions presented in [70, 71, 287].

Sequential Monte Carlo algorithms can be described as sampling methods where on-line observations are used to update the sampling process. They combine ideas from MCMC algorithms and importance sampling to perform on-line Bayesian inference. Our coverage here will be to give a brief description of this family of methods and discuss how to replace their random sampling component by quasi-random sampling.

In our discussion of computer experiments, we first discuss a few proposals for experimental design that naturally lead back to some of the low-discrepancy point sets described in the previous chapters. We then revisit the problem of estimating the sensitivity indices of a function in the context of computer experiments. Our goal here is mostly to establish a few connections between these two fields — computer experiments and quasi–Monte Carlo integration — that should be useful to researchers working in one of these fields who are unfamiliar with the work done in the other field.

All the "quasi–Monte Carlo connections" discussed in this chapter are still at an early stage of study. Our treatment of these topics is meant to give the reader an overview of a few new and exciting possibilities for quasi–Monte Carlo sampling that go beyond the integration applications for which it has been mostly used in the past. Our coverage does not go too far either in depth or in breadth but will hopefully convince the reader of the wide range of problems for which quasi–Monte Carlo sampling can be useful.

8.1 Markov Chain Monte Carlo (MCMC)

In all the examples seen so far in this book, we have assumed that we were able to sample from the distributions of interest. For instance, in financial simulations, to generate Brownian motion paths we simply need to draw observations from the normal distribution, which can easily be done. However, there are several applications — especially those involving *Bayesian inference* — where it is not possible to directly sample from the distribution of interest. For such problems, the idea of MCMC is to cleverly choose a Markov chain whose stationary distribution corresponds to the distribution from which we want to sample. By running simulations of this Markov chain for long enough, one can then construct a sample that approximates the distribution of interest.

A very general way to construct such chains is via what is known as the *Metropolis-Hastings algorithm*, which involves the choice of a *proposal distribution* and then the use of an acceptance-rejection step to converge to the correct distribution. Details are given in the following section. Another popular method is Gibbs sampling, which we will not discuss here because it can be formulated as a special case of Metropolis-Hastings. We refer to [140] for more details and to [290] for a quasi-random Gibbs sampler.

As mentioned above, the underlying Markov chain needs to be run long enough to get a good approximation of the distribution of interest. Determining how long is "long enough" is not obvious, and to circumvent this problem, Propp and Wilson [380] have proposed a way of simulating Markov chains using a *coupling from the past* principle, which allows one to get a sample that has the *exact* distribution. This is what we discuss in Sec. 8.1.2.

Before going further, we introduce the notation that will be used in this section. First, we let $\pi(\mathbf{x})$ denote the distribution from which we want to sample, where $\mathbf{x} \in \mathbb{R}^d$. We let $\{\mathbf{X}_0, \mathbf{X}_1, \ldots\}$ be an ergodic Markov chain whose stationary distribution is given by $\pi(\mathbf{x})$.

A typical problem for which MCMC is useful is Bayesian inference, where one has observed data D, unknown parameters $\boldsymbol{\theta}$, and a model specified by a prior distribution $r(\boldsymbol{\theta})$ and a likelihood distribution $l(D|\boldsymbol{\theta})$. The goal is then to get information about the posterior distribution, which can be written as

$$\pi(\boldsymbol{\theta}|D) = \frac{r(\boldsymbol{\theta})l(D|\boldsymbol{\theta})}{\int r(\boldsymbol{\theta})l(D|\boldsymbol{\theta})d\boldsymbol{\theta}}, \tag{8.1}$$

using Bayes' Theorem. So, in this case, $\mathbf{x} = \boldsymbol{\theta}$ and $\pi(\mathbf{x}) = \pi(\boldsymbol{\theta}|D)$. Because of the integral in the denominator, it is typically impossible to get a closed-form expression for the posterior distribution $\pi(\boldsymbol{\theta}|D)$ given in (8.1). But if we can get a sample from that distribution (or at least from a distribution that approximates it reasonably well), then we can perform inference and get, for example, estimates for the expected value of the parameters $\boldsymbol{\theta}$ given D. MCMC is precisely the tool used to produce such samples.

The general way to use MCMC for inference is to use a *burn-in period* of length M, corresponding to observations $\{\mathbf{x}_t, t \leq M\}$, which we assume have not reached the stationary (or steady-state) distribution, and then approximate

$$\mu(h) = \mathrm{E}_\pi(h(\mathbf{X})) = \int h(\mathbf{x})\pi(\mathbf{x})d\mathbf{x}$$

by

$$\hat{\mu}(h) = \frac{1}{N}\sum_{t=M+1}^{M+N} h(\mathbf{x}_t),$$

where h is some integrable real-valued function defined over $\mathbb{R}^d$. The idea is that if M is large enough, then $\mathbf{X}_{M+1}, \mathbf{X}_{M+2}, \ldots, \mathbf{X}_{M+N}$ are dependent but they each (approximately) follow the stationary distribution $\pi(\cdot)$. Thus by the ergodic theorem, $\hat{\mu}(h)$ converges to $\mu(h)$ almost surely. To get an unbiased variance estimate, one possible approach is to use a *batch means* estimator, where the N observations are grouped into B batches of size N/B. We then form an approximately independent sample $\{Y_1, \ldots, Y_M\}$ by letting

$$Y_i = \frac{1}{N/B}\sum_{t=1}^{N/B} h(\mathbf{X}_{M+(i-1)N/B+t}), \qquad i = 1, \ldots, B.$$

Another possibility is to run a number n of chains $\mathbf{X}_{i,0}, \mathbf{X}_{i,1}, \ldots,$ for $i = 1, \ldots, n$, and then take

$$Y_i = \frac{1}{N}\sum_{t=1}^{N} h(\mathbf{X}_{i,M+t}).$$

Sometimes N is chosen equal to 1, so that no time averaging is done, and the quality of the estimation relies on having a large enough number n of chains. In what follows, unless otherwise stated, we take $M = 0$ (i.e., no burn-in period).

Before explaining the Metropolis-Hastings algorithm, we use a simple example taken from [70] to illustrate the use of MCMC.

Example 8.1. Consider a random walk over the integers $\{1, \ldots, K\}$ with semi-absorbent barriers. That is, here we have a Markov chain with transition probabilities

$$A_{ij} = P(X_t = i | X_{t-1} = j) = \begin{cases} p & \text{if } 1 \leq i = j+1 \leq K \\ 1-p & \text{if } 1 \leq i = j-1 \leq K \\ p & \text{if } i = j = K \\ 1-p & \text{if } i = j = 1 \\ 0 & \text{else.} \end{cases}$$

Stated differently, the transition matrix A for this chain is given by

$$\begin{bmatrix} 1-p & 1-p & 0 & \dots & 0 \\ p & 0 & 1-p & \ddots & \\ 0 & p & 0 & \ddots & \\ \vdots & \ddots & \ddots & & 1-p \\ 0 & & 0 & p & p \end{bmatrix}.$$

The stationary distribution of this Markov chain is described by

$$\pi(k) = c_0 \left(\frac{p}{1-p}\right)^k, \qquad k = 1, \dots, K, \tag{8.2}$$

where c_0 is a normalizing constant. (Problem 8.1 asks you to find its value.) Obviously, in this case one can sample very easily from π. For instance, using inversion, we can proceed as in Fig. 8.1, where $\Pi(i) = \sum_{i=1}^{I} \pi(i)$ and $\Pi(0) = 0$.

```
sample U ~ U(0,1)
return I such that Π(I − 1) ≤ U < Π(I + 1)
```

Fig. 8.1 Using inversion to generate observations from a random walk with semiabsorbent barriers.

As discussed in Chap. 2, the index I can be found by linear search or binary search. Using MCMC to generate samples from this distribution does not make sense in practice, but here is how it would work. First, we need to choose an initial state x_0 and a number N of steps for which we will be running the chain. The chain can then be generated using a random uniform vector $\mathbf{u} = (u_1, \dots, u_N)$ as input, as shown in Fig. 8.2.

In contrast with this artificial example, in typical applications we are often given a description of π that is not defined explicitly as the stationary distribution of a Markov chain, and thus we have to devise an appropriate Markov chain to be used in MCMC. A general way to do this is to use the Metropolis-Hastings algorithm, which we discuss next.

8.1.1 Metropolis-Hastings algorithm

This approach was proposed by Hastings [168] as a generalization to the Metropolis algorithm given in [319]. More recent descriptions can be found, for example, in the texts [140, 386]. It relies on the choice of a *proposal distribution* used to draw a candidate for the next observation of the chain given

```
MCMC(x_0, K, p; u_1, ..., u_N)
for t = 1 to N
    if u_t < p then
        if x_{t-1} < K then
            x_t ← x_{t-1} + 1
        else
            x_t ← x_{t-1}
    else
        if x_{t-1} > 1 then
            x_t ← x_{t-1} - 1
        else
            x_t ← x_{t-1}
```

Fig. 8.2 Simulating a random walk with semiabsorbent barriers over $\{1, \ldots, K\}$.

the current state. The candidate is then accepted with a certain probability. More precisely, let $q(\cdot|\mathbf{x})$ be the proposal distribution. For a candidate $\mathbf{Y}$ generated according to $q(\mathbf{Y}|\mathbf{X}_t)$, it is accepted with probability

$$\alpha(\mathbf{X}_t, \mathbf{Y}) = \min\left(1, \frac{\pi(\mathbf{Y})q(\mathbf{X}_t|\mathbf{Y})}{\pi(\mathbf{X}_t)q(\mathbf{Y}|\mathbf{X}_t)}\right).$$

Hence this *acceptance probability* requires being able to evaluate — at least up to a constant — the target distribution $\pi(\cdot)$.

We illustrate the idea with the following example.

Example 8.2. Consider the random walk with semiabsorbent barriers from Example 8.1. Assume that we are limited in the type of random walk that we can simulate and only have the choice of simulating a symmetric walk. The symmetric random walk can then be used as the proposal distribution within the Metropolis-Hastings algorithm. In that case, we have

$$\alpha(X_t, Y) = \begin{cases} \min\left(1, \frac{1/2}{p}\left(\frac{p}{1-p}\right)^{Y-X_t}\right) & \text{if } Y = X_t + 1 \text{ or } Y = X_t = K \\ \min\left(1, \frac{1/2}{1-p}\left(\frac{p}{1-p}\right)^{Y-X_t}\right) & \text{if } Y = X_t - 1 \text{ or } Y = X_t = 1. \end{cases}$$

Clearly, if $p = 1/2$, then the proposal distribution coincides with the true one, $\alpha(X_t, Y) = 1$, and a proposal is always accepted. If $p > 1/2$, then a "right move" — where $X_t = X_{t-1} + 1$ — is always accepted, while a left one is accepted with probability $1/2p < 1$. This makes sense since the chain used as a proposal is not making enough right moves compared with the true one. Conversely, if $p < 1/2$, then a left move is always accepted, while a right one is accepted only with probability $1/(2(1-p))$.

To prove that the Markov chain produced by the Metropolis-Hastings algorithm has the correct stationary distribution, we need to look at its transition probability, which satisfies

$$\begin{aligned}\phi(\mathbf{X}_{t+1}|\mathbf{X}_t) &= \phi(\mathbf{X}_{t+1}|\mathbf{X}_t, \text{accept})P(\text{accept}) + \phi(\mathbf{X}_{t+1}|\mathbf{X}_t, \text{reject})P(\text{reject}) \\ &= \begin{cases} q(\mathbf{X}_{t+1}|\mathbf{X}_t)\alpha(\mathbf{x}_t, \mathbf{X}_{t+1}) & \text{if } \mathbf{X}_{t+1} \neq \mathbf{X}_t \\ \left(1 - \int q(\mathbf{y}|\mathbf{X}_t)\alpha(\mathbf{X}_t, \mathbf{y})d\mathbf{y}\right) & \text{if } \mathbf{X}_{t+1} = \mathbf{X}_t. \end{cases} \end{aligned} \tag{8.3}$$

We also use the fact that, by definition of $\alpha(\mathbf{X}_t, \mathbf{X}_{t+1})$, if $\pi(\mathbf{Y})q(\mathbf{X}_t|\mathbf{Y}) < \pi(\mathbf{X}_t)q(\mathbf{Y}|\mathbf{X}_t)$, then

$$\alpha(\mathbf{X}_t, \mathbf{Y}) = \frac{\pi(\mathbf{Y})q(\mathbf{X}_t|\mathbf{Y})}{\pi(\mathbf{X}_t)q(\mathbf{Y}|\mathbf{X}_t)} \qquad \text{and} \ \ \alpha(\mathbf{Y}, \mathbf{X}_t) = 1.$$

Therefore, in this case,

$$\alpha(\mathbf{X}_t, \mathbf{Y})\pi(\mathbf{X}_t)q(\mathbf{Y}|\mathbf{X}_t) = \alpha(\mathbf{Y}, \mathbf{X}_t)\pi(\mathbf{Y})q(\mathbf{X}_t|\mathbf{Y}). \tag{8.4}$$

This equality also holds if $\pi(\mathbf{Y})q(\mathbf{X}_t|\mathbf{Y}) \geq \pi(\mathbf{X}_t)q(\mathbf{Y}|\mathbf{X}_t)$. Combining (8.3) and (8.4), we get the *detailed balance* equation/condition

$$\pi(\mathbf{X}_t)\phi(\mathbf{X}_{t+1}|\mathbf{X}_t) = \pi(\mathbf{X}_{t+1})\phi(\mathbf{X}_t|\mathbf{X}_{t+1}). \tag{8.5}$$

If we integrate on both sides of (8.5) with respect to $\mathbf{X}_t$, then we get

$$\int \pi(\mathbf{X}_t)\phi(\mathbf{X}_{t+1}|\mathbf{X}_t)d\mathbf{X}_t = \pi(\mathbf{X}_{t+1}).$$

This equation says that if $\mathbf{X}_t$ is distributed according to $\pi(\cdot)$, then the Markov chain used in the algorithm produces a state $\mathbf{X}_{t+1}$ at time $t+1$ that is also distributed according to $\pi(\cdot)$. Hence the stationary distribution of the chain produced by this algorithm is indeed π. In the case of a continuous distribution $\pi(\cdot)$, a bit more is needed to prove that the chain's distribution will actually converge to $\pi(\cdot)$. We refer the reader to [386, Sect. 7.3] for more information.

To describe the Metropolis-Hastings algorithm in more detail, we assume there is a function $g_{\mathbf{x}} : [0,1)^d \to \mathbb{R}^d$ such that if $\mathbf{u} \sim U([0,1)^d)$, then $\mathbf{y} = g_{\mathbf{x}}(\mathbf{u}) \sim q(\mathbf{y}|\mathbf{x})$. Using this notation, we give in Fig. 8.3 pseudocode describing how to produce N steps of the Metropolis-Hastings algorithm based on an input vector $\mathbf{u}$ of dimension $s = N(d+1)$.

A quasi–Monte Carlo version of this algorithm has been proposed by Owen and Tribble [368]. It uses the concept of a *completely uniformly distributed* sequence, which is reviewed in [288] but goes back to papers by Korobov in the early 1950s.

Definition 8.3. A sequence $u_1, u_2, \ldots \in [0,1]$ is *completely uniformly distributed* (CUD) if, for every integer $d \geq 1$, the points $\mathbf{u}_i = (u_i, \ldots, u_{i+d-1})$ satisfy

$$\lim_{n\to\infty} D^*(P_n) = 0,$$

```
MH(u_1, ..., u_{N(d+1)})
Initialize x_0
for t = 1 to N
    l = (t − 1)(d + 1)
    Y ← g_{x_{t−1}}(u_{l+1}, ..., u_{l+d})
    if u_{l+d+1} ≤ α(x_{t−1}, y) then
        x_t ← y
    else
        x_t ← x_{t−1}
```

Fig. 8.3 Pseudocode describing the Metropolis-Hastings algorithm.

where $P_n = \{\mathbf{u}_1, \ldots, \mathbf{u}_n\}$ and $D^*(\cdot)$ is the star discrepancy defined in Chap. 5.

Note the similarity between the construction P_n used in that definition and the recurrence-based point sets discussed in Chap. 5. In both cases, we construct a multidimensional point set by taking overlapping tuples of a sequence $u_1, u_2, \ldots$ of numbers in $[0, 1]$. The latter case can be viewed as a finite-n version of the above in the sense that we consider a sequence $u_1, u_2, \ldots$ that is periodic with period n and thus contains at most n different values. Hence it can only approximately satisfy the CUD definition. It should also be noted that if the points $\mathbf{u}_i$ are defined by using nonoverlapping (or partially overlapping) tuples, then the star discrepancy of these points still goes to 0 with n if the sequence $u_1, u_2, \ldots$ is CUD [368, Lemma 1].

The quasi–Monte Carlo Metropolis algorithm proposed by Owen and Tribble consists of using the first $N(d+1)$ elements of a CUD sequence in the Metropolis-Hastings algorithm described in Fig. 8.3. Owen and Tribble show that, for chains defined over a finite state-space Ω and under some additional conditions, we have

$$\hat{p}_n(\omega) := \frac{1}{n} \sum_{t=1}^{n} \mathbf{1}_{X_t = \omega} \to \pi(\omega) \text{ as } n \to \infty \text{ for each } \omega \in \Omega.$$

In other words, the observations $X_1, X_2, \ldots$ output by the algorithm are such that the corresponding empirical probability distribution $\hat{p}_n$ converges to the desired one.

In their numerical experiments, Owen and Tribble use approximate CUD sequences based on small LCGs, to which a random shift is added. They effectively use overlapping s-tuples to construct P_n, although they arrange the points in a different order. That is, for an LCG of maximal period of the form

$$x_i = a x_{i-1} \bmod n, \quad u_i = x_i / n \qquad i \geq 1,$$

and in the case $s = d + 1 = 2$, they form the sequence

$$0, 0, u_1, u_2, \ldots, u_{n-1}, u_2, u_3, \ldots, u_{n-1}, u_n, 0, 0, \ldots,$$

add to it (modulo 1) the sequence

$$v_1, v_2, v_1, v_2, v_1, \ldots,$$

where v_1, v_2 are i.i.d. $U(0,1)$, and then take the n nonoverlapping pairs of this sequence of period $2n$. This amounts to using the n points of a randomly shifted Korobov lattice point set based on the generator a but in an order different from the one given by

$$\mathbf{u}_i = \left(\frac{i-1}{n}(1, a) + (v_1, v_2)\right) \bmod 1, i = 1, \ldots, n,$$

and different from the order induced by the LCG within the recurrence-based point set definition, which is given by

$$\mathbf{u}_i = \left(\frac{1}{n}(a^{i-1} \bmod n, a^i \bmod n) + (v_1, v_2)\right) \bmod 1. \tag{8.6}$$

Based on the definition (8.6), they instead use the sequence of points

$$(0, 0), \mathbf{u}_1, \mathbf{u}_3, \ldots, \mathbf{u}_{n-2}, \mathbf{u}_2, \mathbf{u}_4, \ldots, \mathbf{u}_{n-1}.$$

A second quasi–Monte Carlo adaptation of the Metropolis-Hastings algorithm has been proposed in [69]. There, the low-discrepancy sampling is applied in a very different way. It is used to replace the local independent sampling performed within the *multiple-try Metropolis algorithm* proposed in [295]. This algorithm consists in replacing the single trial $\mathbf{Y}_{t+1}$ done at each time step in the Metropolis algorithm by a set of r independent trials $\{\mathbf{Y}_{t+1,1}, \ldots, \mathbf{Y}_{t+1,r}\}$. One of these trials $\mathbf{y}$ is then selected with a probability proportional to its associated *weight function*, given by

$$w(\mathbf{y}, \mathbf{x}_t) = \pi(\mathbf{y}) q(\mathbf{x}_t|\mathbf{y}) \lambda(\mathbf{x}_t, \mathbf{y}),$$

where $\lambda(\cdot, \cdot)$ is a symmetric function to be chosen. The selected proposal $\mathbf{y}$ is accepted with a certain probability p, which must be determined so that the detailed balance condition is preserved. To do so, it is necessary to augment the current state $\mathbf{X}_t = \mathbf{x}$ with a set of $r-1$ states whose distribution depends on $\mathbf{y}$. More precisely, once $\mathbf{y}$ is chosen, we must draw $\mathbf{x}_1^*, \ldots, \mathbf{x}_{r-1}^*$ according to $q(\cdot|\mathbf{y})$, let $\mathbf{x}_r^* = \mathbf{x}$, and define the (generalized) acceptance probability to be

$$p = \min\left\{1, \frac{w(\mathbf{y}_{t+1,1}, \mathbf{x}) + \ldots + w(\mathbf{y}_{t+1,r}, \mathbf{x})}{w(\mathbf{x}_1^*, \mathbf{y}) + \ldots + w(\mathbf{x}_r^*, \mathbf{y})}\right\}.$$

The idea explored in [69] is to replace at each time step the independent sampling used to generate the trials $\{\mathbf{Y}_{t+1,1}, \ldots, \mathbf{Y}_{t+1,r}\}$ by correlated sampling based on some conditional joint density function $\tilde{q}(\mathbf{y}_1, \ldots, \mathbf{y}_r|\mathbf{x})$ whose marginals are precisely given by $q(\mathbf{y}|\mathbf{x})$. One way of getting this conditional joint density function is to choose a randomized low-discrepancy point set $P_r = \{\tilde{\mathbf{u}}_1, \ldots, \tilde{\mathbf{u}}_r\}$ of size r and then generate the sample of trials using

$$\mathbf{y}_1 = g_{\mathbf{x}}(\tilde{\mathbf{u}}_1), \ldots, \mathbf{y}_r = g_{\mathbf{x}}(\tilde{\mathbf{u}}_r).$$

That is, the structure of the point set P_r is used to induce correlation among the trials. The decision to accept or reject $\mathbf{y}$ is still based on a randomly and uniformly drawn number U. It is shown in [69] that the augmented sample $\mathbf{x}_1^*, \ldots, \mathbf{x}_{r-1}^*$ for the current state $\mathbf{x}$ must be generated according to the conditional density $\tilde{q}((\mathbf{x}_1, \ldots, \mathbf{x}_{r-1}|\mathbf{y})|\mathbf{x}_r)$ when this type of correlated sampling is used, so that the detailed balance condition is preserved. When P_r is constructed by taking a deterministic point set $\{\mathbf{u}_1, \ldots, \mathbf{u}_r\}$ and adding a shift $\mathbf{v}$ — by addition modulo 1 or digitally — then it suffices to determine the vector $\mathbf{w} \in [0,1)^s$ such that $\mathbf{x} = g_{\mathbf{y}}(\mathbf{w})$, and then let

$$\mathbf{x}_i = g_{\mathbf{y}}(\mathbf{u}_{i+1} \oplus \mathbf{w}), i = 1, \ldots, r-1,$$

where we assume $\mathbf{u}_1 = \mathbf{0}$, and $\oplus$ corresponds to the operation used to randomize the point set. Hence this form of correlated sampling, based on a low-discrepancy point set, lends itself quite well to this adaptation of the multiple-try Metropolis algorithm.

8.1.2 Exact sampling

As we mentioned before, with MCMC, one needs to simulate the chosen Markov chain for a sufficiently large number of steps in order to get samples that are close enough to the desired distribution $\pi(\cdot)$. Although tests can be done to determine whether we have run the chain for a large enough number of steps (see, for example, [140, Chaps. 3 and 7]), there is something a bit unsatisfying about the fact that this approach does not produce samples that have exactly the desired distribution.

As an alternative to this type of sampling (sometimes called *forward sampling*), Propp and Wilson introduced in 1996 a method called *exact sampling* (also called perfect sampling), which removes the problem of determining for how many steps the chain should be run and produces samples with the correct distribution $\pi(\cdot)$. The idea is to simulate several chains in parallel and use *coupling from the past*. That is, the chains are simulated from some time $-t$ until time 0, with t increased until we go back far enough in time to observe a single common state for all chains at time 0.

To describe this idea in more detail, we assume for now that the Markov chain to be simulated has a finite state-space $\Omega = \{\omega_1, \ldots, \omega_K\}$, and a transition probability $q(\mathbf{y}|\mathbf{x})$, for $\mathbf{x}, \mathbf{y} \in \Omega$. As before, we assume that there exists a function $g_{\mathbf{x}} : [0,1)^d \rightarrow \Omega$ such that if $\mathbf{u} \sim U([0,1)^d)$, then $g_{\mathbf{x}}(\mathbf{u})$ is distributed according to $q(\cdot|\mathbf{x})$. The algorithm as described in [380] also makes use of maps defined as follows. Assume we start K chains at time $t \leq 0$, with one chain starting in each of the K states of Ω. Then we get K paths from time t to time 0 and let the lth path be denoted $\mathbf{X}_t^l, \mathbf{X}_{t+1}^l, \ldots, \mathbf{X}_0^l$. For $t \leq v \leq 0$, define the map $F_t^v : \Omega \rightarrow \Omega$ so that $F_t^v(\omega_l) = \mathbf{X}_v^l$. That is, F_t^v takes as input the initial position of a path at time t and outputs its position at time v. The notation f_t is used to denote the one-step map F_t^{t+1} that determines what happens at time t. This notation is convenient to explain the idea of the coupling from the past approach, which amounts to decreasing t until the map F_t^0 becomes a constant map. Figure 8.4 describes in detail the approach of Propp and Wilson, as explained in [380].

```
ExactSim(u_1, u_2, ...)
t = 0
F_t^0 ← I_K (the identity map over {ω_1, ..., ω_K})
repeat
    t ← t − 1
    for l = 1 to K
        f_t(ω_l) ← g_{ω_l}(u_{−(t+1)d+1}, ..., u_{−(t+1)d+d})
    F_t^0 ← F_{t+1}^0 ∘ f_t
until F_t^0 is constant
return x ← F_t^0(ω_1)
```

Fig. 8.4 Exact sampling algorithm proposed by Propp and Wilson.

Because ultimately our goal is to see how quasi–Monte Carlo sampling can be used within this algorithm, it is important to understand how randomness is used here. As Fig. 8.4 shows, the same random input $\mathbf{u} = (u_1, u_2, \ldots)$ is used for all K chains, and also the same d-dimensional portion,

$$(u_{-(t+1)d+1}, \ldots, u_{-(t+1)d+d}),$$

of that point $\mathbf{u}$ is reused at time t *every time* we go through the repeat loop. The last thing to point out is that since the value of t that will cause all chains to coalesce by time 0 is unknown and unbounded a priori, the total number $d \times (-t)$ of uniform numbers required to perform the algorithm above is random. Hence, exact sampling requires constructions that can handle an unbounded dimension.

The idea of using correlated sampling within the algorithm of Propp and Wilson was first studied in [70, 71]. A quasi–Monte Carlo version of this algorithm was proposed in [287], and further improvements based on the array-RQMC method discussed in Chap. 6 were proposed in [270].

The quasi–Monte Carlo exact sampling proposed in [287] is implemented by first choosing a randomized low-discrepancy point set P_n suitable for dealing with infinite dimensions. Then, each point $\mathbf{u}_i \in P_n$ is used as the input to the algorithm `ExactSim()` described in Fig. 8.4. We thus obtained a sample $\mathbf{x}_1, \ldots, \mathbf{x}_n$, where each $\mathbf{x}_i$ has the desired distribution $\pi(\cdot)$. The whole process can then be repeated using independent randomizations. Numerical experiments reported in [287] show that the samples thus obtained produce approximations

$$\frac{1}{n}\sum_{i=1}^{n} h(\mathbf{x}_i)$$

for $E_\pi(h(\mathbf{X}))$ having less variance than random exact sampling for simple functions h. Examples with continuous state-spaces where exact sampling is applied to Metropolis-Hastings algorithms as in [63] are also given. In this case, the quasi–Monte Carlo versions reduce the variance by factors up to 30 compared with Monte Carlo.

8.2 Sequential Monte Carlo

A very good introduction to sequential Monte Carlo can be found in [87]. Our treatment and notation follow this reference. Sequential Monte Carlo can be used to perform Bayesian inference when the data are accumulated sequentially rather than being given a priori. Hence, inference is performed on-line, with posterior distributions being updated sequentially.

More precisely, here we assume we have an unobserved Markov process $\{\mathbf{X}_t, t = 0, 1, \ldots\}$ with $\mathbf{X}_t \in \Omega$, initial distribution $p_0(\mathbf{X}_0)$, and transition function $q(\mathbf{X}_t|\mathbf{X}_{t-1})$. We also have an observation process $\{\mathbf{Y}_t, t = 1, 2, \ldots\}$ with $\mathbf{Y}_t \in \mathcal{Y}$, where the observations $\mathbf{Y}_1, \ldots, \mathbf{Y}_t$ are conditionally independent given the states $\mathbf{X}_1, \ldots, \mathbf{X}_t$. In addition, we assume a model for $\mathbf{Y}_t$ given $\mathbf{X}_t$ described by a density function $r(\mathbf{y}_t|\mathbf{x}_t)$.

We use the notation $\mathbf{x}_{0:t}$ and $\mathbf{y}_{1:t}$ to denote the sequences $\{\mathbf{x}_0, \ldots, \mathbf{x}_t\}$ and $\{\mathbf{y}_1, \ldots, \mathbf{y}_t\}$, respectively.

The goal is to estimate the posterior distribution $\pi(\mathbf{x}_{0:t}|\mathbf{y}_{1:t})$, expectation of various quantities under that distribution, and also the marginal distribution $p_t(\mathbf{x}_t|\mathbf{y}_{1:t})$ at time t, which is called the *filtering distribution.*

If the model is such that

$$\begin{aligned}
\mathbf{X}_t &= A\mathbf{X}_{t-1} + G\mathbf{a}_t, \\
\mathbf{Y}_t &= H\mathbf{X}_t + \mathbf{b}_t,
\end{aligned}$$

where A, G, and H are matrices and the $\mathbf{a}_t, \mathbf{b}_t$ are independent standard multinormal, then one can use the *Kalman filter* [167, 210] to obtain the exact updated mean and covariance of the posterior distribution. Other types of models also admit analytical solutions, for example when together $(\mathbf{X}_t, \mathbf{Y}_t)$ model a partially observed Markov chain, in which case one can use the *hidden Markov model filter.*

Typically, more complex models are used to represent practical applications, and in such cases it is not possible to obtain analytical expressions for the posterior distribution of interest. Sequential Monte Carlo is meant to be used in such cases. It is based on the idea of generating a set of weighted *particles* $\{(w_i, \mathbf{x}_{i,0:t}), i = 1, \ldots, n\}$, where the weights w_i add up to 1. The information obtained by observing $\mathbf{y}_1, \mathbf{y}_2, \ldots$ is then incorporated sequentially to update the simulation model. The weights are chosen so that the estimator

$$\sum_{i=1}^{n} w_i h(\mathbf{x}_{i,0:t}) \tag{8.7}$$

can be used to approximate expectations of the form

$$\mathrm{E}_\pi(h(\mathbf{x}_{0:t})), \tag{8.8}$$

where $h(\cdot)$ is some integrable function. The purpose of the weights is that in most cases the particles that are generated do not have the correct distribution $\pi(\mathbf{x}_{0:t}|\mathbf{y}_{1:t})$. In such cases, properly chosen weights can be used to produce unbiased (or at least consistent) estimators for (8.8). This is similar to the approach used in importance sampling, with the likelihood ratio acting as a weight in the setting above.

In sequential Monte Carlo methods, most of the time the correct weights — the ones that would make sure (8.7) is an unbiased estimator of (8.8) — are usually known only up to a constant. If we denote them by $\tilde{w}_i$, then the correct (normalized) weights are given by

$$w_i = \frac{\tilde{w}_i}{\sum_{i=1}^{n} \tilde{w}_i}.$$

This is similar to the weighted importance sampling approach discussed in Chap. 4. The following definition, taken from [294], describes a property that such weights should have.

Definition 8.4 ([294]). A set of random draws and weights $\{(w_i, \mathbf{x}_i), i = 1, 2 \ldots\}$ is said to be *properly weighted with respect to the distribution* π if, for any integrable function h, we have

$$\lim_{n \to \infty} \frac{\sum_{i=1}^{n} w_i h(\mathbf{x}_i)}{\sum_{i=1}^{n} w_i} = \mathrm{E}_\pi(h(\mathbf{X})).$$

We now turn to the sequential nature of the algorithms under study in this section. As a first step, we apply Bayes' Theorem and write

$$\pi(\mathbf{x}_{0:t}|\mathbf{y}_{1:t}) = \frac{p(\mathbf{y}_{1:t}|\mathbf{x}_{0:t})p(\mathbf{x}_{0:t})}{\int p(\mathbf{y}_{1:t}|\mathbf{x}_{0:t})p(\mathbf{x}_{0:t})d\mathbf{x}_{0:t}}, \tag{8.9}$$

where we use the same notation $p(\cdot)$ to denote various densities, the arguments inside the parentheses specifying which variables we are considering. Note that the denominator in (8.9) is equal to $p(\mathbf{y}_{1:t})$, which typically cannot be computed in closed form.

We can then derive

$$\begin{aligned}\pi(\mathbf{x}_{0:t+1}|\mathbf{y}_{1:t+1}) &= \frac{p(\mathbf{y}_{1:t+1}|\mathbf{x}_{0:t+1})p(\mathbf{x}_{0:t+1})}{\int p(\mathbf{y}_{1:t+1}|\mathbf{x}_{0:t+1})p(\mathbf{x}_{0:t+1})d\mathbf{x}_{0:t+1}} \\ &= \frac{p(\mathbf{y}_{1:t}|\mathbf{x}_{0:t})r(\mathbf{y}_{t+1}|\mathbf{x}_{t+1})p(\mathbf{x}_{0:t})q(\mathbf{x}_{t+1}|\mathbf{x}_t)}{p(\mathbf{y}_{1:t+1})} \\ &= \frac{\pi(\mathbf{x}_{0:t}|\mathbf{y}_{1:t})r(\mathbf{y}_{t+1}|\mathbf{x}_{t+1})q(\mathbf{x}_{t+1}|\mathbf{x}_t)}{p(\mathbf{y}_{1:t+1}|\mathbf{y}_{1:t})},\end{aligned} \tag{8.10}$$

where for the second equality we used the fact that the observations $\mathbf{y}_t$ are conditionally independent given the states $\mathbf{x}_t$ and for the third equality we used the fact that

$$p(\mathbf{y}_{1:t+1}) = p(\mathbf{y}_{1:t+1}|\mathbf{y}_{1:t})p(\mathbf{y}_{1:t}).$$

Similarly, the filtering distribution can be written recursively as

$$p(\mathbf{x}_t|\mathbf{y}_{1:t}) = \frac{r(\mathbf{y}_t|\mathbf{x}_t)p(\mathbf{x}_t|\mathbf{y}_{1:t-1})}{\int r(\mathbf{y}_t|\mathbf{x}_t)p(\mathbf{x}_t|\mathbf{y}_{1:t-1})d\mathbf{x}_t}. \tag{8.11}$$

Now, with sequential Monte Carlo, the idea is to choose a proposal sampling function $\tilde{q}(\mathbf{x}_t|\mathbf{x}_{t-1}, \mathbf{y}_{1:t})$ from which, at each time t, we generate the next state $\mathbf{x}_{i,t}$ for path i, given $\mathbf{x}_{i,t-1}$ and $\mathbf{y}_{1:t}$. Ideally, one should use $\tilde{q}(\mathbf{x}_t|\mathbf{x}_{t-1}, \mathbf{y}_{1:t}) = \pi(\mathbf{x}_t|\mathbf{y}_t)$, but this is usually impossible. A common choice is to choose the transition function $q(\mathbf{x}_t|\mathbf{x}_{t-1})$, which means that conditioned on the state at time $t-1$, the paths are generated independently from the observation process.

Once the paths $\mathbf{x}_{i,0:t-1}$ are augmented with the next state $\mathbf{x}_{i,t}$ via the proposal $\tilde{q}(\cdot)$, the weights w_i must be adjusted so that the sample paths $\mathbf{x}_{i,0:t}$ are still properly weighted. To determine how this can be done, we see from (8.10) that we should use the recursive weight update

$$\tilde{w}_i = w_i \frac{r(\mathbf{y}_t|\mathbf{x}_{i,t})q(\mathbf{x}_{i,t}|\mathbf{x}_{i,t-1})}{\tilde{q}(\mathbf{x}_{i,t}|\mathbf{x}_{i,0:t-1}, \mathbf{y}_{1:t})}, \quad i = 1, \ldots, n. \tag{8.12}$$

Note that if the transition function $q(\mathbf{x}_t|\mathbf{x}_{t-1})$ is chosen as the proposal function $\tilde{q}$, then the preceding update becomes

$$\tilde{w}_i = w_i \times r(\mathbf{y}_t|\mathbf{x}_{i,t}), \qquad i = 1, \ldots, n.$$

The sequential Monte Carlo method based on this idea is called *sequential importance sampling* (SIS) and is described in Fig. 8.5. As usual, we assume that there exists a function $g(\mathbf{u}; \mathbf{x}_t, \mathbf{y}_{1:t})$ such that if $\mathbf{u} \sim U([0,1)^d)$, then $g(\mathbf{u}; \mathbf{x}_t, \mathbf{y}_{1:t})$ is distributed according to the proposal $\tilde{q}(\cdot|\mathbf{x}_t, \mathbf{y}_{1:t})$. We also assume that all paths are initialized to a common starting point $\mathbf{x}_0$. Thus, an $s = nd$-dimensional point set is required in order to run this algorithm.

```
SeqIS(u_1, ..., u_n)
for i = 1 to n
    x_{i,0} ← x_0
    w_i ← 1/n
for t = 1 to T
    get y_t
    W ← 0
    for i = 1 to n
        x_{i,t} ← g(u_{i,(t-1)d+1}, ..., u_{i,td}; x_{i,t-1}, y_{1:t})
        augment x_{i,0:t-1} with x_{i,t}
        w_i ← w_i × p(y_t|x_{i,t}) q(x_{i,t}|x_{i,t-1}) / q̃(x_{i,t}|x_{i,t-1}, y_{1:t})
        W ← W + w_i
    for i = 1 to n
        w_i ← w_i/W
    // the weighted sample {(w_i, x_{i,0:t}), i = 1, ..., n} can
    // then be used to evaluate, e.g., E_π(g(X_{0:t}))
```

Fig. 8.5 Pseudocode describing the sequential importance sampling approach.

One problem with this approach is that, as t increases, most of the weights tend to get quite small, and then only a small number of paths account for most of the weights. This problem is sometimes referred to as having a small *effective sample size*, which is defined as [88]

$$n^* = \frac{n}{1 + \mathrm{Var}_\pi(w_i)}$$

and estimated as

$$\hat{n}^* = \frac{1}{\sum_{i=1}^n w_i^2}.$$

If we look at the two extreme cases, we see that when the weights w_i are all equal to $1/n$, then $\hat{n}^* = n$, but if one weight is equal to 1 and all the other ones are zero, then $\hat{n}^* = 1$.

One way of circumventing this degeneracy problem is to use a method called the *bootstrap filter*. The idea here is to use resampling — as done in the bootstrap method — at each time step to modify the sample so that only the most likely paths are kept. Paths are resampled according to their associated weights w_i. Figure 8.6 gives the details.

```
BootStrapFilter(x_0; u_1, ..., u_n)
for i = 1 to n
    x_{i,0} ← x_0
    w_i ← 1/n
for t = 1 to T
    W ← 0
    for i = 1 to n
        l ← (t − 1)d + 1
        x̃_{i,t} ← g(u_{i,l}, ..., u_{i,l+d−1}; x_{i,t−1}, y_{1:t})
        x̃_{i,0:t} ← [x_{i,0:t−1}; x̃_{i,t}]
        w_i ← w_i × r(y_t|x_{i,t})q(x_{i,t}|x_{i,t−1})/q̃(x_{i,t}|x_{i,t−1}, y_{1:t})
        W ← W + w_i
    W_0 ← 0
    for i = 1 to n
        w_i ← w_i/W
        W_i ← W_{i−1} + w_i
    // resampling
    for i = 1 to n
        find I such that W_{I−1} < u_{i,dT+t} ≤ W_I
        x_{i,0:t} ← x̃_{I,0:t}
```

Fig. 8.6 Bootstrap filter approach.

Several other variants and generalizations of these approaches have been proposed in the literature. We refer the reader to [87, 294, 386] for more information on this. We will not go much further on these variants, though, since our goal here is just to explain how to replace the "Monte Carlo" part of sequential Monte Carlo by quasi–Monte Carlo.

Our description of both the SIS algorithm and the bootstrap filter explicitly shows how uniform numbers are used to generate the paths. From that point of view, it should be clear how one can apply (randomized) quasi–Monte Carlo instead of Monte Carlo. However, there are some subtle issues arising with the bootstrap filter if quasi–Monte Carlo is used. First, performing the resampling step requires some random numbers. More precisely, in the standard approach described in Fig. 8.6, we need at each time step n uniform numbers in order to perform the resampling step based on the multinomial distribution with parameters $(n, w_1, \ldots, w_n)$. In that figure, we chose to use the last T coordinates of the point set for this purpose. In addition, it is important to realize that the resampling step implies that there is not a one-to-one mapping between points in $P_n = \{(u_{i,1}, \ldots, u_{i,s}),\ i = 1, \ldots, n\}$ and paths $\{\mathbf{x}_{i,0:T}, i = 1, \ldots, n\}$. This is because at time T, due to the resampling mechanism, paths issued from a common "ancestor" will share common initial portions issued from a given point i, while the initial portion of some points in P_n will disappear if they were used to generate a particle that eventually was eliminated. Example 8.5 illustrates this issue.

Example 8.5. Suppose $n = 3$, $d = 2$, and $T = 3$, and that the outcome of the resampling steps done at times 1 and 2 are 1, 1, 2 and 2, 2, 3. Then the three paths obtained and the corresponding coordinates used to generate them are

paths	coordinates
$\mathbf{x}_{1,1}, \mathbf{x}_{2,2}, \mathbf{x}_{1,3}$	$u_{1,1}, u_{1,2}, u_{2,3}, u_{2,4}, u_{1,5}, u_{1,6}$
$\mathbf{x}_{1,1}, \mathbf{x}_{2,2}, \mathbf{x}_{2,3}$	$u_{1,1}, u_{1,2}, u_{2,3}, u_{2,4}, u_{2,5}, u_{2,6}$
$\mathbf{x}_{2,1}, \mathbf{x}_{3,2}, \mathbf{x}_{3,3}$	$u_{2,1}, u_{2,2}, u_{3,3}, u_{3,4}, u_{3,5}, u_{3,6}$.

Related to this, another observation is that there is no obvious way to decide how the points $\{(u_{i,(t-1)(d+1)+1}, \ldots, u_{i,dt}), i = 1, \ldots, n\}$ should be assigned to the newly resampled set of particles $\{\mathbf{x}_{i,0:t-1}, i = 1, \ldots, n\}$ in order to generate the next states conditioned on $\mathbf{x}_{i,t-1}$. Equivalently, one must decide how the paths should be ordered after the resampling step is performed. In [354], the above-mentioned assignment is done at random (i.e., using a random permutation). Also, the uniform numbers required for the resampling step are simply taken as i.i.d. uniform numbers and are thus independent from the ones used to generate the states $\mathbf{x}_{i,t}$.

Since the point set used in [354] is a randomly shifted Korobov lattice and is therefore dimension-stationary, this corresponds to using the Latin supercube sampling method discussed in Chap. 6, with T blocks of size d based on T copies of a d-dimensional (randomly shifted) Korobov point set and then a block of size $T-1$ based on Monte Carlo sampling. That is, the underlying point set used in the code described in Fig. 8.6 has its ith point given by

$$\tilde{\mathbf{u}}_i = (\tilde{u}^1_{i,1}, \ldots, \tilde{u}^1_{i,d}, \tilde{u}^2_{\pi_1[i],1}, \ldots, \tilde{u}^2_{\pi_1[i],d}, \tilde{u}^3_{\pi_2[i],1}, \ldots, \tilde{u}^3_{\pi_2[i],d}, \ldots,$$
$$\tilde{u}^T_{i,\pi_{T-1}[i],1}, \ldots, \tilde{u}^T_{\pi_{T-1}[i],d}, w_{i,Td+1}, \ldots, w_{i,dT+T-1}),$$

where

$$\tilde{u}^l_{i,j} = (u_{i,j} + v^l_j) \bmod 1, \qquad i = 1, \ldots, n, j = 1, \ldots, d, l = 1, \ldots, T,$$

and the numbers v^l_j are i.i.d. $U(0,1)$, $\mathbf{u}_i$ is the ith point of the d-dimensional Korobov point set, and the numbers $w_{i,j}$ used for the resampling step are i.i.d. $U(0,1)$. Clearly, one could also use the $(d+1)$th coordinates of a $(d+1)$-dimensional Korobov point set to perform the resampling step.

An interesting idea would be to try using array-RQMC for sequential quasi–Monte Carlo. That is, one could choose a $(d+1)$-dimensional low-discrepancy point set P_n and a way to order the states $\mathbf{x}_{i,t}$. Then, at time t, the order induced by $\{\mathbf{x}_{i,t-1}, i = 1, \ldots, n\}$ can be used to assign the points $\mathbf{u}_i$ of a randomized version of P_n — independent from the one used at other time steps — to the resampling step and the generation of the next state $\mathbf{x}_{i,t}$.

Although the resampling step avoids the degeneracy problem that can occur in sequential importance sampling, it has some disadvantages, too. The main problem is that this step introduces additional variability in the simulated paths. A possible remedy is to perform *residual resampling* [294], whereby instead of performing a completely random resampling step, each path i is chosen deterministically $m_i = \lfloor nw_i \rfloor$ times, and the remaining $n - (m_1 + \ldots + m_n)$ draws are done at random, based on the adjusted weights proportional to

$$nw_i - m_i, \qquad i = 1, \ldots, n.$$

Note that in this case the resampling step only requires $n - (m_1 + \ldots + m_n)$ uniform numbers. Other authors have even suggested ways of doing the resampling step that only require one uniform number [386, p. 555], a method called *systematic resampling*. In fact, here the n uniform numbers required to perform the resampling step are chosen to be

$$u_i = \frac{i-1}{n} + v, \qquad i = 1, \ldots, n,$$

where $v \sim U(0, 1/n)$. Hence this amounts to using a one-dimensional randomly shifted lattice point set.

In addition to the reference [354] mentioned above, other papers that discuss the use of quasi–Monte Carlo within bootstrap filters are [117, 376].

We conclude this section with a simple example that illustrates the use of randomized quasi–Monte Carlo sampling within the two sequential Monte Carlo methods that we discussed.

Example 8.6. Consider a symmetric two-dimensional random walk where the step sizes are normally distributed. That is,

$$\mathbf{x}_t = \mathbf{x}_{t-1} + \xi_t, \qquad t \geq 1,$$

where ξ_t is a standard bivariate normal with marginal variances c and $\mathbf{x}_0 = 0$. Suppose that only a noisy observation of the position of $\mathbf{x}_t$ can be done at each time step. That is, $\mathbf{y}_t = \mathbf{x}_t + \epsilon_t$, where ϵ_t is a standard bivariate normal with marginal variances σ^2 and is recorded at each time $t = 1, 2, \ldots$. The goal is to get an estimate of the position $\mathbf{x}_t$ at time t given the observations $\mathbf{y}_1, \mathbf{y}_2, \ldots, \mathbf{y}_t$ gathered so far.

In this case, one could use the Kalman filter to derive exact expressions for the updated mean and variance of $\mathbf{x}_t$ given $\mathbf{y}_1, \ldots, \mathbf{y}_t$ at each time step. In Figs. 8.7 and 8.8, we show how to use sequential importance sampling and the bootstrap filter, respectively. In both cases, we assume the transition function $q(\mathbf{x}_t|\mathbf{x}_{t-1})$ is used as the proposal $\tilde{q}$.

```
RW-SIS(u_1, ..., u_n)
for i = 1 to n
    x_{i,0} ← 0
    w_i ← 1/n
for t = 1 to T
    W ← 0
    for i = 1 to n
        x_{i,t} ← x_{i,t-1} + √c(Φ^{-1}(u_{i,2t-1}), Φ^{-1}(u_{i,2t}))
        w_i ← w_i × exp(-||y_t - x_{i,t}||^2/2c)
        W ← W + w_i
    for i = 1 to n
        w_i ← w_i/W
// estimate of x_T
μ̂ ← 0
for i = 1 to n
    μ̂ ← μ̂ + w_i × h(x_{i,T})
return(μ̂)
```

Fig. 8.7 Using sequential importance sampling for the two-dimensional random walk.

```
RW-BootStFil(u_1, ..., u_n)
for i = 1 to n
    x_{i,0} ← 0
    w_i ← 1/n
for t = 1 to T
    W ← 0
    for i = 1 to n
        x̃_{i,t} ← x_{i,t-1} + √c(Φ^{-1}(u_{i,3t-2}), Φ^{-1}(u_{i,3t-1}))
        x̃_{i,0:t} ← x̃_{i,0:t-1}; x̃_{i,t}
        w_i ← exp(-||y_t - x̃_{i,t}||^2/2c)
        W ← W + w_i
    W_0 ← 0
    for i = 1 to n
        w_i ← w_i/W
        W_i ← W_{i-1} + w_i
    // estimate of E_π(h(x_t))
    μ_t ← 0
    for i = 1 to n
        μ_t ← μ_t + w_i × h(x_{i,t})
    // resampling
    for i = 1 to n
        find I such that W_{I-1} < u_{i,3t} ≤ W_I
        x_{i,0:t} ← x̃_{I,0:t}
    // can reorder the samples here
```

Fig. 8.8 Bootstrap filter for a simple two-dimensional random walk example.

In the pseudocode for the bootstrap filter, when we say "can reorder the samples here" on the last line, we are referring to the comment made previously about the possibility of choosing a random permutation to assign the newly chosen paths to the points $\mathbf{u}_i$. Also, in the code for the bootstrap filter, we perform the inference step before the resampling, as advised in [294, Sect. 2.4].

8.3 Computer experiments

Computer experiments [392, 393, 472] is an area that has a lot in common with stochastic simulation and quasi–Monte Carlo methods. This methodology can be used to study complex systems for which true physical experimentation would be too costly. When these systems can be modeled as stochastic processes, one uses stochastic simulation to perform inference on the measures of interest. We have seen several such examples in this book so far. Instead, computer experiments deal with systems where the output measures of interest, called *responses*, are determined in a very complex way, but usually deterministically, by several input variables, called *factors*. Furthermore, there is typically some level of uncertainty associated with the values taken by these factors.

In a computer experiment, the response corresponding to a certain choice of factors is obtained by running a computer program that implements a model of the system. As we just mentioned, this model is usually assumed to be deterministic (i.e., the program will output the same response if the same set of factors is used). The model can thus be represented as a function $f : \mathbb{R}^d \to \mathbb{R}$ that takes as input the values $x_1, \ldots, x_d$ of the factors and outputs the response $y = f(x_1, \ldots, x_d)$. To complete the model, the range of possible values for the factors must be determined, possibly also with their probability distribution over this range. Typically, for factors x_j that are controllable, we choose a range $[L_j, H_j]$ giving the possible values for x_j, and for inference purposes we simply assume a uniform distribution over this range. Factors that are not controllable might be modeled differently. For instance, in [12] the authors study the problem of circuit design in electrical engineering. There the factors are divided into two categories: (i) 20 adjustable engineering variables for the sizes of electrically active devices (such as transistors) and (ii) 16 factors representing variability due to manufacturing noise. An output measure of interest in this case is the time delay for propagation of signals through the circuit. In this model, the 20 controllable factors are each assumed to take a value within a specified range $[L_j, H_j]$, but the 16 uncontrollable factors are assumed to have a multivariate normal distribution.

Before going further, we discuss a simpler example often found in the computer experiments literature, which is the *borehole function problem* presented in [328].

Example 8.7. The problem here is to study the flow of water through a borehole that is drilled from the ground surface through two aquifers. The output measure of interest is the flow rate y through the borehole in m^3/yr, and there are eight factors determining this quantity, which are listed below along with their range of possible values.

$x_1 = r_w =$ radius of the borehole in $[0.05, 0.15]$,
$x_2 = r =$ radius of influence in $[100, 50000]$,
$x_3 = T_u =$ transmissivity of upper aquifer in $[63070, 115600]$,
$x_4 = H_u =$ potentiometric head of upper aquifer in $[990, 1110]$,
$x_5 = T_l =$ transmissivity of lower aquifer in $[63.1, 116]$,
$x_6 = H_l =$ potentiometric head of lower aquifer $[700, 820]$,
$x_7 = L =$ length of borehole in $[1120, 1680]$,
$x_8 = K_w =$ hydraulic conductivity of borehole in $[9855, 12045]$.

These eight factors determine the response y in the following way:

$$y = f(x_1, \ldots, x_8) = \frac{2\pi x_3(x_4 - x_6)}{\ln(x_2/x_1)\left(1 + \frac{2x_7x_3}{\ln(x_2/x_1)x_1^2x_8} + \frac{x_3}{x_5}\right)}. \quad (8.13)$$

Even if this function can be written in a compact form and evaluated very quickly on a computer, when we look at (8.13) it is not easy to determine how each factor is influencing the response. Computer experiments techniques can thus be used to get useful information on this function.

In general, the systems under study are very complex, and we do not necessarily attempt to write out explicitly the function that describes how the computer program transforms the factors into the response. Instead, we rely on the computer program to gather information on this function and use it to perform a number of tasks of interest. Generally speaking, these tasks attempt to better understand the relationship between the factors and the response.

The first task might be sensitivity analysis, where we try to determine which of the factors are the most important and how the response is affected by changes in the values of the factors. Answering this question can in turn suggest which factors need to be estimated the most accurately. The second task might be to identify a *surrogate function* that approximates reasonably well the more complicated one under study. That is, given some values for the factors, the surrogate should output a response close to the one output by the computer program. Usually, the goal is to find a surrogate that can be evaluated rather easily and is therefore less costly to work with than the computer program, which for complex systems could require a lot of computation time. Once we have a good approximation, then other tasks can be conducted more easily such as optimization. That is, one might be interested in determining which values for the factors provide the optimal response according to some appropriate optimality criterion.

Now, for all these tasks, one needs to query the computer program that implements the function under study at a certain number of well-chosen values for the factors. This task is often referred to as *experimental design* and provides a first connection with quasi–Monte Carlo. A second connection can be established when looking at methods used to perform sensitivity analysis in the context of computer experiments. Indeed, the "global sensitivity indices" that were defined in Chap. 6 have also been used in the context of computer experiments, and recent work has been done in this area to devise efficient methods to estimate these indices. These two topics are the ones we chose to discuss in this section. It is clear that many other connections between quasi-random sampling and computer experiments could lead to interesting advances for either field. For example, there have been several papers written recently investigating the idea of using low-discrepancy point sets to construct *approximations* for functions rather than simply estimating their integrals (see, for example, [79, 190] and the references therein). We refer the reader to [110, 296] for more on these connections.

Experimental design and low-discrepancy point sets

For a computer experiment with d factors, a d-dimensional space needs to be sampled. Often, each factor can take values in a certain finite range that is a subset of the real numbers. By rescaling these ranges appropriately, we can assume the sampling space in the unit cube $[0, 1]^d$. This assumption is made explicitly in [355, 417].

We now discuss different methods that have been used to choose a *design* $P_n = \{\mathbf{u}_1, \ldots, \mathbf{u}_n\}$ containing the n vectors at which the function f will be evaluated. We start with the two extremes, which are (i) to use random sampling (i.e., take P_n as a set of i.i.d. vectors in $[0, 1)^d$) and (ii) to use a 2^d-*factorial design*, where for each factor j we select two possible values u_j^l and u_j^h and then evaluate f at each of the 2^d possible combinations of the form

$$\{(u_1^{x(1)}, \ldots, u_d^{x(d)}), x(j) \in \{l, h\}, j = 1, \ldots, d\}.$$

One could obviously extend this to an N^d-factorial design, where for example each factor takes each of the N possible values $\{0, 1/N, \ldots, (N-1)/N\}$, much like in the rectangle rule for integration described in Chap. 1. It is clear that designs like this require the total number of sample points n to be much too large for moderate values of d. On the other hand, a completely random sample might fail to appropriately sample the factors. Similarly to researchers working on Monte Carlo methods who have come up with quasi–

Monte Carlo counterparts in order to avoid this property of random sampling, people working in computer experiments have devised improved sampling mechanisms, typically referred to as *space-filling designs* in that area.

A first step in that direction is *Latin hypercube sampling* (LHS), which was discussed in Chap. 6. For the sake of completeness, we recall this construction using notation that will be useful in the forthcoming discussion. With LHS one makes sure that each factor is evaluated exactly once in each interval of the form $[(i-1)/n, i/n)$, for $i = 1, \ldots, n$, over the n evaluations of f that are performed. This goal is achieved by taking

$$u_{i,j} = \frac{\tilde{A}_{i,j} - 1}{n} + w_{i,j}, \tag{8.14}$$

where the variables $w_{i,j}$ are i.i.d. $U(0, 1/n)$, $A_{i,j} = i$ for all $j = 1, \ldots, d$, and $\tilde{A}$ is obtained by applying random i.i.d. permutations of $[1, \ldots, n]$ to each of the d columns of A. In practice, for $d > 1$, it is equivalent to taking the first permutation to be the identity, and the remaining permutations are then drawn randomly.

Using this description, it is clear that for each factor there will be exactly one value in each interval of the form $[(i-1)/n, i/n)$ among the n trial vectors used.

In the description above, we used an $n \times d$ matrix A of the form

$$A = \begin{bmatrix} 1 & 1 & \ldots & 1 \\ 2 & 2 & \ldots & 2 \\ \vdots & \vdots & \vdots & \vdots \\ n & n & \ldots & n \end{bmatrix} \tag{8.15}$$

to describe the design used by LHS. This matrix can be viewed as the deterministic structure that underlies LHS and is often called a *sampling plan*. That is, if in LHS all permutations were given by the identity, then the value for $A_{i,j}$ would tell us that, for the ith design point, we will use a value in the $(A_{i,j})$th cell of $[0, 1)$, given by $[(i-1)/n, i/n)$, for the jth factor. Hence, if no permutations were used in LHS, we would then have

$$\mathbf{u}_i \in \left[\frac{i-1}{n}, \frac{i}{n}\right)^d .$$

Obviously, this is not very good since it means all n points $\mathbf{u}_1, \ldots, \mathbf{u}_n$ fall within a distance — taken in the sup norm sense — of $1/n$ of the main diagonal in the unit cube $[0, 1)^d$. Using random permutations allows the points to be better distributed in the unit cube while preserving the one-dimensional stratification. Figure 8.9 illustrates the effect of the permutations on a small example.

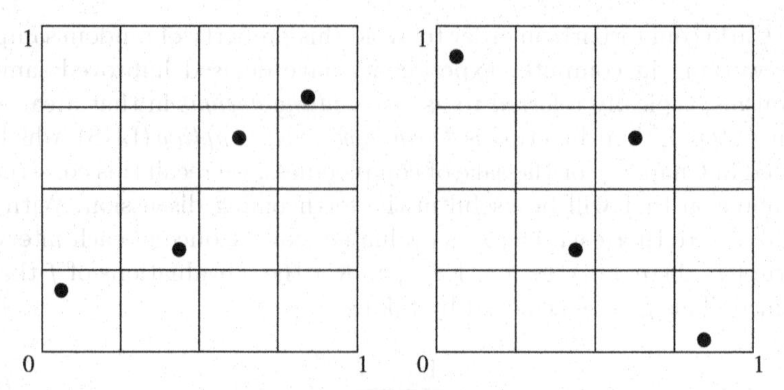

Fig. 8.9 Stratified design with no permutations (left) and with permutation $\pi_2 = [4231]$ as in LHS (right).

We will be using this matrix A to describe a generalization of LHS based on *orthogonal arrays*, as discussed in [356]. This matrix will also be convenient to explain methods that are used for sensitivity analysis. In addition, this description is helpful for handling slightly more general setups than the one chosen here, where we assumed each factor had been rescaled to the interval $[0,1]$. Sometimes authors prefer to work with real-valued factors $X_1, \ldots, X_d$ assumed to be independent and each having a marginal pdf $\varphi_j(x)$ for $j = 1, \ldots, d$. In that context, the choice of design is usually done in two steps: (1) produce a set of n vectors

$$\{(x_{i,1}, \ldots, x_{i,d}), i = 1, \ldots, n\}$$

according to some sampling method, where $x_{i,j}$ is distributed according to φ_j for each $i = 1, \ldots, n$, and each $j = 1, \ldots, d$; and (2) use a sampling plan A, possibly modified with permutations, in order to define the design

$$\{(x_{A[i,1],1}, \ldots, x_{A[i,d],d}), i = 1, \ldots, n\}.$$

(Here we use the notation $A[i,j]$ instead of $A_{i,j}$ to avoid double subscripts.)

This more general framework can, however, be converted to the previous one, where the goal is to construct a good design over $[0,1]^d$. Example 8.8 illustrates this idea, which refers back to the integration versus simulation formulation discussed throughout this book.

Example 8.8. Suppose $d = 2$, and X_1, X_2 are assumed to be independent and exponentially distributed random variables with mean β. Using inversion, we can obtain such variables using

$$X = -\beta \ln(1 - U).$$

The following steps describe how to use LHS sampling to generate a set of n inputs $\{(x_{i,1}, x_{i,2}), i = 1, \ldots, n\}$. First generate n independent uniform vectors $\mathbf{u}_i \in [0,1)^d$, $i = 1, \ldots, n$. Then produce two stratified samples over $[0, 1)$ as

$$\left\{w_{i,j} = \frac{i-1}{n} + \frac{u_{j,1}}{n}, i = 1, \ldots, n\right\} \text{ for } j = 1, 2,$$

and let

$$x_{i,j} = -\beta \ln(1 - u_{i,j}), \qquad i = 1, \ldots, n, j = 1, 2.$$

This completes Step (1) as indicated above. Now consider the sampling plan A for LHS given in (8.15) with $d = 2$ columns. Generate one random uniform permutation π of $[1, \ldots, n]$, and use it to permute the second column of A. That is, A becomes

$$\tilde{A} = \begin{bmatrix} 1 & \pi(1) \\ 2 & \pi(2) \\ \vdots & \vdots \\ n & \pi(n) \end{bmatrix}.$$

Then the LHS sample is given by

$$\{(x_{\tilde{A}[i,1],1}, x_{\tilde{A}[i,2],2}), i = 1, \ldots, n\}$$

or, equivalently,

$$\{(x_{i,1}, x_{\pi(i),2}), i = 1, \ldots, n\}.$$

Going back to the general application of LHS, its superiority over Monte Carlo shows up when we measure the variability of the estimator

$$\hat{\mu}_{\text{lhs}} = \frac{1}{n}\sum_{i=1}^{n} f(\mathbf{u}_i)$$

as an approximation for the mean output value

$$I(f) = \int_{[0,1)^d} f(\mathbf{u}) d\mathbf{u}. \tag{8.16}$$

In [313], the authors prove a result that, translated in our setup, is as follows.

Theorem 8.9 ([313]). *If f is monotonic in each of its arguments, then* $\text{Var}(\hat{\mu}_{\text{lhs}}) \leq \text{Var}(\hat{\mu}_{\text{mc}})$.

The proof of this theorem relies on results from [275], which were also used to prove a similar theorem for antithetic variates, as discussed in Chap. 4. More results on the variance of the LHS estimator are given in [355, 427]. In [427], the following theorem is given, which uses the concept of ANOVA decomposition described in Chap. 6.

Theorem 8.10. *If f is square-integrable, then*

$$\mathrm{Var}(\hat{\mu}_{\mathrm{lhs}}) = \frac{1}{n} \sum_{J \subseteq \{1,\dots,s\}, |J|>1} \sigma_J^2 + o(1/n).$$

Neglecting the $o(1/n)$ term, this result implies that for a function whose effective dimension in the superposition sense is 1, the variance of the LHS estimator is negligible. Said differently, the result implies that the one-dimensional ANOVA components are very well integrated by LHS: Their corresponding variance terms $\sigma_{\{j\}}^2$ get "knocked out" of the variance expression, with only some residual components that are absorbed in the $o(1/n)$ term. This holds because the one-dimensional projections of the LHS point set are stratified along each axis, as we mentioned earlier.

A natural way of trying to "knock out" more terms in the variance is to consider generalized versions of LHS where higher-dimensional projections of P_n are well stratified. Readers of this book might immediately think of (t, k, s)-nets as a way of achieving that. In the computer experiments community, people have also looked at *orthogonal arrays* [356, 355], which turn out to be closely connected to digital nets, as we explain below.

With LHS, the sampling plan A described in (8.15) is given by d identical columns containing the elements from 1 to n. Hence, if we look at two columns, we only get n of the possible n^2 pairs of the form $\{(i,j), 1 \leq i,j \leq n\}$. Correspondingly, this means that, without the permutations, any two-dimensional projection of the LHS point set would have its points close to the main diagonal. If A is built more carefully — not merely by padding the same column d times — we can avoid this behavior. This is the idea of an orthogonal array, which we now define.

Definition 8.11. An $n \times d$ matrix A with elements in $\{1, \dots, q\}$ is called an *orthogonal array (OA) of strength* $\tau \leq d$ if any τ columns of A form an $n \times \tau$ matrix in which each of the q^τ possible rows appears the same number $\lambda = (n/q^\tau)$ of times. The array A is then denoted $OA(n, d, q, \tau)$. The *maximal strength* of the OA is the largest value of τ for which A is an OA of strength τ.

It is clear that n must be a multiple of a power of q in order for this definition to make sense. For example, with LHS, $q = n$ and the sampling plan used is an $OA(n, d, n, 1)$. Lists of OAs of different strengths can be found on the Internet; for example, in the databases [488, 502].

To use (8.14) with an OA, we must first redefine the variables $w_{i,j}$, so that they are now i.i.d. $U(0, 1/q)$ instead of $U(0, 1/n)$, for $i = 1, \dots, n$, $j = 1, \dots, d$. This is because we are now possibly using bigger cells for the stratification as $q \leq n$. If we then take $\tilde{A}$ to be an $OA(n, d, q, \tau)$ in (8.14), with the elements in each column randomly permuted, and divide by q instead of n, we get a *randomized orthogonal array* estimator $\hat{\mu}_{\mathrm{roa}}$, given by

$$\hat{\mu}_{\mathrm{roa}} = \frac{1}{n} \sum_{i=1}^{n} f(\mathbf{u}_i), \tag{8.17}$$

where

$$u_{i,j} = \frac{\tilde{A}_{i,j} - 1}{q} + w_{i,j}, \qquad i = 1, \ldots, n, j = 1, \ldots, d.$$

The following example illustrates how to construct such an estimator in a simple case.

Example 8.12. Consider an $OA(4, 3, 2, 2)$ given by

$$\begin{bmatrix} 1\,1\,1 \\ 1\,2\,2 \\ 2\,1\,2 \\ 2\,2\,1 \end{bmatrix}.$$

We can use this OA to produce a sample $\mathbf{u}_1, \ldots, \mathbf{u}_4$ to be used in (8.17) as follows. First, generate two random permutations π_2, π_3 of $[1, 2, 3, 4]$, and then construct $\tilde{A}$ by permuting the second and third columns of the OA by π_2 and π_3, respectively. For instance, if $\pi_2 = [1\,4\,2\,3]$ and $\pi_3 = [2\,1\,3\,4]$, then the OA becomes

$$\tilde{A} = \begin{bmatrix} 1\,1\,2 \\ 1\,2\,1 \\ 2\,2\,2 \\ 2\,1\,1 \end{bmatrix},$$

and we have

$$u_{i,j} = \frac{\tilde{A}_{i,j} - 1}{2} + w_{i,j},$$

where the variables $w_{i,j}$ are i.i.d. $U(0, 1/2)$. The sample P_n obtained is such that if we consider the projection $P_n(\{1, 2\})$, then we have one observation randomly distributed within each cell of the form

$$[j_1/2, (j_1 + 1)/2) \times [j_2/2, (j_2 + 1)/2),$$

where $j_1, j_2 \in \{0, 1\}$. The same is true for the two other two-dimensional projections, $P_n(\{1, 3\})$ and $P_n(\{2, 3\})$. However, there is no guarantee that in one dimension we will have one observation in each cell of the form $[j/4, (j+1)/4)$ for $j = 0, \ldots, 3$. We can only say that there will be two observations in each cell of the form $[j/2, (j + 1)/2)$ for $j = 0, 1$.

Consider a modified "midpoint rule" version $\tilde{\mu}_{\text{roa}}$ of this estimator where $w_{i,j} = 1/2q$ for each $i = 1, \ldots, n$ and $j = 1, \ldots, s$. That is, instead of having a point randomly distributed within each cell, it is placed in the center. Figure 8.10 illustrates the idea.

If the maximal strength of the OA is τ, then the corresponding estimator has a variance approximately given by [358, 355]

$$\frac{1}{n} \sum_{J:|J|>\tau} \sigma_J^2. \tag{8.18}$$

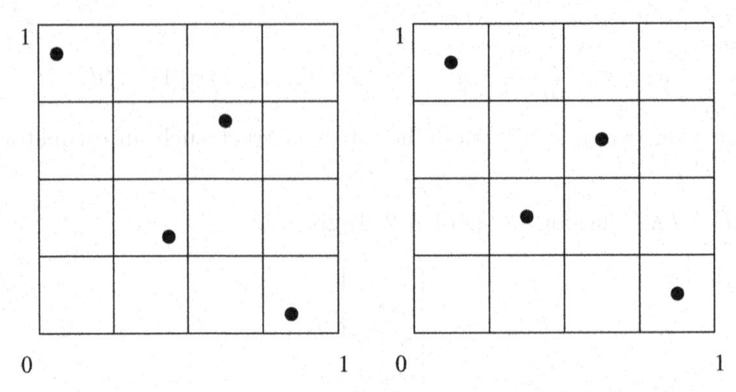

Fig. 8.10 Sample based on stratification (left) versus midpoint rule (right).

Instead of discussing (8.18) further, we focus on an example where we can give more precise information on the quality of the approximation above. For an $OA(q^2, d, q, 2)$, it can be shown that [355]

$$\text{Var}(\tilde{\mu}_{\text{roa}}) = \frac{1}{n} \sum_{I:|I|>2} \sigma_I^2 \times (1 + \epsilon(n)),$$

where $\epsilon(n) \in O(n^{-1/2})$ and $n = q^2$.

A few remarks are in order here. First, the midpoint rule version of the randomized OA estimator is biased, and this bias can be shown to be in $O(q^{-2})$, or equivalently in $O(n^{-2/\tau})$. Hence, to make sure that the bias does not dominate the variance in the MSE of this estimator, we must have $\tau \leq 3$. Another remark is that an OA of strength $\tau > 1$ produces a point set P_n with good τ-dimensional projections. But, for $s < \tau$, the s-dimensional projections are not that good because they stratify these subspaces in a number of cells smaller than n. For instance, the midpoint rule version of an $OA(q^2, d, q, 2)$ has very good two-dimensional projections of the form

$$\left\{ \left(\frac{i-1}{q} + \frac{1}{2q}, \frac{j-1}{q} + \frac{1}{2q} \right), 1 \leq i, j \leq q \right\},$$

but the one-dimensional projections are projecting q points on each coordinate of the form

$$\frac{i-1}{q} + \frac{1}{2q}, \qquad i = 1, \ldots, q.$$

Hence this construction is not fully projection-regular. If instead of the midpoint rule we use uniform draws within the cells, as discussed initially when we defined the estimator $\hat{\mu}_{\text{roa}}$, then at least the point set obtained is fully projection-regular with probability 1. But the stratification done on

s-dimensional projections for $s < \tau$ is still over a number of cells smaller than n.

A related construction proposed by Tang for computer experiments has the advantage of retaining the one-dimensional "maximal" stratification performed by LHS [433]. Tang calls it an *OA-based Latin hypercube design.* It works as follows. First choose an $OA(q^2, d, q, 2)$ and permute the order of each column independently. Then, in each column, replace the q occurrences of the symbol $j \in \{1, \ldots, q\}$ by a random permutation of $[(j-1)q+1, \ldots, jq]$. Call the matrix obtained B. Then define $P_n = \{\mathbf{u}_i, i = 1, \ldots, n\}$ by

$$u_{ij} = \frac{B_{ij} - 1}{n} + w_{ij}, j = 1, \ldots, d,$$

where the variables w_{ij} are i.i.d. $U(0, 1/n)$. It is easy to see that the point set obtained is in fact an $(0, 2, s)$-net in base q.

The idea above avoids the problem of points projecting onto each other that is encountered by the OA-based design discussed previously. This feature of OAs is also the reason why we cannot say that an OA-based design of strength τ is an $(0, \tau, d)$-net in base q. This is because the corresponding point set P_n is not $(q_1, \ldots, q_d)$-equidistributed when each q_l is 0 except one, which is equal to τ. However, if P_n is a (t, k, s)-net in base b and we define A so that

$$A_{i,j} = \lfloor q \times u_{i,j} \rfloor + 1,$$

then A is an $OA(b^k, s, b, k)$.

Starting from the concept of orthogonal arrays, several other combinatorial connections can be established. For instance, OAs can be related to families of hash functions and linear codes [428]. Also, OAs can be generalized to a concept called an *ordered (or generalized) orthogonal array* [31, 96, 305], which is more closely related to (t, k, s)-nets than OAs are.

More recently, another connection between (t, k, s)-nets and computer experiments was made [32]. More precisely, one of the ideas explored in that paper is to use scrambled nets to construct designs for computer experiments. However, instead of working with cubic cells, hyper-rectangles are used to allow a better sampling of the (not necessarily uniform) distribution of the input variables. More connections are discussed in [296].

Revisiting sensitivity indices estimation in the context of computer experiments

As we mentioned above, one of the goals of computer experiments is to perform sensitivity analysis, and more precisely to determine which factors are the most important in determining the response. The concept of ANOVA decomposition can be used for this purpose. In particular, recall what we defined as the *global sensitivity indices* in Chap. 6 following the terminology of [9]. They are the quantities

$$S_I = \frac{\sigma_I^2}{\sigma^2}$$

indicating the proportion of the variance of f explained by the ANOVA component f_I. We have discussed in Chap. 6 methods that can be used to estimate these indices S_I. Here we will focus on the method used by Sobol' and his collaborators in [9, 419, 417] and explain it using the concept of a sampling plan described above so that we can tie this back to current work in this area.

If we focus on the task of estimating quantities of the form $\sigma_{\{j\}}^2$ for $j = 1, \ldots, d$, then the idea of these authors amounts to using a sampling plan of the form

$$\left[\begin{array}{cccc} 1 & 1 & \ldots & 1 \\ 2 & 2 & \ldots & 2 \\ \vdots & \vdots & & \vdots \\ n & n & \ldots & n \\ \hline 1 & n+1 & \ldots & n+1 \\ 2 & n+2 & \ldots & n+2 \\ \vdots & \vdots & & \vdots \\ n & 2n & \ldots & 2n \\ \hline & & \vdots & \\ \hline n+1 & n+1 & \ldots & 1 \\ n+2 & n+2 & \ldots & 2 \\ \vdots & \vdots & & \vdots \\ 2n & 2n & \ldots & 2n \end{array}\right].$$

This $n(d+1) \times d$ plan is obtained by drawing, for each of the d factors, an i.i.d. sample of $2n$ values labeled from 1 to $2n$. To estimate $\sigma_{\{j\}}^2$, we use the first n rows and then the $(j+1)$th block of n rows sequentially pairing each of the rows in the two blocks and then summing the product of the value of f evaluated at each input in the pair. That is, we form the estimator

$$\hat{\sigma}_{\{j\}}^2 = \frac{1}{n} \sum_{i=1}^{n} f(u_{i,j} \mathbf{u}_{i,-j}) f(u_{i,j}, \mathbf{u}_{n+i,-j}) - \hat{\mu}^2, \qquad (8.19)$$

where $-j = \{1, \ldots, d\} \backslash \{j\}$, and

$$\hat{\mu} = \frac{1}{2n} \sum_{i=1}^{2n} f(\mathbf{u}_i).$$

For each term appearing in the sum (8.19), we see that the jth factor is always evaluated at the same value $u_{i,j}$. This approach is called the *substituted-columns plan* in the literature.

One criticism of the substituted-columns plan approach is that the "sampling efficiency" of the plan is not maximal. This concept refers to the number of degrees of freedom of the estimator for $\sigma^2_{\{j\}}$ divided by the total number of runs (or function evaluations) used by the design. In the setting above, this number is

$$\frac{n}{n(d+1)} = \frac{1}{d+1}$$

since each estimate of $\sigma^2_{\{j\}}$ is based on n independent sample values, while we perform a total of $n(d+1)$ function evaluations.

Consequently, other authors have proposed alternative sampling plans that have a higher sampling efficiency [329]. We discuss one such proposal here, called the *permuted-columns plan*, where one starts with an i.i.d. sample of n draws for each factor. The permuted column plan is an $na \times d$ matrix obtained by generating a groups of $d-1$ permutations of $[1,\ldots,n]$ that are used to *scramble* the original sample, where $a > 0$ is an integer to be chosen. That is, the $(ln+k)$th row of the sampling plan is defined as

$$k, \pi_{l+1,2}(k)\ldots, \pi_{l+1,d}(k),$$

where the permutations $\pi_{l+1,j}$ of $[1,\ldots,n]$ are independent. By definition, each value between 1 and n appears exactly a times in each column. Let $\{r_j(i,1),\ldots,r_j(i,a)\}$ be the set of row numbers in the sampling plan where the jth column is equal to i. To estimate $\sigma^2_{\{j\}}$, we can then use the estimator

$$\hat{S}_j = \frac{1}{n}\sum_{i=1}^{n}\sum_{l=1}^{a}\frac{(f(u_{i,j}, \mathbf{u}_{r_j(i,l),-j}) - \hat{f}_{i,j})^2}{a-1},$$

where

$$\hat{f}_{i,j} = \frac{1}{a}\sum_{l=1}^{a} f(u_{i,j}, \mathbf{u}_{r_j(i,l),-j}).$$

The advantage of this estimator over that of the substituted-columns plan is that the function is evaluated at $a \times n$ points in total and each estimator $\sigma^2_{\{j\}}$ is based on a total of $a \times n$ evaluations. However, the problem is that this estimator is in general biased because, unlike the substituted-columns plan estimator, the vectors $\{\mathbf{u}_{r_j(i,l),-j}, i = 1,\ldots,n\}$ are not necessarily independent since some repetitions may occur through the permutations. For instance, suppose $n = d = 3$, $a = 2$, and that we have

$$\pi_{1,2} = [3,2,1],$$
$$\pi_{2,2} = [3,1,2],$$
$$\pi_{1,3} = [2,1,3],$$
$$\pi_{2,3} = [3,2,1].$$

Then we get the sampling plan

$$\begin{bmatrix} 1 & 3 & 2 \\ 2 & 2 & 1 \\ 3 & 1 & 3 \\ 1 & 3 & 3 \\ 2 & 1 & 2 \\ 3 & 2 & 1 \end{bmatrix}.$$

Hence, in that case, for $j = 1$, if we look at the two rows for which the first column has a 1, we get

$$\begin{bmatrix} 1 & 3 & 2 \\ 1 & 3 & 3 \end{bmatrix},$$

and thus we reuse the same value twice for the second factor. Consequently, the two random quantities $f(u_{1,1}, u_{3,2}, u_{2,3})$ and $f(u_{1,1}, u_{3,2}, u_{3,3})$ used to compute an estimate of the variance of f given $u_1 = u_{1,1}$ are not independent. One possibility to avoid this problem is to use a *balanced incomplete block design* [329]. We will not discuss this idea further here, and we refer the reader to [329] for more information and to [111] for connections between balanced incomplete block designs and quasi–Monte Carlo concepts.

The reader may recall from Chap. 6 that in addition to the approach of Sobol' and his collaborators, we also discussed a method for estimating the global sensitivity indices based on function approximation and quasi-regression. More generally, this approach can be used to construct an approximation for the function under study in computer experiments. As discussed at the beginning of this section, finding a surrogate that is more amenable to certain tasks such as optimization is often of interest. Hence, if one uses the approach of [286] for the purpose of estimating the global sensitivity indices, the benefit is that we get "for free" an approximation for f as well.

Problems

The problems in this chapter are simply meant to get the reader familiar with some of the underlying tools used in the statistical techniques discussed in this chapter.

8.1. Find the value of the normalizing constant c_0 in (8.2) that ensures we indeed have a density function.

8.2. Verify that the stationary distribution of the Markov chain described in Example 8.1 satisfies (8.2).

8.3. Implement the code shown in Fig. 8.2 to simulate a random walk with semiabsorbent barriers, with $K = 20$ and (i) $N = 100$, (ii) $N = 1000$, and (iii) N =10,000. Graph in each case the histogram depicting the sample x_1 to x_N, and compare it with the true distribution as described in (8.2).

8.4. Is the Markov chain produced by the Metropolis-Hastings algorithm time-reversible? Explain.

8.5. Consider a proposal distribution $q(\mathbf{y}|\mathbf{x})$ given by a bivariate normal density with mean $\mathbf{x}$ and covariance matrix given by

$$\begin{bmatrix} 2 & 1 \\ 1 & 2 \end{bmatrix}.$$

Following our discussion in Subsect. 8.1.1 of the multiple-try Metropolis algorithm based on correlated sampling, write a computer program that correctly generates a sample of $r = 8$ trials based on this distribution and using a randomly shifted Korobov point set with $n = r = 8$ points and multiplier $a = 5$ and then generates the augmented sample $\mathbf{x}_1^*, \ldots, \mathbf{x}_{r-1}^*$.

8.6. Explain how a Kalman filter could be used to determine the posterior distribution of the model discussed in Example 8.6.

8.7. Show that (8.11) holds.

8.8. Show that if $\{(w_i, \mathbf{x}_{i,0:t}), i = 1, \ldots, n\}$ is a properly weighted sample, then

$$\{(\hat{w}_i, \mathbf{x}_{i,0:t+1}), i = 1, \ldots, n\},$$

with $\hat{w}_i$ a rescaled version of (8.12) and $\mathbf{x}_{i,0:t+1}$ being obtained by augmenting $\mathbf{x}_{i,0:t}$ with $\mathbf{x}_{i,t+1}$ drawn from $\tilde{q}(\cdot|\mathbf{x}_{i,0:t}, \mathbf{y}_{1:t})$ (as in the SIS approach discussed on p. 314), is also a properly weighted sample.

8.9. Show that the OA-based Latin hypercube design proposed by Tang in [433] is a $(0, 2, d)$-net in base q.

8.10. Show that, as stated on p. 329, if P_n is a (t, k, s)-net in base b and we define A so that

$$A_{i,j} = \lfloor q \times u_{i,j} \rfloor + 1,$$

then A is an $OA(b^k, s, b, k)$.

8.11. Describe the sampling plan that would be used to construct the $d - 1$ estimators of the form $\hat{\sigma}^2_{\{1,j\}}$ for $j = 2, \ldots, d$.

8.12. Consider a randomized low-discrepancy point set $P_n = \{\mathbf{u}_1, \ldots, \mathbf{u}_n\}$ in $[0, 1)^d$, where $\mathbf{u}_i \sim U([0, 1)^d$ for each $i = 1, \ldots, n$. Construct an unbiased estimator for the variance of f based on P_n.

Appendix A
Review of Algebra

The purpose of this appendix is to provide the background on algebra — rings, fields, and, in particular, polynomial rings and formal Laurent series — required to understand the concepts discussed in this work. We refer the reader to [94, 291, 390] for further information.

Definition A.1. A *group* is an ordered pair $(G, *)$, where G is a set and $*$ is a binary operation on G satisfying the following axioms:

(1) Associativity: $(a * b) * c = a * (b * c)$ for all $a, b, c \in G$.
(2) Identity: There exists an element $e \in G$ — called an identity — such that for all $a \in G$, $a * e = e * a = a$.
(3) Inverse: For each $a \in G$, there exists an element a^{-1} — called the inverse of a — such that $a * a^{-1} = e$.

If, in addition, we have that $a * b = b * a$ for all $a, b \in G$ (commutativity), then the group is said to be *Abelian*.

Note: Rather than referring to a group as an ordered pair, we can also say that G *is a group under* $*$.

Example A.2. The set $\mathbb{Z}$ of integers is a group under the operation $+$ with $e = 0$ and $a^{-1} = -a$.

Definition A.3. A *ring* is a set together with two binary operations $+$ and $\times$ (called addition and multiplication) satisfying:

(1) $(R, +)$ is an abelian group.
(2) $\times$ is associative.
(3) Distributivity: For all $a, b, c \in R$, we have $(a + b) \times c = (a \times c) + (b \times c)$ and $a \times (b + c) = (a \times b) + (a \times c)$.

The ring is said to be *commutative* if multiplication is commutative. The ring is said to have an *identity* (or to *contain a 1*) if there is an element $1 \in R$ such that $1 \times a = a \times 1 = a$ for all $a \in R$.

Note: We often write ab instead of $a \times b$ and denote the additive identity of R by 0 and the additive inverse of a by $-a$.

Example A.4. The set $\mathbb{Z}$ of integers is a ring under the usual addition and multiplication operations. It is in fact a commutative ring with identity, which is the integer 1.

Example A.5. Consider the set of equivalence classes over $\mathbb{Z}$ under the modulo n operation. That is, consider the set $\mathbb{Z}_n = \{[0], [1], \ldots, [n-1]\}$, where

$$[a] = \{m \in \mathbb{Z} : m = a \bmod n\}$$

for $a = 0, 1, \ldots, n-1$ is called the congruence class of a mod n. Define addition and multiplication over this set as follows:

$$[a] + [b] = [a+b] \text{ and } [a][b] = [ab].$$

Then $\mathbb{Z}_n$ is a commutative ring with identity, which is given by [1], and is called the *ring of integers modulo n*. We often identify the sets $[i]$ with the integer i and think of $\mathbb{Z}_n$ as the set $\{0, \ldots, n-1\}$ equipped with addition and multiplication modulo n.

Definition A.6. Given a ring R, a *polynomial $f(z)$ over R* is a sequence

$$f(z) = (c_0, c_1, \ldots, c_n, 0, 0, \ldots)$$

with $c_i \in R$ for all i and $c_i = 0$ for all $i > n = \deg(f)$. We denote by $R[z]$ the set of polynomials over R. We define addition and multiplication on $R[z]$ by

$$(c_0, c_1, \ldots) + (d_0, d_1, \ldots) = (c_0 + d_0, c_1 + d_1, \ldots)$$

and

$$(c_0, c_1, \ldots)(d_0, d_1, \ldots) = (e_0, e_1, \ldots),$$

where $e_0 = c_0 d_0$, $e_1 = c_0 d_1 + c_1 d_0$, and, in general,

$$e_k = \sum_{i,j:i+j=k} c_i d_j.$$

We define the zero polynomial by $(0, 0, \ldots)$ and denote it by 0; similarly, we denote $(1, 0, 0, \ldots)$ by 1. Then $R[z]$ is a commutative ring with identity.

Example A.7. Consider the polynomial ring $\mathbb{Z}_2[z]$ and two of its elements: $f(z) = 1 + z$ and $g(z) = 1 + z^2$. Then $f(z) + g(z) = z + z^2$ and $f(z)g(z) = 1 + z + z^2 + z^3$.

Definition A.8. A *field F* is a commutative ring with an identity in which each element $a \in F$ has a multiplicative inverse. That is, for each $a \in F$, there exists $a^{-1} \in F$ such that $aa^{-1} = 1$. This means each element $a \in F$ is a *unit*.

Example A.9. If n is prime, then $\mathbb{Z}_n$ is a field.

Example A.10. The set $\mathbb{R}$ of real numbers is a field under the usual addition and multiplication rules.

Notation. For b a positive integer, the notation $\mathbb{F}_b$ is used to denote the (Galois) field with b elements. When b is prime, we thus have the correspondence $\mathbb{F}_b = \mathbb{Z}_b$.

A polynomial ring over a field F is denoted $F[z]$. The advantage of working with a field when defining a polynomial ring is that we have a division algorithm. That is, if we choose a nonzero polynomial $g(z) \in F[z]$, then, for any $f(z) \in F[z]$, we can find polynomials $q(z), r(z) \in F[z]$ such that $f(z) = q(z)g(z) + r(z)$, where $\deg(r) < \deg(g)$. This in turn can be used to determine the gcd of two polynomials $f(z)$ and $g(z)$ over $F[z]$, which is defined as follows [390].

Definition A.11. For a field F and two polynomials $f(z), g(z) \in F[z]$, we have that $\gcd(f(z), g(z))$ is given by a polynomial $d(z) \in F[z]$ such that (i) $d(z)$ divides $f(z)$ and $g(z)$; (ii) if $c(z)$ is any common divisor of $f(z)$ and $g(z)$, then $c(z)$ divides $d(z)$; and (iii) $d(z)$ is monic (its leading coefficient is the identity).

Next, we have the following definition.

Definition A.12. A polynomial $f(z) \in F[z]$ is said to be *irreducible over F* if $\deg(f) > 0$ and $f(z) = g(z)q(z)$ with $g(z), r(z) \in F[z]$ can hold only if either q or g is a constant polynomial.

In a certain sense, irreducible polynomials play the same role as prime numbers. Another important concept is the following.

Definition A.13. The *residue class ring* $F[z]/(f(z))$ is the set $\{[r(z)] : \deg(r) < \deg(f)\}$, where

$$\begin{aligned}[r(z)] &= \{p(z) \in F[z] : p(z) = r(z) \bmod f(z)\} \\ &= \{p(z) \in F[z] : p(z) = q(z)f(z) + r(z) \text{ for some } q(z) \in F[z]\}.\end{aligned}$$

It can be shown that the relation $\bmod f(z)$ is an equivalence relation over $F[z]$, so that each $g(z) \in F[z]$ belongs in exactly one set $[r(z)]$.

For instance, if $b = 2$ and $f(z) = z^7 + z^3 + 1$, then

$$z^8 \bmod (z^7 + z^3 + 1) = z^8 - z(z^7 + z^3 + 1) = z^4 + z.$$

Theorem A.14. *The residue class ring* $F[z]/(f(z))$ *is a field if and only if* $f(z)$ *is irreducible over* F.

For a special class of irreducible polynomials $f(z)$, the field $F[z]/(f(z))$ has the following particularly useful representation.

Definition A.15. A *primitive polynomial* $f(z) \in \mathbb{F}_b[z]$ is an irreducible polynomial for which the set $\{z^k \bmod f(z), k = 0, \ldots, b^d - 1\}$ is equal to the set of all polynomials in $\mathbb{F}_b[z]$ with degree less than $d = \deg(f(z))$.

Hence, if $f(z)$ is a primitive polynomial of degree d, then the elements of $\mathbb{F}_b[z]/(f(z))$ can be identified with the powers z^k for $k = 0, \ldots, b^d - 1$.

Formal Laurent series

The field of formal Laurent series plays an important role in the definition of several families of digital nets and sequences. It is thus important to define this concept.

A useful analogy is to think of the field of formal Laurent series (in a given base b) as the field of real numbers. In this context, we view the ring $\mathbb{F}_b[z]$ of polynomials over $\mathbb{F}_b$ as the "integers". Similarly, we consider the ring of polynomial quotients of the form $f(z)/g(z)$, where $f(z), g(z)$ are polynomials in $\mathbb{F}_b[z]$, and view these quotients as the "rational numbers". Then, the field of formal Laurent series over $\mathbb{F}_b[z]$ is defined as the set $\mathbb{F}_b((z^{-1}))$ of elements L of the form

$$L = \sum_{r=w}^{\infty} a_r z^{-r},$$

where the coefficients a_r are in $\mathbb{F}_b$. In particular, quotients $f(z)/g(z)$ of polynomials can be expressed as formal Laurent series. Some examples will be given shortly.

At this point, it is useful to mention that when we consider expansions coming from ratios of polynomials of the form

$$\frac{f(z)}{g(z)} = \sum_{r=w} a_r z^{-r}, \tag{A.1}$$

where $g(z)$ is a monic polynomial of degree e, and the degree of $f(z)$ is no larger than e, then $w = 1$ in (A.1), and the coefficients $a_1, a_2, \ldots$ can be shown to follow a recurrence whose characteristic polynomial is $g(z)$ [337, p. 65]. The role of $f(z)$ is to initialize this recurrence. An example follows.

Example A.16. Let $b = 2$ and consider $g(z) = z^3 + z + 1$. Then we have that

$$\frac{1}{z^3 + z + 1} = a_1 z^{-1} + a_2 z^{-2} + a_3 z^{-3} + \ldots.$$

Rearranging, we have that

$$1 = a_1 z^2 + a_2 z + (a_1 + a_3) + (a_1 + a_2 + a_4) z^{-1} + (a_2 + a_3 + a_5) z^{-2} + \ldots,$$

which means we must find coefficients $a_1, a_2, \ldots$ that satisfy

$$\begin{aligned} a_1 &= 0, \\ a_2 &= 1, \\ a_1 + a_3 &= 1, \\ a_1 + a_2 + a_4 &= 0, \\ a_2 + a_3 + a_5 &= 0, \end{aligned}$$

and so on. Hence we get $a_1 = a_2 = 0$, $a_3 = 1$, and then the next coefficient a_r follows the recurrence $a_r = a_{r-2} + a_{r-3}$, implying that $a_4 = 0$, $a_5 = 1$, $a_6 = 1$, $a_7 = 1$, and so on. Therefore we have that

$$\frac{1}{1+z+z^3} = z^{-3} + z^{-5} + z^{-6} + z^{-7} + \ldots.$$

If instead we wish to compute for instance

$$\frac{1+z}{1+z+z^3},$$

then this means the initial conditions are now $a_1 = 0$, $a_2 = 1$, $a_1 + a_3 = 1$, so that

$$\frac{1+z}{1+z+z^3} = z^{-2} + z^{-3} + z^{-4} + z^{-7} + \ldots.$$

Alternatively, we can simply compute

$$\frac{1+z}{1+z+z^3}$$

as

$$\begin{aligned} \frac{1}{1+z+z^3} + \frac{z}{1+z+z^3} &= z^{-3} + z^{-5} + z^{-6} + z^{-7} + \ldots \\ &\quad + z^{-2} + z^{-4} + z^{-5} + z^{-6} + \ldots \\ &= z^{-2} + z^{-3} + z^{-4} + z^{-7} + \ldots. \end{aligned}$$

Appendix B
Error and Variance Analysis for Halton Sequences

In this appendix, we extend to Halton sequences results that are known for digital nets and that were mentioned in Chap. 6.

Consider a *scrambled* Halton sequence based on nonsingular lower-triangular generating matrices in respective bases $b_1, \ldots, b_s$. That is, the jth coordinate of the ith point of that sequence is given by

$$u_{ij} = \sum_{r=1}^{\infty} \sum_{l=1}^{\infty} c^j_{r,l} a_{b_j,l}(i),$$

where $c^j_{r,l}$ is the entry on the rth row and lth column of the jth generating matrix C_j, and $a_{b_j,l}(i)$ is the lth digit in the base b expansion of i,

$$i = \sum_{l=1}^{\infty} a_{b_j,l}(i) b_j^{l-1}.$$

For simplicity, assume the bases $b_1, \ldots, b_s$ are the first s primes.

Consider the point set $P_n = \{\mathbf{u}_1, \ldots, \mathbf{u}_n\}$ in $[0,1)^s$ formed by the first

$$n = b_1^{k_1} \ldots b_s^{k_s}$$

points of this sequence, where the k_j are positive integers. Furthermore, for the following analysis, we are making the assumption that each coordinate $u_{i,j}$ is determined by only k_j digits in base b_j. Equivalently, and since the generating matrices are assumed to be lower-triangular, this means we are using a generating matrix C_j with k_j rows and k_j columns for $j = 1, \ldots, s$.

The dual space of this point set P_n is defined as

$$\mathcal{C}_s^* = \{\mathbf{h} \in \mathbb{N}_0^s : C_j^{\mathrm{T}} \cdot (h_j)_{k_j} = \mathbf{0} \text{ for all } j = 1, \ldots, s\},$$

where the notation $C_j^{\mathrm{T}} \cdot (h_j)_{k_j}$ means we only consider the first k_j digits in the expansion of h_j in base b_j. That is,

$$C_j^{\mathrm{T}} \cdot (h_j)_{k_j} = \begin{pmatrix} \sum_{l=1}^{k_j} C_{j,l,1} h_{j,l-1} \\ \vdots \\ \sum_{l=1}^{k_j} C_{j,l,k_j} h_{j,l-1} \end{pmatrix},$$

where the coefficients $h_{j,l}$ come from

$$h_j = \sum_{l=0}^{\infty} h_{j,l} b_j^l.$$

(An equivalent definition — to be used later — is to pad C_j with zeros and make it have an infinite number of rows, so that we can then compute the untruncated product $C_j^{\mathrm{T}} \cdot h_j$.)

For $\mathbf{h} \in \mathbb{N}_0^s$, define the multibase Walsh basis function

$$\phi_{\mathbf{h}}^{(b_1,\ldots,b_s)}(\mathbf{u}) = e^{2\pi \mathrm{i} \sum_{j=1}^{s} (h_j \cdot u_j)/b_j}$$

where $\mathrm{i} = \sqrt{-1}$,

$$h_j \cdot u_j = \sum_{l=1}^{\infty} h_{j,l-1} u_{j,l} \in \mathbb{Z}_{b_j},$$

and the coefficients $u_{j,l}$ come from the decomposition of u_j in base b_j. That is,

$$u_j = \sum_{l=1}^{\infty} u_{j,l} b_j^{-l}.$$

Now, for a real-valued function f defined over $[0,1)^s$ and $\mathbf{h} \in \mathbb{N}_0^s$, define the Walsh coefficient

$$\tilde{f}^{(b_1,\ldots,b_s)}(\mathbf{h}) = \int_{[0,1)^s} f(\mathbf{u}) \phi_{-\mathbf{h}}^{(b_1,\ldots,b_s)}(\mathbf{u}) d\mathbf{u}.$$

Proposition B.1. *Let P_n be the first n points of a scrambled Halton sequence based on generating matrices C_j of size $k_j \times k_j$ with elements in $\mathbb{Z}_{b_j}$ for $j = 1, \ldots, s$, where $n = b_1^{k_1} \ldots b_s^{k_s}$ and each $k_j \in \mathbb{N}$. Then, for a given $\mathbf{h} \in \mathbb{N}_0^s$, we have*

$$\frac{1}{n} \sum_{i=1}^{n} \phi_{\mathbf{h}}^{(b_1,\ldots,b_s)}(\mathbf{u}_i) = \begin{cases} 1 & \text{if } \mathbf{h} \in C_s^* \\ 0 & \text{otherwise.} \end{cases}$$

Proof: We can write

$$h_j \cdot u_j = h_j^{\mathrm{T}} (\tilde{C}_j \cdot \mathbf{x}_{i,j}),$$

where $\tilde{C}_j$ is an $\infty \times n_j$ matrix described by

$$\tilde{C}_{j,l,r} = \begin{cases} C_{j,l,r} & \text{if } l \leq k_j, r \leq k_j \\ 0 & \text{otherwise.} \end{cases}$$

That is, $\tilde{C}_j$ is an $\infty \times n_j$ matrix obtained by padding C_j with zeros to fill up the extra dimensions and where

$$n_j = \lceil \log_{b_j} n \rceil$$

is the number of digits in base b_j required for the decomposition of the indices $i = 0, \ldots, n-1$ in that base. The vector $\mathbf{x}_{i,j}$ is an n_j-dimensional vector containing the expansion of i in base b_j. That is,

$$\mathbf{x}_{i,j} = (x_{i,j,0}, x_{i,j,1}, \ldots, x_{i,j,n_j})^{\mathrm{T}},$$

where the $x_{i,j,l}$ are such that

$$i - 1 = \sum_{l=0}^{n_j} x_{i,j,l} b_j^l.$$

Now, if $\mathbf{h}$ is in the dual $\mathcal{C}_s^*$, then $h_j^{\mathrm{T}} \cdot \tilde{C}_j = \mathbf{0}^{\mathrm{T}}$ for each $j = 1, \ldots, s$, by definition. Hence, in that case $h_j \cdot u_j = 0$, and thus $\phi_{\mathbf{h}}^{(b_1,\ldots,b_s)}(\mathbf{u}_i) = 1$ for each i, from which the first part of the result follows.

If $\mathbf{h}$ is not in $\mathcal{C}_s^*$, then for at least one j we have

$$\mathbf{y}_j := \tilde{C}_j^{\mathrm{T}} \cdot h_j \neq \mathbf{0}.$$

Let

$$\mathcal{J} = \{j : \mathbf{y}_j \neq \mathbf{0}\}$$

and

$$\tilde{n} = \prod_{j \in \mathcal{J}} b_j \geq 2.$$

Now consider an index j for which $\mathbf{y}_j \neq \mathbf{0}$. Observe that from the definition of $\tilde{C}_j$ we have that

$$h_j \cdot u_j = h_j^{\mathrm{T}}(\tilde{C}_j \cdot \mathbf{x}_{i,j}) = (\mathbf{y}_j)_{k_j} \cdot (\mathbf{x}_{i,j})_{k_j}. \tag{B.1}$$

That is, we only need to consider the first k_j components of $\mathbf{y}_j$ and $\mathbf{x}_{i,j}$. When $i-1$ runs from 0 to $n-1$, these first k_j components of $\mathbf{x}_{i,j}$ take each of the $b_j^{k_j}$ values in $\mathbb{Z}_{b_j}^{k_j}$ a total of $n/b_j^{k_j}$ times. Since $\mathbf{y}_j \neq \mathbf{0}$, the product (B.1) takes each of the b_j values in $\mathbb{Z}_{b_j}$ a total of n/b_j times. In addition, since the b_j are primes, the $(k_1 + \ldots + k_s)$-dimensional vector

$$(\mathbf{x}_{i,1}| \ldots |\mathbf{x}_{i,s})$$

takes each value in $\mathbb{Z}_{b_1}^{k_1} \times \ldots \times \mathbb{Z}_{b_s}^{k_s}$ exactly once as i goes from 1 to n. Hence the sum

$$\sum_{j \in \mathcal{J}} (h_j \cdot u_j)/b_j \bmod 1$$

takes each value of the form $w/\tilde{n}$ for $w \in \{0, \dots, \tilde{n}-1\}$ a total of

$$\prod_{j \in \mathcal{J}} b_j^{k_j - 1} \prod_{j \notin \mathcal{J}} b_j^{k_j}$$

times, which is equal to $n/\tilde{n}$.

Since for any positive integer b we have that

$$\sum_{v=0}^{b-1} e^{2\pi \mathrm{i}(v/b)} = 0,$$

this proves the result (use this with $b = \tilde{n}$).

With this result, it becomes easy to prove the following ones.

Proposition B.2. *Let P_n be defined as above. Then, for a function f such that*

$$\sum_{\mathbf{h} \in \mathbb{N}_0^s} |\tilde{f}^{(b_1, \dots, b_s)}(\mathbf{h})| < \infty,$$

the integration error is given by

$$\frac{1}{n} \sum_{i=1}^{n} f(\mathbf{u}_i) - I(f) = \sum_{0 \neq \mathbf{h} \in C_s^*} \tilde{f}^{(b_1, \dots, b_s)}(\mathbf{h}).$$

Proof. We can write

$$\begin{aligned} \frac{1}{n} \sum_{i=1}^{n} f(\mathbf{u}_i) &= \frac{1}{n} \sum_{i=1}^{n} \sum_{\mathbf{h} \in \mathbb{N}_0^s} \phi_{\mathbf{h}}^{(b_1, \dots, b_s)}(\mathbf{u}_i) \tilde{f}^{(b_1, \dots, b_s)}(\mathbf{h}) \\ &= \frac{1}{n} \sum_{\mathbf{h} \in \mathbb{N}_0^s} \tilde{f}^{(b_1, \dots, b_s)}(\mathbf{h}) \sum_{i=1}^{n} \phi_{\mathbf{h}}^{(b_1, \dots, b_s)}(\mathbf{u}_i) \\ &= \sum_{\mathbf{h} \in C_s^*} \tilde{f}^{(b_1, \dots, b_s)}(\mathbf{h}), \end{aligned}$$

where the change of order in the summation — done from the first to the second line — is allowed because we assumed that the Walsh coefficients of f converge absolutely. Since $I(f) = \tilde{f}^{(b_1, \dots, b_s)}(\mathbf{0})$, the result follows.

Proposition B.3. *Let P_n be defined as above, and consider the set $\tilde{P}_n$ obtained by performing a random digital shift in the multibase $(b_1, \dots, b_s)$. Then, for a square-integrable function f, the variance of the estimator $\hat{\mu}$ based on $\tilde{P}_n$ is given by*

$$\mathrm{Var}(\hat{\mu}) = \sum_{0 \neq \mathbf{h} \in C_s^*} |\tilde{f}^{(b_1, \dots, b_s)}(\mathbf{h})|^2.$$

Proof. Let $\mathbf{w} \sim U([0,1)^s)$. The randomized point set $\tilde{P}_n = \{\tilde{\mathbf{u}}_1, \ldots, \tilde{\mathbf{u}}_n\}$ is defined by

$$\tilde{\mathbf{u}}_i = S^{(b_1,\ldots,b_s)}(\mathbf{u}_i, \mathbf{w}), i = 1, \ldots, n,$$

where $S^{(b_1,\ldots,b_s)}(\mathbf{u}_i, \mathbf{w})$ has its jth component given by

$$\tilde{u}_{i,j} = \sum_{l=1}^{\infty} (u_{i,j,l} + w_{j,l}) b_j^{-l},$$

where addition is performed in $\mathbb{Z}_{b_j}$.

We first write

$$\hat{\mu} = g(\mathbf{w}) := \frac{1}{n} \sum_{i=1}^{n} f(\tilde{\mathbf{u}}_i).$$

We can then use Parseval's identity [124]

$$\mathrm{Var}(\hat{\mu}) = \mathrm{Var}(g(\mathbf{w})) = \sum_{\mathbf{0} \neq \mathbf{h} \in \mathbb{N}_0^s} |\tilde{g}(\mathbf{h})|^2 \tag{B.2}$$

because the multibase Walsh basis functions form an orthonormal set (which follows from the fact that each of the j components is a standard Walsh basis function). Moreover,

$$\begin{aligned}
\tilde{g}(\mathbf{h}) &= \int g(\mathbf{w}) \phi_{-\mathbf{h}}(\mathbf{w}) d\mathbf{w} \\
&= \frac{1}{n} \int_{[0,1)^s} \sum_{i=1}^{n} f(S^{(b_1,\ldots,b_s)}(\mathbf{u}_i, \mathbf{w})) \phi_{-\mathbf{h}}(\mathbf{w}) d\mathbf{w} \\
&= \frac{1}{n} \sum_{i=1}^{n} \int_{[0,1)^s} f(\tilde{\mathbf{u}}_i) \phi_{-\mathbf{h}}(M^{(b_1,\ldots,b_s)}(\tilde{\mathbf{u}}_i, \mathbf{u}_i)) d\tilde{\mathbf{u}}_i \\
&= \frac{1}{n} \sum_{i=1}^{n} \phi_{\mathbf{h}}(\mathbf{u}_i) \int_{[0,1)^s} f(\tilde{\mathbf{u}}_i) \phi_{-\mathbf{h}}(\tilde{\mathbf{u}}_i) d\tilde{\mathbf{u}}_i \\
&= \frac{1}{n} \sum_{i=1}^{n} \phi_{\mathbf{h}}(\mathbf{u}_i) \tilde{f}(\mathbf{h}) \\
&= \begin{cases} \tilde{f}(\mathbf{h}) & \text{if } \mathbf{h} \in C_s^* \\ 0 & \text{otherwise,} \end{cases}
\end{aligned} \tag{B.3}$$

where $M^{(b_1,\ldots,b_s)}(\tilde{\mathbf{u}}_i, \mathbf{u}_i)$ has its jth component given by

$$\sum_{l=1}^{\infty} (\tilde{u}_{i,j,l} - u_{i,j,l}) b_j^{-l}$$

and where the subtraction is done in $\mathbb{Z}_{b_j}$. Substituting (B.3) in (B.2), we get the desired result.

References

1. M. Abramowitz and I. A. Stegun, editors. *Handbook of Mathematical Functions with Formulas, Graphs and Mathematical Tables*, volume 55 of Applied Mathematics. National Bureau of Standards, Washington D.C., 1964.
2. P. Acworth, M. Broadie, and P. Glasserman. A comparison of some Monte Carlo and quasi-Monte Carlo techniques for option pricing. In P. Hellekalek and H. Niederreiter, editors, *Monte Carlo and Quasi-Monte Carlo Methods in Scientific Computing*, volume 127 of Lecture Notes in Statistics, pages 1–18. Springer-Verlag, New York, 1997.
3. R. J. Adler. *An Introduction to Continuity, Extrema, and Related Topics for General Gaussian Processes*, volume 12 of IMS Lecture Notes–Monograph Series. Institute of Mathematical Statistics, Hayward, CA, 1990.
4. F. Åkesson and J. P. Lehoczy. Path generation for quasi-Monte Carlo simulation of mortgage-backed securities. *Management Science*, 46:1171–1187, 2000.
5. J. An and A. B. Owen. Quasi-regression. *Journal of Complexity*, 17:588–607, 2001.
6. I. J. Andréasson. Combinations of antithetic methods in simulation. Technical Report NA 72.49, Royal Institute of Technology, Stockholm, 1972.
7. I. J. Andréasson and G. Dahlquist. Groups of antithetic transformations in simulation. Technical Report NA 72.57, Royal Institute of Technology, Stockholm, 1972.
8. I. A. Antonov and V. M. Saleev. An economic method of computing LP_τ-sequences. *USSR Computational Mathematics and Mathematical Physics*, 19:252–256, 1980.
9. G. E. B. Archer, A. Saltelli, and I. M. Sobol'. Sensitivity measures, ANOVA-like techniques and the use of bootstrap. *Journal of Statistical Computation and Simulation*, 58:99–120, 1997.
10. P. Artzner, F. Delbaen, J. M. Eber, and D. Heath. Coherent risk measures. *Mathematical Finance*, 9:203–228, 1999.
11. J. Arvo et al. *State of the Art in Monte Carlo Ray Tracing for Realistic Image Synthesis*. ACM SIGGRAPH 2001 Course 29. ACM, New York, 2001.
12. R. Aslett, R. J. Buck, S. G. Duvall, J. Sacks, and W. J. Welch. Circuit optimization via sequential computer experiments: Design of an output buffer. *Applied Statistics*, 47:31–48, 1998.
13. S. Asmussen. Conjugate processes and the simulation of ruin problems. *Stochastic Processes and Their Applications*, 20:213–229, 1985.
14. S. Asmussen. Ruin probabilities expressed in terms of storage process. *Advances in Applied Probability*, 20:913–916, 1988.
15. S. Asmussen, K. Binswanger, and B. Højgaard. Rare events simulation for heavy-tailed distributions. *Bernoulli*, 6:303–322, 2000.

16. S. Asmussen and R. Rubinstein. Complexity properties of steady-state rare events simulation in queueing models. In J. Dshalalow, editor, *Advances in Queueing: Theory, Methods, and Open Problems*, pages 429–462. CRC Press, Boca Raton, FL, 1995.
17. D. I. Asotsky, E. E. Myshetskaya, and I. M. Sobol'. The average dimension of a multidimensional function for quasi-Monte Carlo estimates of an integral. *Computational Mathematics and Mathematical Physics*, 46:2061–2067, 2006.
18. E. Atanassov. On the discrepancy of the Halton sequences. *Mathematica Balkanica*, 18:15–32, 2004.
19. E. Atanassov and M. K. Durchova. Generating and testing the modified Halton sequences. In I. Dimov, I. Lirkov, S. Margenov, and Z. Zlatev, editors, *Numerical Methods and Applications, 5th International Conference, NMA 2002, Borovets, Bulgaria, August 20-24, 2002*, volume 2542 of Lecture Notes in Computer Science, pages 91–98. Springer-Verlag, Berlin, 2002.
20. M. Avellaneda. Minimum-entropy calibration of asset-pricing models. *International Journal of Theoretical and Applied Finance*, 1(4):447–472, 1998.
21. A. N. Avramidis, K. W. Bauer, Jr., and J. R. Wilson. Simulation of stochastic activity networks using path control variates. *Journal of Naval Research*, 38:183–201, 1991.
22. A. N. Avramidis and J. R. Wilson. Integrated variance reduction strategies for simulation. *Operations Research*, 44:327–346, 1996.
23. A. N. Avramidis and J. R. Wilson. Correlation-induction techniques for estimating quantiles in simulation experiments. *Operations Research*, 46(4):574–591, 1998.
24. K. I. Babenko. Approximation by trigonometric polynomials in a certain class of periodic functions of several variables. *Soviet Mathematics Doklady*, 1:672–675, 1960.
25. K. I. Babenko. Approximation of periodic functions of several variables by trigonometric polynomials. *Soviet Mathematics Doklady*, 1:513–516, 1960.
26. K. G. Beauchamp. *Applications of Walsh and Related Functions.* Academic Press, London, 1984.
27. J. Beck and W. W. L. Chen. *Irregularities of Distribution.* Cambridge University Press, Cambridge, 1987.
28. R. Bellman. *Adaptive Control Processes: A Guided Tour.* Princeton University Press, Princeton, NJ, 1961.
29. H. Ben Ameur, P. L'Ecuyer, and C. Lemieux. Variance reduction of Monte Carlo and randomized quasi-Monte Carlo estimators for stochastic volatility models in finance. In P. A. Farrington and H. B. Nemhard, editors, *Proceedings of the 1999 Winter Simulation Conference*, pages 336–343. IEEE Press, Piscataway, NJ, 1999.
30. W. A. Beyer, R. B. Roof, and D. Williamson. The lattice structure of multiplicative congruential pseudo-random vectors. *Mathematics of Computation*, 25(114):345–363, 1971.
31. J. Bierbrauer, Y. Edel, and W. Ch. Schmid. Coding-theoretic constructions for (t, m, s)-nets and ordered orthogonal arrays. *Journal of Combinatorial Designs*, 10:403–418, 2002.
32. D. Bingham and D. Mease. Latin hyperrectangle sampling for computer experiments. *Technometrics*, 48:467–477, 2006.
33. F. Black and M. Scholes. The pricing of options and corporate liabilities. *Journal of Political Economy*, 81:637–654, 1973.
34. N. Bolia and S. Juneja. Function-approximation-based perfect control variates for pricing American options. In N. Steiger and M. E. Kuhl, editors, *Proceedings of the 2005 Winter Simulation Conference*, pages 1876–1883. IEEE Press, Piscataway, NJ, 2005.
35. I. Borosh and H. Niederreiter. Optimal multipliers for pseudo-random number generation by the linear congruential method. *Bit*, 23:115–129, 1983.
36. G. E. P. Box and M. E. Muller. A note on the generation of random normal deviates. *Annals of Mathematical Statistics*, 29:610–611, 1958.

37. P. Boyle. Options: A Monte Carlo approach. *Journal of Financial Economics*, 4:323–338, 1977.
38. P. Boyle, M. Broadie, and P. Glasserman. Monte Carlo methods for security pricing. *Journal of Economic Dynamics and Control*, 21(8–9):1267–1321, 1997.
39. P. Boyle, A. W. Kolkiewicz, and K. S. Tan. An improved simulation method for pricing high-dimensional American derivatives. *Mathematics and Computers in Simulation*, 62:315–322, 2003.
40. P. Boyle, Y. Lai, and K. S. Tan. Pricing options using lattice rules. *North American Actuarial Journal*, 9:50–76, 2005.
41. E. Braaten and G. Weller. An improved low-discrepancy sequence for multidimensional quasi-Monte Carlo integration. *Journal of Computational Physics*, 33:249–258, 1979.
42. G. Brassard and P. Bratley. *Fundamentals of Algorithmics*. Prentice-Hall, Englewood Cliffs, NJ, 1996.
43. P. Bratley and B. L. Fox. Algorithm 659: Implementing Sobol's quasirandom sequence generator. *ACM Transactions on Mathematical Software*, 14(1):88–100, 1988.
44. P. Bratley, B. L. Fox, and H. Niederreiter. Implementation and tests of low-discrepancy sequences. *ACM Transactions on Modeling and Computer Simulation*, 2:195–213, 1992.
45. P. Bratley, B. L. Fox, and L. E. Schrage. *A Guide to Simulation*, second edition. Springer-Verlag, New York, 1987.
46. V. Brazauskas, B. L. Jones, M. L. Puri, and R. Zitikis. Estimating conditional tail expectations with actuarial applications in view. *Journal of Statistical Planning and Inference*, 138:3590–3604, 2007.
47. M. Broadie and P. Glasserman. Estimating security price derivatives using simulation. *Management Science*, 42:269–285, 1996.
48. M. Broadie and P. Glasserman. A stochastic mesh method for pricing high-dimensional American options. Manuscript, 1997.
49. M. Broadie and Ö. Kaya. Exact simulation of option greeks under stochastic volatility and jump-diffusion models. In R. G. Ingalls, M. D. Rossetti, J. S. Smith, and B. A Peters, editors, *Proceedings of the 2004 Winter Simulation Conference*, pages 1607–1615. IEEE Press, Piscataway, NJ, 2004.
50. J. M. Burt and M. B. Garman. Conditional Monte Carlo: A simulation technique for stochastic network analysis. *Management Science*, 18:207–217, 1972.
51. R. E. Caflisch, W. Morokoff, and A. B. Owen. Valuation of mortgage-backed securities using Brownian bridges to reduce effective dimension. *The Journal of Computational Finance*, 1(1):27–46, 1997.
52. R. E. Caflisch and B. Moskowitz. Modified Monte Carlo methods using quasi-random sequences. In H. Niederreiter and P. J.-S. Shiue, editors, *Monte Carlo and Quasi-Monte Carlo Methods in Scientific Computing*, volume 106 of Lecture Notes in Statistics, pages 1–16. Springer-Verlag, New York, 1995.
53. J. Carriere. Valuation of early-exercise price of options using simulations and nonparametric regression. *Insurance: Mathematics and Economics*, 19:19–30, 1996.
54. J. W. S. Cassels. *An Introduction to the Geometry of Numbers*. Classics in Mathematics. Springer-Verlag, Berlin, 1997. Corrected reprint of the 1971 edition.
55. C. S. Chang, P. Heidelberger, and P. Shahabuddin. Fast simulation of packet loss rates in a shared buffer communications switch. *ACM Transactions on Modeling and Computer Simulation*, 5(4):306–325, 1995.
56. S. K. Chaudhary. *Acceleration of Monte Carlo methods using low discrepancy sequences*. PhD thesis, UCLA, 2004.
57. R. C. H. Cheng. The use of antithetic variates in computer simulations. *Journal of the Operational Research Society*, 33:229–237, 1982.
58. H. Chi, M. Mascagni, and T. Warnock. On the optimal Halton sequence. *Mathematics and Computers in Simulation*, 70:9–21, 2005.

59. E. Clément, D. Lamberton, and P. Protter. An analysis of a least squares regression method for American option pricing. *Finance and Stochastics*, 6:449–471, 2002.
60. W. G. Cochran. *Sampling Techniques*, second edition. John Wiley and Sons, New York, 1977.
61. J. H. Conway and N. J. A. Sloane. *Sphere Packings, Lattices and Groups*, third edition, volume 290 of Grundlehren der Mathematischen Wissenschaften. Springer-Verlag, New York, 1999.
62. R. Cools, F. Y. Kuo, and D. Nuyens. Constructing embedded lattice rules for multivariate integration. *SIAM Journal on Scientific Computing*, 28(6):2162–2188, 2006.
63. J. N. Corcoran and R. L. Tweedie. Perfect sampling from independent Metropolis-Hastings chains. *Journal of Statistical Planning and Inference*, 104(2):297–314, 2002.
64. RAND Corporation. *A Million Random Digits with 100,000 Normal Deviates*. The Free Press, Glencoe, IL, 1955.
65. R. Couture and P. L'Ecuyer. On the lattice structure of certain linear congruential sequences related to AWC/SWB generators. *Mathematics of Computation*, 62(206):798–808, 1994.
66. R. Couture and P. L'Ecuyer. Lattice computations for random numbers. *Mathematics of Computation*, 69(230):757–765, 2000.
67. R. Couture, P. L'Ecuyer, and S. Tezuka. On the distribution of k-dimensional vectors for simple and combined Tausworthe sequences. *Mathematics of Computation*, 60(202):749–761, S11–S16, 1993.
68. R. R. Coveyou and R. D. MacPherson. Fourier analysis of uniform random number generators. *Journal of the ACM*, 14:100–119, 1967.
69. R. V. Craiu and C. Lemieux. Acceleration of the multiple-try metropolis algorithm using antithetic and stratified sampling. *Statistics and Computing*, 17:109–120, 2007.
70. R. V. Craiu and X.-L. Meng. Antithetic coupling for perfect sampling. In E. I. George, editor, *Bayesian Methods with Applications to Science, Policy, and Official Statistics (Selected Papers from ISBA 2000)*, pages 99–108. Eurostat, Luxembourg, 2000.
71. R. V. Craiu and X.-L. Meng. Multi-process parallel antithetic coupling for forward and backward Markov chain Monte Carlo. *Annals of Statistics*, 33:661–697, 2005.
72. R. Cranley and T. N. L. Patterson. Randomization of number theoretic methods for multiple integration. *SIAM Journal on Numerical Analysis*, 13(6):904–914, 1976.
73. P. Davis and P. Rabinowitz. *Methods of Numerical Integration*, second edition. Academic Press, New York, 1984.
74. D. C. Dembeck. Dynamic numerical integration using randomized quasi-Monte Carlo methods. Master's thesis, University of Calgary, 2003.
75. L. Devroye. *Non-uniform Random Variate Generation*. Springer-Verlag, New York, 1986.
76. J. Dick. The construction of extensible polynomial lattice rules with small weighted star discrepancy. *Mathematics of Computation*, 76:2077–2085, 2007.
77. J. Dick. Explicit constructions of quasi-Monte Carlo rules for the numerical integration of high dimensional periodic functions. *SIAM Journal on Numerical Analysis*, 45:2141–2176, 2007.
78. J. Dick. Walsh spaces containing smooth functions and quasi-Monte Carlo rules of arbitrary high order. *SIAM Journal on Numerical Analysis*, 46:1519–1553, 2008.
79. J. Dick, P. Kritzer, and F. Y. Kuo. Approximation of functions using digital nets. In A. Keller, S. Heinrich, and H. Niederreiter, editors, *Monte Carlo and Quasi-Monte Carlo Methods 2006*, pages 275–298. Springer, New York, 2008.
80. J. Dick, P. Kritzer, F. Pillichshammer, and W. Ch. Schmid. On the existence of higher order polynomial lattices based on a generalized figure of merit. *Journal of Complexity*, 23:581–593, 2007.
81. J. Dick, F. Y. Kuo, F. Pillichshammer, and I. H. Sloan. Construction algorithms for polynomial lattice rules for multivariate integration. *Mathematics of Computation*, 74:1895–1921, 2005.

82. J. Dick and H. Niederreiter. On the exact t-value of some standard low-discrepancy sequences. *Journal of Complexity*, 2008. In press.
83. J. Dick, F. Pillichshammer, and B. J. Waterhouse. The construction of good extensible Korobov rules. *Computing*, 79:79–91, 2007.
84. J. Dick, F. Pillichshammer, and B. J. Waterhouse. The construction of good extensible rank-1 lattices. *Mathematics of Computation*, 77:2345–2373, 2008.
85. U. Dieter. How to calculate shortest vectors in a lattice. *Mathematics of Computation*, 29(131):827–833, 1975.
86. S. A. R. Disney and I. H. Sloan. Lattice integration rules of maximal rank formed by copying rank 1 rules. *SIAM Journal on Numerical Analysis*, 29:566–577, 1992.
87. A. Doucet, N. de Freitas, and N. Gordon, editors. *Sequential Monte Carlo Methods in Practice.* Springer-Verlag, New York, 2001.
88. A. Doucet, S. Godsill, and C. Andrieu. On sequential Monte Carlo sampling methods for Bayesian filtering. *Statistics and Computing*, 10:197–208, 2000.
89. M. Drmota and R. F. Tichy. *Sequences, Discrepancies and Applications*, volume 1651 of Lecture Notes in Mathematics. Springer-Verlag, New York, 1997.
90. J.-C. Duan, G. Gauthier, and J.-G. Simonato. Asymptotic distribution of the EMS option price estimator. *Management Science*, 47(8):1122–1132, 2001.
91. J.-C. Duan and J.-G. Simonato. Empirical martingale simulation for asset prices. *Management Science*, 44:1218–1233, 1998.
92. D. Duffie. *Dynamic Asset Pricing Theory*, second edition. Princeton University Press, Princeton, NJ, 1996.
93. D. Duffie and P. Glynn. Efficient Monte Carlo simulation for security prices. *The Annals of Applied Probability*, 5(4):897–905, 1995.
94. D. S. Dummit and R. M. Foote. *Abstract Algebra*, second edition. John Wiley and Sons, New York, 1999.
95. R. Eckhardt. Stan Ulam, John von Neumann and the Monte Carlo method. *Los Alamos Science*, pages 131–143, 1987. Special Issue.
96. Y. Edel and J. Bierbrauer. Construction of digital nets from bch-codes. In H. Niederreiter, P. Hellekalek, G. Larcher, and P. Zinterhof, editors, *Monte Carlo and Quasi-Monte Carlo Methods 1996*, volume 127 of Lecture Notes in Statistics, pages 221–231. Springer-Verlag, New York, 1997.
97. B. Efron. Bootstrap methods: Another look at the jackknife. *Annals of Statistics*, 7:1–26, 1979.
98. B. Efron. *The Jackknife, the Bootstrap and Other Resampling Plans*, volume 38 of CBMS-NSF Regional Conference Series in Applied Mathematics. SIAM, Philadelphia, 1982.
99. B. Efron and C. Stein. The jackknife estimator of variance. *Annals of Statistics*, 9:586–596, 1981.
100. B. Efron and R. J. Tibshirani. *An Introduction to the Bootstrap*. Chapman and Hall, New York, 1993.
101. S. M. T. Ehrlichman and S. G. Henderson. American options from MARS. In B. Lawson, J. Liu, F. Perrone, and F. Wieland, editors, *Proceedings of the 2006 Winter Simulation Conference*, pages 719–726. IEEE Press, Piscataway, NJ, 2006.
102. J. Eichenauer-Herrmann. Inversive congruential pseudorandom numbers: A tutorial. *International Statistical Reviews*, 60:167–176, 1992.
103. R. J. Elliott, L. Chan, and T. K. Siu. Option pricing and Esscher transform under regime switching. *Annals of Finance*, 1:423–432, 2005.
104. P. Embrechts. Copulas: A personal view. *Journal of Risk and Insurance*, 2009. To appear.
105. K. Entacher. Bad subsequences of well-known linear congruential pseudorandom number generators. *ACM Transactions on Modeling and Computer Simulation*, 8(1):61–70, 1998.

106. K. Entacher and B. Hechenleitner. A parallel search for good lattice points using lll-spectral tests. *Journal of Computational and Applied Mathematics*, 189:424–441, 2006.
107. K. Entacher, P. Hellekalek, and P. L'Ecuyer. Quasi-Monte Carlo node sets from linear congruential generators. In H. Niederreiter and J. Spanier, editors, *Monte Carlo and Quasi-Monte Carlo Methods 1998*, pages 188–198. Springer, Berlin, 2000.
108. K. Entacher, G. Laimer, H. Röck, and A. Uhl. Normalization of the spectral test in high dimensions. *Monte Carlo Methods and Applications*, 10:265–272, 2004.
109. K. Entacher, T. Schell, and A. Uhl. Bad lattice points. *Computing*, 75:281–295, 2005.
110. K.-T. Fang. Some applications of quasi-Monte Carlo methods in statistics. In K.-T. Fang, F. J. Hickernell, and H. Niederreiter, editors, *Monte Carlo and Quasi-Monte Carlo Methods 2000*, pages 10–26. Springer, New York, 2001.
111. K.-T. Fang, Y. Tang, and J. Yin. Lower bounds for wrap-around L_s-discrepancy and constructions of symmetrical uniform designs. *Journal of Complexity*, 21:757–771, 2005.
112. H. Faure. Discrépance des suites associées à un système de numération (en dimension s). *Acta Arithmetica*, 41:337–351, 1982.
113. H. Faure. Good permutations for extreme discrepancy. *Journal of Number Theory*, 42(1):47–56, 1992.
114. H. Faure. Selection criteria for (random) generation of digital $(0, s)$-sequences. In H. Niederreiter and D. Talay, editors, *Monte Carlo and Quasi-Monte Carlo Methods 2004*, pages 113–126. Springer, New York, 2006.
115. H. Faure and C. Lemieux. Generalized Halton sequence in 2008: A comparative study. Manuscript, 2008.
116. H. Faure and S. Tezuka. Another random scrambling of digital (t, s) sequences. In K.-T. Fang, F. J. Hickernell, and H. Niederreiter, editors, *Monte Carlo and Quasi-Monte Carlo Methods 2000*, pages 242–256. Springer, New York, 2001.
117. P. Fearnhead. Using random quasi-Monte Carlo within particle filters, with application to financial time series. *Journal of Computational and Graphical Statistics*, 14:751–769, 2005.
118. U. Fincke and M. Pohst. Improved methods for calculating vectors of short length in a lattice, including a complexity analysis. *Mathematics of Computation*, 44:463–471, 1985.
119. G. S. Fishman. Multiplicative congruential random number generators with modulus 2^β: An exhaustive analysis for $\beta = 32$ and a partial analysis for $\beta = 48$. *Mathematics of Computation*, 54(189):331–344, 1990.
120. G. S. Fishman. *Monte Carlo: Concepts, Algorithms, and Applications.* Springer Series in Operations Research. Springer-Verlag, New York, 1996.
121. G. S. Fishman. *A First Course in Monte Carlo.* Duxbury Press, Belmont, CA, 2005.
122. G. S. Fishman and B. D. Huang. Antithetic variates revisited. *Communications of the ACM*, 26:964–971, 1983.
123. G. S. Fishman and L. S. Moore III. An exhaustive analysis of multiplicative congruential random number generators with modulus $2^{31} - 1$. *SIAM Journal on Scientific and Statistical Computing*, 7(1):24–45, 1986.
124. G. B. Folland. *Fourier Analysis and Its Applications.* Wadsworth and Brooks, Pacific Grove, CA, 1992.
125. G. E. Forsythe and R. A. Leibler. Matrix inversion by a Monte Carlo method. *Mathematical Tables and Other Aids to Computation*, 4(31):127–129, 1950.
126. B. L. Fox. Generation of random samples from the Beta and F distributions. *Technometrics*, 5:269–270, 1963.
127. B. L. Fox. Implementation and relative efficiency of quasirandom sequence generators. *ACM Transactions on Mathematical Software*, 12:362–376, 1986.
128. B. L. Fox. *Strategies for Quasi-Monte Carlo.* Kluwer Academic, Boston, 1999.
129. B. L. Fox and P. W. Glynn. Computing Poisson probabilities. *Communications of the ACM*, 31:440–445, 1988.

130. E. W. Frees and E. A. Valdez. Understanding relationships using copulas. *North American Actuarial Journal*, 2:1–25, 1998.
131. R. Freivalds. Fast probabilistic algorithms. In *Proceedings of the 8th Symposium on the Mathematical Foundations of Computer Science*, volume 74 of Lecture Notes in Computer Science. Springer-Verlag, Berlin, 1979.
132. J. H. Friedman and M. H. Wright. A nested partitioning procedure for numerical integration. *ACM Transactions on Mathematical Software*, 7:76–92, 1981.
133. M. C. Fu. Optimization via simulation: A review. *Annals of Operations Research*, 53:199–248, 1994.
134. M. C. Fu, S. B. Laprise, D. B. Madan, Y. Su, and R. Wu. Pricing American options: A comparison of Monte Carlo simulation approaches. *Journal of Computational Finance*, 2:49–74, 1999.
135. M. Fushimi and S. Tezuka. The k-distribution of generalized feedback shift register pseudorandom numbers. *Communications of the ACM*, 26(7):516–523, 1983.
136. C. Genest. Frank's family of bivariate distributions. *Biometrika*, 74:549–555, 1987.
137. J. E. Gentle. *Random Number Generation and Monte Carlo Methods*, second edition. Springer, New York, 2003.
138. H. Gerber and E. Shiu. Option pricing by Esscher transforms. *Transactions of the Society of Actuaries*, 46:99–140, 1994.
139. M. C. Giles. Multi-level Monte Carlo path simulation. *Operations Research*, 56:607–617, 2008.
140. W. R. Gilks, S. Richardson, and D. J. Spiegelhalter. *Markov Chain Monte Carlo in practice.* Chapman and Hall/CRC, Boca Raton, FL, 1998.
141. H. S. Gill and C. Lemieux. A search for extensible Korobov rules. *Journal of Complexity*, 23:603–613, 2007.
142. D. Gillespie. Exact stochastic simulation of coupled chemical reactions. *Journal of Physical Chemistry*, 81:2340–2361, 1977.
143. D. Gillespie. Approximate accelerated stochastic simulation of chemically reacting systems. *Journal of Chemical Physics*, 115:1716–1733, 2001.
144. P. Glasserman. *Gradient Estimation via Perturbation Analysis.* Kluwer Academic, Norwell, MA, 1991.
145. P. Glasserman. *Monte Carlo Methods in Financial Engineering*, volume 53 of Application of Mathematics – Stochastic Modelling and Applied Probability. Springer, New York, 2004.
146. P. Glasserman, P. Heidelberger, and P. Shahabuddin. Asymptotically optimal importance sampling and stratification for pricing path dependent options. *Journal of Mathematical Finance*, 9(2):117–152, 1999.
147. P. Glasserman, P. Heidelberger, and P. Shahabuddin. Importance sampling and stratification for value-at-risk. In Y. S. Abu-Mostafa, B. LeBaron, A. W. Lo, and A. S. Weigend, editors, *Computational Finance 1999 (Proceedings of the Sixth International Conference on Computational Finance)*. MIT Press, Cambridge, MA, 1999.
148. P. Glasserman, P. Heidelberger, and P. Shahabuddin. Variance reduction techniques for estimating value-at-risk. *Management Science*, 46:1349–1364, 2000.
149. P. Glasserman, P. Heidelberger, P. Shahabuddin, and T. Zajic. Multilevel splitting for estimating rare event probabilities. *Operations Research*, 47(4):585–600, 1999.
150. P. Glasserman and B. Yu. Simulation for American options: Regression now or regression later. In H. Niederreiter, editor, *Monte Carlo and Quasi-Monte Carlo Methods 2002*, pages 213–226. Springer, New York, 2004.
151. P. Glasserman and B. Yu. Large sample properties of weighted Monte Carlo estimators. *Operations Research*, 53:298–312, 2005.
152. P. W. Glynn. Likelihood ratio gradient estimation: An overview. In *Proceedings of the 1987 Winter Simulation Conference*, pages 366–375. IEEE Press, Piscataway, NJ, 1987.
153. P. W. Glynn. Efficiency improvement techniques. *Annals of Operations Research*, 53:175–197, 1994.

154. P. W. Glynn. Importance sampling for Monte Carlo estimation of quantiles. In *Proceedings of the Second International Workshop on Mathematical Methods in Stochastic Simulation and Experimental Design*, pages 180–185. St. Petersburg University Press, St. Petersburg, Russia, 1996.
155. P. W. Glynn and R. Szechtman. Some new perspectives on the method of control variates. In K.-T. Fang, F. J. Hickernell, and H. Niederreiter, editors, *Monte Carlo and Quasi-Monte Carlo Methods 2000*, pages 27–49. Springer-Verlag, Berlin, 2002.
156. P. W. Glynn and M. Torres. Nonparametric estimation of tail probabilities for the single-server queue. In P. Glasserman, K. Sigman, and D. D. Yao, editors, *Stochastic Networks: Stability and Rare Events*, volume 117 of Lecture Notes in Statistics, pages 109–138. Springer, New York, 1996.
157. P. W. Glynn and W. Whitt. The asymptotic efficiency of simulation estimators. *Operations Research*, 40:505–520, 1992.
158. M. Goresky and A. Klapper. Efficient multiple-with-carry random number generators with maximal period. *ACM Transactions on Modeling and Computer Simulation*, 13:310–321, 2003.
159. A. Grube. Mehrfach rekursiv-erzeugte Pseudo-Zufallszahlen. *Zeitschrift für angewandte Mathematik und Mechanik*, 53:T223–T225, 1973.
160. S. Haber. Parameters for integrating periodic functions of several variables. *Mathematics of Computation*, 41:115–129, 1983.
161. J. H. Halton. On the efficiency of certain quasi-random sequences of points in evaluating multi-dimensional integrals. *Numerische Mathematik*, 2:84–90, 1960.
162. J. M. Hammersley. Conditional Monte Carlo. *Journal of the ACM*, 3:73–76, 1956.
163. J. M. Hammersley. Monte Carlo methods for solving multivariable problems. *Annals of the New York Academy of Sciences*, 86:844–874, 1960.
164. J. M. Hammersley and D. C. Handscomb. A new Monte Carlo technique: Antithetic variates. *Proceedings of the Cambridge Philosophical Society*, 52:449–475, 1956.
165. J. M. Hammersley and D. C. Handscomb. *Monte Carlo Methods.* Methuen, London, 1964.
166. M. R. Hardy, R. K. Freeland, and M. C. Till. Validation of long-term equity return models for equity-linked guarantees. *North American Actuarial Journal*, 10:28–47, 2006.
167. P. J. Harrison and C. F. Stevens. Bayesian forecasting (with discussion). *Journal of the Royal Statistical Society, Series B*, 38:205–247, 1976.
168. W. K. Hastings. Monte Carlo sampling methods using Markov chains and systems. *Biometrika*, 57:97–109, 1970.
169. M. B. Haugh and L. Kogan. Pricing American options: A duality approach. *Operations Research*, 52:258–270, 2004.
170. S. Heinrich. Efficient algorithms for computing the L_2 discrepancy. *Mathematics of Computation*, 65:1621–1633, 1996.
171. P. Hellekalek. General discrepancy estimates: The Walsh function system. *Acta Arithmetica*, 67:209–218, 1994.
172. P. Hellekalek. On the assessment of random and quasi-random point sets. In P. Hellekalek and G. Larcher, editors, *Random and Quasi-Random Point Sets*, volume 138 of Lecture Notes in Statistics, pages 49–108. Springer, New York, 1998.
173. P. Hellekalek and H. Leeb. Dyadic diaphony. *Acta Arithmetica*, 80:187–196, 1997.
174. P. Hellekalek and H. Niederreiter. The weighted spectral test: Diaphony. *ACM Transactions on Modeling and Computer Simulation*, 8(1):43–60, 1998.
175. S. G. Henderson and B. L. Nelson, editors. *Elsevier Handbooks in Operations Research and Management Science: Simulation*, volume 13. Elsevier Science, Amsterdam, 2006.
176. T. Hesterberg. *Advances in importance sampling.* PhD thesis, Statistics Department, Stanford University, 1988.
177. T. Hesterberg. Control variates and importance sampling for efficient bootstrap simulations. *Statistics and Computing*, 6:147–157, 1996.

178. T. Hesterberg and B. L. Nelson. Control variates for probability and quantile estimation. *Management Science*, 44:1295–1312, 1998.
179. S. L. Heston. A closed-form solution for options with stochastic volatility with applications to bond and currency options. *Review of Financial Studies*, 6:327–343, 1993.
180. F. J. Hickernell. Quadrature error bounds with applications to lattice rules. *SIAM Journal on Numerical Analysis*, 33:1995–2016, 1996.
181. F. J. Hickernell. A generalized discrepancy and quadrature error bound. *Mathematics of Computation*, 67:299–322, 1998.
182. F. J. Hickernell. Lattice rules: How well do they measure up? In P. Hellekalek and G. Larcher, editors, *Random and Quasi-Random Point Sets*, volume 138 of Lecture Notes in Statistics, pages 109–166. Springer, New York, 1998.
183. F. J. Hickernell. Obtaining $o(n^{-2+\epsilon})$ convergence for lattice quadrature rules. In K.-T. Fang, F. J. Hickernell, and H. Niederreiter, editors, *Monte Carlo and Quasi-Monte Carlo Methods 2000*, pages 274–289. Springer, New York, 2001.
184. F. J. Hickernell and H. S. Hong. Computing multivariate normal probabilities using rank-1 lattice sequences. In G. H. Golub, S. H. Lui, F. T. Luk, and R. J. Plemmons, editors, *Proceedings of the Workshop on Scientific Computing (Hong Kong)*, pages 209–215. Springer-Verlag, Singapore, 1997.
185. F. J. Hickernell and H. S. Hong. The asymptotic efficiency of randomized nets for quadrature. *Mathematics of Computation*, 68:767–791, 1999.
186. F. J. Hickernell, H. S. Hong, P. L'Ecuyer, and C. Lemieux. Extensible lattice sequences for quasi-Monte Carlo quadrature. *SIAM Journal on Scientific Computing*, 22:1117–1138, 2001.
187. F. J. Hickernell, C. Lemieux, and A. B. Owen. Control variates for quasi-Monte Carlo. *Statistical Science*, 20:1–31, 2005.
188. F. J. Hickernell and H. Niederreiter. The existence of good extensible rank-1 lattices. *Journal of Complexity*, 19:286–300, 2003.
189. F. J. Hickernell and X. Wang. The error bounds and tractability of quasi-Monte Carlo methods in infinite dimension. *Mathematics of Computation*, 71:1641–1661, 2001.
190. F. J. Hickernell and H. Woźniakowski. Integration and approximation in arbitrary dimensions. *Advances in Computational Mathematics*, 12:25–58, 2000.
191. E. Hlawka. Funktionen von beschränkter variation in der theorie der gleichverteilung. *Annali di Matematica Pura ed Applicata*, 54:325–333, 1961.
192. Y.-C. Ho and X.-R. Cao. *Discrete-Event Dynamic Systems and Perturbation Analysis*. Kluwer Academic, Norwell, MA, 1991.
193. W. Hoeffding. A class of statistics with asymptotically normal distributions. *Annals of Mathematical Statistics*, 19:293–325, 1948.
194. H. S. Hong and F. J. Hickernell. Algorithm 823: Implementing scrambled digital sequences. *ACM Transactions on Mathematical Software*, 29:95–109, 2003.
195. W. Hörmann and J. Leydold. Importance sampling to accelerate the convergence of quasi-Monte Carlo. Technical Report 49, Department of Statistics and Mathematics,Wirtschaftuniversität Wien, February 2007.
196. W. Hörrmann, J. Leydold, and G. Derflinger. *Automatic Nonuniform Random Variate Generation*. Springer-Verlag, New York, 2003.
197. L. K. Hua and Y. Wang. *Applications of Number Theory to Numerical Analysis*. Springer, Berlin, 1981.
198. J. Hull. *Options, Futures, and Other Derivative Securities*, sixth edition. Prentice-Hall, Englewood Cliffs, NJ, 2006.
199. J. Hull and A. White. The pricing of options on assets with stochastic volatilities. *Journal of Finance*, 42:281–300, 1987.
200. J. Imai and K. S. Tan. Enhanced quasi-Monte Carlo methods with dimension reduction. In E. Yücesan and C.-H. Chen, editors, *Proceedings of the 2002 Winter Simulation Conference*, pages 1502–1510. IEEE Press, Piscataway, NJ, 2002.

201. J. Imai and K. S. Tan. An accelerating quasi-Monte Carlo method for option pricing under the generalized hyperbolic Lévy process. To appear, 2008.
202. P. Jäckel. *Monte Carlo Methods in Finance.* Wiley, New York, 2002.
203. J. Jaffari and M. Anis. On efficient Monte Carlo-based statistical static timing analysis of digital circuits. In J. Roychowdhury and L. Scheffer, editors, *Proceedings of the IEEE International Conference on Computer Aided Design (IEEE-ICCAD) 2008.* IEEE Press, New York, 2008.
204. F. James. A review of pseudorandom number generators. *Computer Physics Communications*, 60:329–344, 1990.
205. T. Jiang and A. B. Owen. Quasi-regression with shrinkage. *Mathematics and Computers in Simulation*, 62:231–241, 2003.
206. X. Jin and A. X. Zhang. Reclaiming quasi-Monte Carlo efficiency in portfolio value-at-risk simulation through Fourier transform. *Management Science*, 52:925–938, 2006.
207. S. Joe and F. Y. Kuo. Remark on Algorithm 659: Implementing Sobol's quasirandom sequence generator. *ACM Transactions on Mathematical Software*, 29(1):49–57, 2003.
208. S. Joe and F. Y. Kuo. Constructing Sobol' sequences with better two-dimensional projections. *SIAM Journal on Scientific Computing*, 30:2635–2654, 2008.
209. H. Kahn. Use of different Monte Carlo sampling techniques. In H. Meyer, editor, *Symposium on Monte Carlo Methods*, pages 146–190. John Wiley and Sons, New York, 1956.
210. R. E. Kalman. A new approach to linear filtering and prediction problems. *Transactions of the ASME Journal of Basic Enginnering, Series D*, 82:35–45, 1960.
211. M. H. Kalos and P. A. Whitlock. *Monte Carlo Methods.* Wiley–Interscience, New York, 1986.
212. I. Karatzas and S. Shreve. *Brownian Motion and Stochastic Calculus*, second edition. Springer-Verlag, New York, 1988.
213. A. Keller. *Quasi-Monte Carlo methods for photorealistic image synthesis.* PhD thesis, Universität Kaiserlautern, 1997.
214. A. Keller. Stratification by rank-1 lattices. In H. Niederreiter, editor, *Monte Carlo and Quasi-Monte Carlo Methods 2002*, pages 299–314. Springer, New York, 2004.
215. A. G. Z. Kemna and A. C. F. Vorst. A pricing method for options based on average asset values. *Journal of Banking and Finance*, 14:113–129, 1990.
216. W. J. Kennedy, Jr. and J. E. Gentle. *Statistical Computing.* Dekker, New York, 1980.
217. J. P. C. Kleijnen. *Statistical Techniques in Simulation, Part. 1.* Dekker, New York, 1974.
218. J. P. C. Kleijnen. *Statistical Techniques in Simulation, Part. 2.* Dekker, New York, 1975.
219. P. E. Kloeden and E. Platen. *Numerical Solution of Stochastic Differential Equations.* Springer-Verlag, Berlin, 1992.
220. D. E. Knuth. *The Art of Computer Programming, Volume 2: Seminumerical Algorithms*, second edition. Addison-Wesley, Reading, MA, 1981.
221. D. E. Knuth. *The Art of Computer Programming, Volume 2: Seminumerical Algorithms*, third edition. Addison-Wesley, Reading, MA, 1998.
222. L. Kocis and W. J. Whiten. Computational investigations of low-discrepancy sequences. *ACM Transactions on Mathematical Software*, 23(2):266–294, 1997.
223. J. F. Koksma. Een algemeene stelling uit de theorie der gelikmatige verdeeling modulo 1. *Mathematica B (Zutphen)*, 11:7–11, 1942/1943.
224. N. M. Korobov. The approximate computation of multiple integrals. *Doklady Akademii Nauk SSSR*, 124:1207–1210, 1959. In Russian.
225. P. Kritzer. Improved upper bounds on the star discrepancy of (t, m, s)-nets and (t, s)-sequences. *Journal of Complexity*, 22:336–347, 2006.
226. P. Kritzer and F. Pillichshammer. Constructions of general polynomial lattices for multivariate integration. *Bulletin of the Australian Mathematical Society*, 76:93–110, 2007.

227. P. Kritzer and F. Pillichshammer. The weighted dyadic diaphony of digital sequences. In A. Keller, S. Heinrich, and H. Niederreiter, editors, *Monte Carlo and Quasi-Monte Carlo Methods 2006*, pages 549–560. Springer, New York, 2008.
228. L. Kuipers and H. Niederreiter. *Uniform Distribution of Sequences*. John Wiley and Sons, New York, 1974.
229. F. Y. Kuo. Component-by-component constructions achieve the optimal rate of convergence for multivariate integration in weighted Korobov and Sobolev spaces. *Journal of Complexity*, 19:301–320, 2003.
230. F. Y. Kuo and S. Joe. Component-by-component constructions of good lattice rules with a composite number of points. *Journal of Complexity*, 18:943–976, 2002.
231. F. Y. Kuo and S. Joe. Component-by-component construction of good intermediate-rank lattice rules. *SIAM Journal on Numerical Analysis*, 41:1465–1486, 2003.
232. F. Y. Kuo, I. H. Sloan, and H. Woźniakowski. Periodization strategy may fail in high dimensions. *Numerical Algorithms*, 46:369–391, 2007.
233. F. Y. Kuo, W. T. M. Dunsmuir I. H. Sloan, M. P. Wand, and R. S. Womersley. Quasi-Monte Carlo for highly structured generalized response models. *Methodology and Computing in Applied Probability*, 10:239–275, 2008.
234. H. J. Kushner and D. S. Clark. *Stochastic Approximation Methods for Constrained and Unconstrained Systems*, volume 26 of Applied Mathematical Sciences. Springer-Verlag, New York, 1978.
235. Y. Lai and K.S. Tan. Simulation of nonlinear portfolio value-at-risk by Monte Carlo and quasi-Monte Carlo methods. In M. Holder, editor, *Financial Engineering and Applications 2006*. ACTA Press, Cambridge, 2006.
236. D. P. Landau and K. Binder. *A Guide to Monte Carlo Simulations in Statistical Physics*, second edition. Cambridge University Press, Cambridge, 2005.
237. G. Larcher. Digital point sets: Analysis and applications. In P. Hellekalek and G. Larcher, editors, *Random and Quasi-Random Point Sets*, volume 138 of Lecture Notes in Statistics, pages 167–222. Springer, New York, 1998.
238. G. Larcher, A. Lauss, H. Niederreiter, and W. Ch. Schmid. Optimal polynomials for (t, m, s)-nets and numerical integration of multivariate Walsh series. *SIAM Journal on Numerical Analysis*, 33(6):2239–2253, 1996.
239. G. Larcher, H. Niederreiter, and W. Ch. Schmid. Digital nets and sequences constructed over finite rings and their application to quasi-Monte Carlo integration. *Monatshefte für Mathematik*, 121(3):231–253, 1996.
240. G. Larcher and C. Traunfellner. The numerical integration of Walsh series. *Mathematics of Computation*, 63:277–291, 1994.
241. S. S. Lavenberg, T. L. Moeller, and P. D. Welch. Statistical results on multiple control variables with application to queueing network simulation. *Operations Research*, 30(1):182–202, 1982.
242. S. S. Lavenberg and P. D. Welch. A perspective on the use of control variables to increase the efficiency of Monte Carlo simulations. *Management Science*, 27:322–335, 1981.
243. A. M. Law and W. D. Kelton. *Simulation Modeling and Analysis*, third edition. McGraw-Hill, New York, 2000.
244. K. M. Lawrence, A. Mahalanabis, G. L. Mullen, and W. Ch. Schmid. Construction of digital (t, m, s)-nets from linear codes. In S. D. Cohen and H. Niederreiter, editors, *Finite Fields and Applications*, volume 233 of Lecture Notes Series of the London Mathematical Society, pages 189–208. Cambridge University Press, Cambridge, 1996.
245. C. Lécot and S. Ogawa. Quasirandom walk methods. In K.-T. Fang, F. J. Hickernell, and H. Niederreiter, editors, *Monte Carlo and Quasi-Monte Carlo Methods 2000*, pages 63–85. Springer, New York, 2001.
246. C. Lécot and B. Tuffin. Quasi-Monte Carlo methods for estimating transient measures of discrete time Markov chains. In H. Niederreiter, editor, *Monte Carlo and Quasi-Monte Carlo Methods 2002*, pages 329–343. Springer, New York, 2004.

247. P. L'Ecuyer. Efficiency improvement via variance reduction. In J. D. Tew, S. Manivannan, D. A. Sadowski, and A. F. Seila, editors, *Proceedings of the 1994 Winter Simulation Conference*, pages 122–132. IEEE Press, Piscataway, NJ, 1994.
248. P. L'Ecuyer. Uniform random number generation. *Annals of Operations Research*, 53:77–120, 1994.
249. P. L'Ecuyer. Maximally equidistributed combined Tausworthe generators. *Mathematics of Computation*, 65(213):203–213, 1996.
250. P. L'Ecuyer. Bad lattice structures for vectors of non-successive values produced by some linear recurrences. *INFORMS Journal on Computing*, 9(1):57–60, 1997.
251. P. L'Ecuyer. Random number generators and empirical tests. In P. Hellekalek, G. Larcher, H. Niederreiter, and P. Zinterhof, editors, *Monte Carlo and Quasi-Monte Carlo Methods in Scientific Computing*, volume 127 of Lecture Notes in Statistics, pages 124–138. Springer, New York, 1998.
252. P. L'Ecuyer. Good parameters and implementations for combined multiple recursive random number generators. *Operations Research*, 47(1):159–164, 1999.
253. P. L'Ecuyer. Tables of linear congruential generators of different sizes and good lattice structure. *Mathematics of Computation*, 68(225):249–260, 1999.
254. P. L'Ecuyer. Tables of maximally equidistributed combined LFSR generators. *Mathematics of Computation*, 68(225):261–269, 1999.
255. P. L'Ecuyer. Software for uniform random number generation:distinguishing the good and the bad. In B. A. Peters, J. S. Smith, D. J. Medeiros, and M. W. Rohrer, editors, *Proceedings of the 2001 Winter Simulation Conference*, pages 95–105. IEEE Press, Pistacaway, NJ, 2001.
256. P. L'Ecuyer. Polynomial integration lattices. In H. Niederreiter, editor, *Monte Carlo and Quasi-Monte Carlo Methods 2002*, pages 73–98. Springer, New York, 2004.
257. P. L'Ecuyer. Random number generation. In S. G. Henderson and B. L. Nelson, editors, *Elsevier Handbooks in Operations Research and Management Science: Simulation*, chapter 3, pages 55–81. Elsevier Science, Amsterdam, 2006.
258. P. L'Ecuyer and Y. Champoux. Estimating small cell-loss ratios in ATM switches via importance sampling. *ACM Transactions on Modeling and Computer Simulation*, 11:76–105, 2001.
259. P. L'Ecuyer and R. Couture. An implementation of the lattice and spectral tests for multiple recursive linear random number generators. *INFORMS Journal on Computing*, 9(2):206–217, 1997.
260. P. L'Ecuyer, V. Demers, and B. Tuffin. Rare-event, splitting and quasi-Monte Carlo. *ACM Transactions on Modeling and Computer Simulation*, 17:1–45, 2006.
261. P. L'Ecuyer and J. Granger-Piché. Combined generators with components from different families. *Mathematics and Computers in Simulation*, 62:395–404, 2003.
262. P. L'Ecuyer and P. Hellekalek. Random number generators: Selection criteria and testing. In P. Hellekalek and G. Larcher, editors, *Random and Quasi-Random Point Sets*, volume 138 of Lecture Notes in Statistics, pages 223–265. Springer, New York, 1998.
263. P. L'Ecuyer, C. Lécot, and B. Tuffin. Randomized quasi-Monte Carlo simulation of Markov chains with an ordered state space. In H. Niederreiter and D. Talay, editors, *Monte Carlo and Quasi-Monte Carlo Methods 2004*, pages 331–342. Springer-Verlag, New York, 2006.
264. P. L'Ecuyer and C. Lemieux. Variance reduction via lattice rules. *Management Science*, 46(9):1214–1235, 2000.
265. P. L'Ecuyer and C. Lemieux. Recent advances in randomized quasi-Monte Carlo methods. In M. Dror, P. L'Ecuyer, and F. Szidarovszki, editors, *Modeling Uncertainty: An Examination of Stochastic Theory, Methods, and Applications*, pages 419–474. Kluwer Academic Publishers, Boston, 2002.
266. P. L'Ecuyer, L. Meliani, and J. Vaucher. SSJ: A framework for stochastic simulation in Java. In E. Yücesan and C.-H. Chen, editors, *Proceedings of the 2002 Winter Simulation Conference*, pages 234–242. IEEE Press, 2002.

267. P. L'Ecuyer and F. Panneton. A new class of linear feedback shift register generators. In J. A. Joines, R. R. Barton, K. Kang, and P. A. Fishwick, editors, *Proceedings of the 2000 Winter Simulation Conference*, pages 690–696. IEEE Press, Pistacaway, NJ, 2000.
268. P. L'Ecuyer and F. Panneton. Construction of equidistributed generators based on linear recurrences modulo 2. In K.-T. Fang, F. J. Hickernell, and H. Niederreiter, editors, *Monte Carlo and Quasi-Monte Carlo Methods 2000*, pages 318–330. Springer, New York, 2002.
269. P. L'Ecuyer and F. Panneton. Random number generators based on linear recurrences modulo 2. In C.Alexopoulos, D. Goldsman, and J. R. Wilson, editors, *Advancing the Frontiers of Simulation: A Festschrift in Honor of George S. Fishman.* Springer, New York, 2008. Forthcoming.
270. P. L'Ecuyer and C. Sanvido. Coupling from the past with randomized quasi-Monte Carlo. Manuscript.
271. P. L'Ecuyer and R. Simard. On the performance of birthday spacings tests for certain families of random number generators. *Mathematics and Computers in Simulation*, 55:139–148, 2001.
272. P. L'Ecuyer, R. Simard, E. J. Chen, and W. D. Kelton. An object-oriented random-number package with many long streams and substreams. *Operations Research*, 50(6):1073–1075, 2002.
273. P. L'Ecuyer, R. Simard, and S. Wegenkittl. Sparse serial tests of uniformity for random number generators. *SIAM Journal on Scientific Computing*, 24:652–668, 2002.
274. H. Leeb and S. Wegenkittl. Inversive and linear congruential pseudorandom number generators in empirical tests. *ACM Transactions on Modeling and Computer Simulation*, 7(2):272–286, 1997.
275. E. L. Lehmann. Some concepts of dependence. *Annals of Mathematical Statistics*, 37:1137–1153, 1966.
276. D. H. Lehmer. Mathematical methods in large scale computing units. *Annals of the Computation Laboratory of Harvard University*, 26:141–146, 1951.
277. C. Lemieux. A comparison of copy rules and Korobov rules. Yellow Series Research Paper No. 836, Department of Mathematics and Statistics, University of Calgary, 2004.
278. C. Lemieux. Randomized quasi-Monte Carlo methods: A tool for improving the efficiency of simulations in finance. In R. G. Ingalls, M. D. Rossetti, J. S. Smith, and B. A Peters, editors, *Proceedings of the 2004 Winter Simulation Conference*, pages 1565–1573. IEEE Press, 2004.
279. C. Lemieux, M. Cieslak, and K. Luttmer. RandQMC user's guide: A package for randomized quasi-Monte Carlo methods in C. Technical Report 2002-712-15, Department of Computer Science, University of Calgary, 2002.
280. C. Lemieux and J. La. A study of variance reduction techniques for American option pricing. In N. Steiger and M. E. Kuhl, editors, *Proceedings of the 2005 Winter Simulation Conference*, pages 1884–1891. IEEE Press, Piscataway, NJ, 2005.
281. C. Lemieux and P. L'Ecuyer. Lattice rules for the simulation of ruin problems. In H. Szczerbicka, editor, *Proceedings of the 1999 European Simulation Multiconference*, volume 2, pages 533–537. The Society for Computer Simulation, Ghent, Belgium, 1999.
282. C. Lemieux and P. L'Ecuyer. A comparison of Monte Carlo, lattice rules and other low-discrepancy point sets. In H. Niederreiter and J. Spanier, editors, *Monte Carlo and Quasi-Monte Carlo Methods 1998*, pages 326–340. Springer, Berlin, 2000.
283. C. Lemieux and P. L'Ecuyer. Using lattice rules for variance reduction in simulation. In J. A. Joines, R. R. Barton, K. Kang, and P. A. Fishwick, editors, *Proceedings of the 2000 Winter Simulation Conference*, pages 509–516. IEEE Press, Piscataway, NJ, 2000.

284. C. Lemieux and P. L'Ecuyer. Selection criteria for lattice rules and other low-discrepancy point sets. *Mathematics and Computers in Simulation*, 55:139–148, 2001.
285. C. Lemieux and P. L'Ecuyer. Randomized polynomial lattice rules for multivariate integration and simulation. *SIAM Journal on Scientific Computing*, 24(5):1768–1789, 2003.
286. C. Lemieux and A. B. Owen. Quasi-regression and the relative importance of the ANOVA components of a function. In K.-T. Fang, F. J. Hickernell, and H. Niederreiter, editors, *Monte Carlo and Quasi-Monte Carlo Methods 2000*, pages 331–344. Springer, New York, 2001.
287. C. Lemieux and P. Sidorsky. Exact sampling with highly-uniform point sets. *Mathematical and Computer Modelling*, 43:339–349, 2006.
288. M. B. Levin. Discrepancy estimates of completely uniformly distributed and pseudo-random number sequences. *International Mathematics Research Notices*, 22:1231–1251, 1999.
289. T. G. Lewis and W. H. Payne. Generalized feedback shift register pseudorandom number algorithm. *Journal of the ACM*, 20(3):456–468, 1973.
290. J. G. Liao. Variance reduction in Gibbs sampler using quasi-random numbers. *Journal of Computational and Graphical Statistics*, 3:253–266, 1998.
291. R. Lidl and H. Niederreiter. *Introduction to Finite Fields and Their Applications*, revised edition. Cambridge University Press, Cambridge, 1994.
292. D. V. Lindley. The theory of queues with a single server. *Proceedings of the Cambridge Philosophical Society*, 43:277–289, 1952.
293. J. S. Liu. *Monte Carlo Strategies in Scientific Computing*. Springer-Verlag, New York, 2001.
294. J. S. Liu and R. Chen. Sequential Monte Carlo methods for dynamic systems. *Journal of the American Statistical Association*, 93:1032–1044, 1998.
295. J. S. Liu, F. Liang, and W. H. Wong. The use of multiple-try method and local optimization in Metropolis sampling. *Journal of the American Statistical Association*, 95:121–134, 2000.
296. M. Q. Liu and F. J. Hickernell. Experimental designs using digital nets with small number of points. In H. Niederreiter and D. Talay, editors, *Monte Carlo and Quasi-Monte Carlo Methods 2004*, pages 343–354. Springer-Verlag, New York, 2006.
297. R. Liu and A. B. Owen. Estimating mean dimensionality of analysis of variance decompositions. *Journal of the American Statistical Association*, 101:712–721, 2006.
298. F. A. Longstaff and E. S. Schwartz. Valuing American options by simulations: A simple least-squares approach. *Review of Financial Studies*, 14(1):113–147, 2001.
299. D. B. Madan, P. Carr, and E. C. Chang. Tha variance gamma process and option pricing. *European Finance Review*, 2:79–105, 1998.
300. D. Maisonneuve. Recherche et utilisation des bons treillis. Programmation et résultats numériques. In S. K. Zaremba, editor, *Application de la théorie des nombres à l'analyse numérique*, pages 121–201. Academic Press, New York, 1972.
301. G. Marsaglia. Random variables and computers. In J. Koseznik, editor, *Information Theory, Statistical Decision Functions, Random Processes: Transactions of the Third Prague Conference*, pages 499–510. Czechoslovak Academy of Sciences, Prague, 1962.
302. G. Marsaglia. Random numbers fall mainly in the planes. *Proceedings of the National Academy of Sciences of the United States of America*, 60:25–28, 1968.
303. G. Marsaglia. A current view of random number generators. In L. Billard, editor, *Computer Science and Statistics: The Interface*, pages 3–10. Elsevier Science Publishers, Amsterdam, 1985.
304. G. Marsaglia and A. Zaman. A new class of random number generators. *The Annals of Applied Probability*, 1:462–480, 1991.
305. W. J. Martin and D. R. Stinson. Association schemes for ordered orthogonal arrays and (t,m,s)-nets. *Canadian Journal of Mathematics*, 51:326–346, 1999.
306. M. Mascagni and H. Chi. On the scrambled Halton sequence. *Monte Carlo Methods and Applications*, 10:435–442, 2004.

307. J. Matoušek. On the L_2-discrepancy for anchored boxes. *Journal of Complexity*, 14:527–556, 1998.
308. J. Matoušek. *Geometric Discrepancy*. Springer, Berlin, 1999.
309. M. Matsumoto and Y. Kurita. Twisted GFSR generators. *ACM Transactions on Modeling and Computer Simulation*, 2(3):179–194, 1992.
310. M. Matsumoto and T. Nishimura. Mersenne Twister: A 623-dimensionally equidistributed uniform pseudo-random number generator. *ACM Transactions on Modeling and Computer Simulation*, 8(1):3–30, 1998.
311. M. Matsumoto, I. Wada, A. Kuramoto, and H. Ashihara. Common defects in initializing pseudorandom number generators. *ACM Transactions Modeling Comp. Simulation*, 17(4):15, 2007.
312. D. J. S. Mayor and H. Niederreiter. A new construction of (t, s)-sequences and some improved bounds on their quality parameters. *Acta Arithmetica*, 128:177–191, 2007.
313. M. D. Mckay, R. J. Beckman, and W. J. Conover. A comparison of three methods for selecting values of input variables in the analysis of output from a computer code. *Technometrics*, 21:239–245, 1979.
314. D. L. McLeish. *Monte Carlo Simulation and Finance*. John Wiley and Sons, Hoboken, NJ, 2005.
315. A. J. McNeil, R. Frey, and P. Embrechts. *Quantitative Risk Management: Concepts, Techniques, Tools*. Princeton Series in Finance. Princeton University Press, Princeton, NJ, 2005.
316. R. Merton. The theory of rational option pricing. *Bell Journal of Economics and Management Science*, 4:141–183, 1973.
317. R. C. Merton. Option pricing when the underlying stock returns are discontinuous. *Journal of Financial Economics*, 3:125–144, 1976.
318. N. Metropolis. The beginning of the Monte Carlo method. *Los Alamos Science*, 15:125–130, 1987.
319. N. Metropolis, A. W. Rosenbluth, M. N. Rosenbluth, A. H. Teller, and E. Teller. Equation of state calculations by fast computing machines. *Journal of Chemical Physics*, 21:1087–1092, 1953.
320. N. Metropolis and S. M. Ulam. The Monte Carlo method. *Journal of the American Statistical Association*, 44:335–341, 1949.
321. H. Meyer, editor. *Symposium on Monte Carlo Methods*. John Wiley and Sons, New York, 1956.
322. F. Michaud. Estimating the probability of ruin for variable premiums by simulation. *Astin Bulletin*, 26:93–105, 1996.
323. G. Miller. Riemann's hypothesis and tests for primality. *Journal of Computer and System Sciences*, 13(3):300–317, 1976.
324. B. Moro. The full Monte. *Risk*, 8:57–58, February 1995.
325. H. Morohosi and M. Fushimi. A practical approach to the error estimation of quasi-Monte Carlo integration. In H. Niederreiter and J. Spanier, editors, *Monte Carlo and Quasi-Monte Carlo Methods 1998*, pages 377–390. Springer-Verlag, Berlin, 2000.
326. W. J. Morokoff and R. E. Caflisch. Quasi-random sequences and their discrepancies. *SIAM Journal on Scientific Computing*, 15:1251–1279, 1994.
327. W. J. Morokoff and R. E. Caflisch. Quasi-Monte Carlo simulation of random walks in finance. In H. Niederreiter, P. Hellekalek, G. Larcher, and P. Zinterhof, editors, *Monte Carlo and Quasi-Monte Carlo Methods 1996*, volume 127 of Lecture Notes in Statistics, pages 340–352. Springer-Verlag, New York, 1997.
328. M. D. Morris, T. J. Mitchell, and D. Ylvisaker. Bayesian design and analysis of computer experiments: Use of derivatives in surface prediction. *Technometrics*, 35:243–255, 1993.
329. M. D. Morris, L. M. Moore, and M. D. McKay. Sampling plans based on balanced incomplete block designs for evaluating the importance of computer model inputs. *Journal of Statistical Planning and Inference*, 136:3203–3220, 2006.

330. R. Měch. *Modeling and simulation of the interaction of plants with the environment using L-systems and their extensions.* PhD thesis, University of Calgary, 1997.
331. B. L. Nelson. Control-variate remedies. *Operations Research*, 38:974–992, 1990.
332. J. Von Neumann. Various techniques used in connection with random digits. *U.S. National Bureau of Standards Applied Mathematics Series*, 12:36–38, 1951.
333. H. Niederreiter. Quasi-Monte Carlo methods and pseudorandom numbers. *Bulletin of the American Mathematical Society*, 84(6):957–1041, 1978.
334. H. Niederreiter. Multidimensional numerical integration using pseudorandom numbers. *Mathematical Programming Study*, 27:17–38, 1986.
335. H. Niederreiter. Point sets and sequences with small discrepancy. *Monatshefte für Mathematik*, 104:273–337, 1987.
336. H. Niederreiter. Low discrepancy and low dispersion sequences. *Journal of Number Theory*, 30:51–70, 1988.
337. H. Niederreiter. Remarks on nonlinear congruential pseudorandom numbers. *Metrika*, 35:321–328, 1988.
338. H. Niederreiter. Low-discrepancy point sets obtained by digital constructions over finite fields. *Czechoslovak Mathematical Journal*, 42:143–166, 1992.
339. H. Niederreiter. *Random Number Generation and Quasi-Monte Carlo Methods*, volume 63 of SIAM CBMS-NSF Regional Conference Series in Applied Mathematics. SIAM, Philadelphia, 1992.
340. H. Niederreiter. The existence of good extensible polynomial lattice rules. *Monatshefte für Mathematik*, 139:295–307, 2003.
341. H. Niederreiter. Nets, (t, s)-sequences and codes. In A. Keller, S. Heinrich, and H. Niederreiter, editors, *Monte Carlo and Quasi-Monte Carlo Methods 2006*, pages 83–100. Springer, New York, 2008.
342. H. Niederreiter and F. Özbudak. Low-discrepancy sequences using duality and global function fields. *Acta Arithmetica*, 130:79–97, 2007.
343. H. Niederreiter and G. Pirsic. Duality for digital nets and its applications. *Acta Arithmetica*, 97:173–182, 2001.
344. H. Niederreiter and I. Shparlinski. Recent advances in the theory of nonlinear pseudorandom number generators. In K.-T. Fang, F. J. Hickernell, and H. Niederreiter, editors, *Monte Carlo and Quasi-Monte Carlo Methods 2000*, pages 86–102. Springer, New York, 2001.
345. H. Niederreiter and C. Xing. Low-discrepancy sequences obtained from algebraic function fields over finite fields. *Acta Arithmetica*, 72:281–298, 1995.
346. H. Niederreiter and C. Xing. Low-discrepancy sequences and global function fields with many rational places. *Finite Fields and Their Applications*, 2:241–273, 1996.
347. H. Niederreiter and C. Xing. The algebraic-geometry approach to low-discrepancy sequences. In P. Hellekalek, G. Larcher, H. Niederreiter, and P. Zinterhof, editors, *Monte Carlo and Quasi-Monte Carlo Methods 1996*, volume 127 of Lecture Notes in Statistics, pages 139–160. Springer-Verlag, New York, 1997.
348. S. Ninomiya and S. Tezuka. Toward real-time pricing of complex financial derivatives. *Applied Mathematical Finance*, 3:1–20, 1996.
349. R. E. Odeh and J. O. Evans. Algorithm AS 70: Percentage points of the normal distribution. *Applied Statistics*, 23:96–97, 1974.
350. B. Øksendal. *Stochastic Differential Equations: An Introduction with Applications*, third edition. Springer-Verlag, New York, 1992.
351. G. Ökten. A probabilistic result on the discrepancy of a hybrid-Monte Carlo sequence and applications. *Monte Carlo Methods and Applications*, 2:255–270, 1996.
352. G. Ökten. Applications of a hybrid-Monte Carlo sequence to option pricing. In H. Niederreiter and J. Spanier, editors, *Monte Carlo and Quasi-Monte Carlo Methods 1998*, pages 391–406. Springer, Berlin, 2000.
353. G. Ökten, B. Tuffin, and V. Burago. A central limit theorem and improved error bounds for a hybrid Monte Carlo sequence with applications in computational finance. *Journal of Complexity*, 22:435–458, 2006.

354. D. Ormoneit, C. Lemieux, and D. J. Fleet. Lattice particle filters. In D. Koller and J. Breese, editors, *Proceedings of the 17th Conference on Uncertainty in Artificial Intelligence*, pages 395–402. Morgan Kaufmann, San Francisco, CA, 2001.
355. A. B. Owen. Orthogonal arrays for computer experiments, integration and visualization. *Statistica Sinica*, 2:439–452, 1992.
356. A. B. Owen. Lattice sampling revisited: Monte Carlo variance of means over randomized orthogonal arrays. *Annals of Statistics*, 22:930–945, 1994.
357. A. B. Owen. Randomly permuted (t, m, s)-nets and (t, s)-sequences. In H. Niederreiter and P. J.-S. Shiue, editors, *Monte Carlo and Quasi-Monte Carlo Methods in Scientific Computing*, volume 106 of Lecture Notes in Statistics, pages 299–317. Springer-Verlag, New York, 1995.
358. A. B. Owen. Monte Carlo variance of scrambled equidistribution quadrature. *SIAM Journal on Numerical Analysis*, 34(5):1884–1910, 1997.
359. A. B. Owen. Scrambled net variance for integrals of smooth functions. *Annals of Statistics*, 25(4):1541–1562, 1997.
360. A. B. Owen. Latin supercube sampling for very high-dimensional simulations. *ACM Transactions on Modeling and Computer Simulation*, 8(1):71–102, 1998.
361. A. B. Owen. Scrambling Sobol and Niederreiter-Xing points. *Journal of Complexity*, 14:466–489, 1998.
362. A. B. Owen. Necessity of low effective dimension. Manuscript, 2002.
363. A. B. Owen. The dimension distribution and quadrature test functions. *Statistica Sinica*, 13:1–17, 2003.
364. A. B. Owen. Quasi-Monte Carlo sampling. In H. W. Jensen, editor, *Monte Carlo Ray Tracing: SIGGRAPH 2003 Course 44*, pages 69–88. ACM, New York, 2003.
365. A. B. Owen. Variance and discrepancy with alternative scramblings. *ACM Transactions on Modeling and Computer Simulation*, 13:363–378, 2003.
366. A. B. Owen. Multidimensional variation for quasi-Monte Carlo. In Jianqing Fan and Gang Li, editors, *International Conference on Statistics in honour of Professor Kai-Tai Fang's 65th birthday*, pages 49–74. World Scientific Publications, Hackensack, NJ, 2005.
367. A. B. Owen and D. A. Tavella. Scrambled nets for value-at-risk calculations. In S. Grayling, editor, *VAR Understanding and Applying Value-At-Risk*, pages 257–273. Risk Publications, London, 1997.
368. A. B. Owen and S. D. Tribble. A quasi-Monte Carlo Metropolis algorithm. *Proceedings of the National Academy of Sciences*, 102(25):8844–8849, 2005.
369. G. Pagès. Functional quantization for pricing derivatives. Technical Report 5392, INRIA, 2004.
370. H. H. Panjer, editor. *Financial Economics: With Applications to Investments, Insurance, and Pensions.* The Actuarial Foundation, Schaumburg, IL, 1998.
371. F. Panneton and P. L'Ecuyer. Infinite-dimensional highly-uniform point sets defined via linear recurrences in $\mathbb{F}_{2^w}$. In H. Niederreiter and D. Talay, editors, *Monte Carlo and Quasi-Monte Carlo Methods 2004*, pages 419–430. Springer, New York, 2006.
372. F. Panneton, P. L'Ecuyer, and M. Matsumoto. Improved long-period random number generators based on linear recurrences modulo 2. *ACM Transactions on Mathematical Software*, 32(1):1–16, 2006.
373. A. Papageorgiou. The Brownian bridge does not offer a consistent advantage in quasi-Monte Carlo integration. *Journal of Complexity*, 18(1):171–186, 2002.
374. A. Papageorgiou and J. Traub. Beating Monte Carlo. *Risk*, 9:63–65, June 1996.
375. S. Paskov and J. Traub. Faster valuation of financial derivatives. *Journal of Portfolio Management*, 22:113–120, 1995.
376. V. Philomin, R. Duraiswami, and L. Davis. Quasi-random sampling for condensation. In D. Vernon, editor, *Proceedings of the European Conference on Computer Vision, Part II*, volume 1843 of Lecture Notes in Computer Science, pages 139–149. Springer, New York, 2000.

377. G. Pirsic. A software implementation of Niederreiter-Xing sequences. In K.-T. Fang, F. J. Hickernell, and H. Niederreiter, editors, *Monte Carlo and Quasi-Monte Carlo Methods 2000*, pages 434–445. Springer, New York, 2001.
378. G. Pirsic and W. Ch. Schmid. Calculation of the quality parameter of digital nets and application to their construction. *Journal of Complexity*, 17:827–839, 2001.
379. W. H. Press, S. A. Teukolsky, W. T. Vetterling, and B. P. Flannery. *Numerical Recipes in C.* Cambridge University Press, Cambridge, 1992.
380. J. G. Propp and D. B. Wilson. Exact sampling with coupled Markov chains and applications to statistical mechanics. *Random Structures and Algorithms*, 9(1–2):223–252, 1996.
381. M. O. Rabin. Probabilistic algorithms for primality testing. *Journal of Number Theory*, 12:128–138, 1980.
382. M. M. Rao. *Stochastic Processes: Inference Theory.* Mathematics and Its Applications. Kluwer Academic Publishers, Dordrecht, 2000.
383. R. D. Richtmyer. On the evaluation of definite integrals and a quasi-Monte Carlo method based on properties of algebraic numbers. Technical Report LA-1342, Los Alamos Scientific Laboratory, 1951.
384. B. D. Ripley. The lattice structure of pseudo-random number generators. *Proceedings of the Royal Society of London, Series A*, 389:197–204, 1983.
385. H. Robbins and S. Monro. A stochastic approximation method. *Annals of Mathematical Statistics*, 22:400–407, 1951.
386. C. P. Robert and G. Casella. *Monte Carlo Statistical Methods*, second edition. Springer Texts in Statistics. Springer, New York, 2005.
387. L. C. G. Rogers. Monte Carlo valuation of American options. *Mathematical Finance*, 12:271–286, 2002.
388. S. M. Ross. *Introduction to Probability Models*, fifth edition. Academic Press, New York, 1993.
389. S. M. Ross. *Simulation*, fourth edition. Elsevier Academic Press, New York, 2006.
390. J. Rotman. *Galois Theory*, second edition. Springer, New York, 1998.
391. R. Y. Rubinstein. *Simulation and the Monte Carlo Method.* John Wiley and Sons, New York, 1981.
392. J. Sacks, S. B. Schiller, and W. J. Welch. Designs for computer experiments. *Technometrics*, 31:41–47, 1989.
393. J. Sacks, W. J. Welch, T. J. Mitchell, and H. P. Wynn. Design and analysis of computer experiments. *Statistical Science*, 4:409–423, 1989.
394. C.-E. Särndal, B. Swensson, and J. Wretman. *Model Assisted Survey Sampling.* Springer, New York, 1992.
395. W. Ch. Schmid. Shift-nets: A new class of binary digital (t, m, s)-nets. In P. Hellekalek, G. Larcher, H. Niederreiter, and P. Zinterhof, editors, *Monte Carlo and Quasi-Monte Carlo Methods in Scientific Computing*, volume 127 of Lecture Notes in Statistics, pages 369–381. Springer-Verlag, New York, 1997.
396. W. Ch. Schmid. Improvements and extensions of the "Salzburg Tables" by using irreducible polynomials. In H. Niederreiter and J. Spanier, editors, *Monte Carlo and Quasi-Monte Carlo Methods 1998*, pages 436–447. Springer, Berlin, 2000.
397. W. Ch. Schmid. Projections of digital nets and sequences. *Mathematics and Computers in Simulation*, 55:239–248, 2001.
398. W. Ch. Schmid and R. Schürer. Shift-nets and Salzburg tables: Power computing in number-theoretical numerics. In E. Efinger and A. Uhl, editors, *Scientific Computing in Salzburg – Festschrift on the Occasion of Peter Zinterhof's 60th Birthday*, pages 175–184. Österreichische Computer Gesellschaft, Vienna, 2005.
399. W. M. Schmidt. Irregularities of distribution. vii. *Acta Arithmetica*, 21:45–50, 1972.
400. R. Schürer and W. Ch. Schmid. MinT: A database for optimal net parameters. In H. Niederreiter and D. Talay, editors, *Monte Carlo and Quasi-Monte Carlo Methods 2004*, pages 457–469. Springer, Berlin, 2006.

401. R. J. Serfling. *Approximation Theorems for Mathematical Statistics.* Wiley, New York, 1980.
402. J. E. H. Shaw. A quasirandom approach to Bayesian statistics. *Annals of Statistics*, 16:895–914, 1988.
403. A. Sidi. A new variable transformation for numerical integration. In H. Brass and G. Hämmerlin, editors, *Numerical Integration IV*, volume 112 of Internationl Series on Numerical Mathematics, pages 359–373. Birkhäuser, Basel, 1993.
404. A. Sklar. Fonctions de répartition à n dimensions et leurs marges. *Publications de l'Institut de Statistique de l'Université de Paris*, 8:229–231, 1959.
405. M. M. Skriganov. Coding theory and uniform distributions. Technical report, Steklov Mathematical Institute, St. Petersburg, 1998.
406. I. H. Sloan. QMC integration – beating intractability by weighting the coordinate directions. In K.-T. Fang, F. J. Hickernell, and H. Niederreiter, editors, *Monte Carlo and Quasi-Monte Carlo Methods 2000*, pages 103–123. Springer, New York, 2001.
407. I. H. Sloan and S. Joe. *Lattice Methods for Multiple Integration.* Clarendon Press, Oxford, 1994.
408. I. H. Sloan and P. J. Kachoyan. Lattice methods for multiple integration: Theory, error analysis and examples. *SIAM Journal on Numerical Analysis*, 24:116–128, 1987.
409. I. H. Sloan, F. Y. Kuo, and S. Joe. Constructing randomly shifted lattice rules in weighted Sobolev spaces. *SIAM Journal on Numerical Analysis*, 40:1650–1665, 2002.
410. I. H. Sloan, F. Y. Kuo, and S. Joe. On the step-by-step construction of quasi-Monte Carlo integration rules that achieve strong tractability error bounds in weighted Sobolev spaces. *Mathematics of Computation*, 71:1609–1640, 2002.
411. I. H. Sloan and A. V. Rezstov. Component-by-component construction of good lattice rules. *Mathematics of Computation*, 71:263–273, 2002.
412. I. H. Sloan and L. Walsh. A computer search of rank 2 lattice rules for multidimensional quadrature. *Mathematics of Computation*, 54:281–302, 1990.
413. I. H. Sloan and H. Woźniakowski. When are quasi-Monte Carlo algorithms efficient for high dimensional integrals? *Journal of Complexity*, 14:1–33, 1998.
414. I. H. Sloan and H. Woźniakowski. Tractability of multivariate integration for weighted Korobov classes. *Journal of Complexity*, 17:697–721, 2001.
415. I. M. Sobol'. On the distribution of points in a cube and the approximate evaluation of integrals. *USSR Computational Mathematics and Mathematical Physics*, 7:86–112, 1967.
416. I. M. Sobol'. *Multidimensional Quadrature Formulas and Haar Functions.* Nauka, Moskow, 1969. In Russian.
417. I. M. Sobol. Sensitivity estimates for nonlinear mathematical models. *Mathematical Modeling and Computer Experiments*, 1:407–414, 1993. Published in Russian in 1990.
418. I. M. Sobol'. *A Primer for the Monte Carlo Method.* CRC Press, Boca Raton, FL, 1994.
419. I. M. Sobol'. Global sensitivity indices for nonlinear mathematical models and their Monte Carlo estimates. *Mathematics and Computers in Simulation*, 55:271–280, 2001.
420. I. M. Sobol' and Y. L. Levitan. The production of points uniformly distributed in a multidimensional cube. Technical Report Preprint 40, Institute of Applied Mathematics, USSR Academy of Sciences, 1976. In Russian.
421. I. M. Sobol' and Y. L. Levitan. On the use of variance reducing multipliers in Monte Carlo computations of a global sensitivity index. *Computer Physics Communications*, 117:52–61, 1999.
422. I. M. Sobol', V. I. Turchaninov, Y. L. Levitan, and B. V. Shukhman. Quasirandom sequence generators. Technical report, Keldysh Institute of Applied Mathematics, 1992.
423. J. Spanier. A new family of estimators for random walk problems. *Journal of the Institute for Mathematics and Its Applications*, 23:1–31, 1979.

424. J. Spanier and E. M. Gelbard. *Monte Carlo Principles and Neutron Transport Problems.* Addison-Wesley, Reading, MA, 1969.
425. J. Spanier and E. H. Maize. Quasi-random methods for estimating integrals using relatively small samples. *SIAM Review*, 36:18–44, 1994.
426. A. Speight. A multilevel approach to control variates. Manuscript, 2007.
427. M. Stein. Large sample properties of simulations using Latin hypercube sampling. *Technometrics*, 29:143–151, 1987.
428. D. R. Stinson. Combinatorial techniques for univeral hashing. *Journal of Computer and System Sciences*, 48:337–346, 1994.
429. O. Strauch and Š. Porubský. *Distribution of Sequences: A Sampler.* Peter Lang Publishing Group, Frankfurt am Main, 2005.
430. Y. Su and M. C. Fu. Importance sampling in derivative securities pricing. In J. A. Joines, R. R. Barton, K. Kang, and P. A. Fishwick, editors, *Proceedings of the 2000 Winter Simulation Conference*, pages 587–596. IEEE Press, Piscataway, NJ, 2000.
431. P. R. Tadikamalla. Computer generation of gamma random variables ii. *Communications of the ACM*, 21:925–928, 1978.
432. A. Tajima, S. Ninomiya, and S. Tezuka. On the anomaly of ran1() in Monte Carlo pricing of financial derivatives. In J. Charnes and D. Morrice, editors, *Proceedings of the 1996 Winter Simulation Conference*, pages 360–366. IEEE Press, Piscataway, NJ, 1996.
433. B. Tang. Orthogonal array-based Latin hypercubes. *Journal of the American Statistical Association*, 88:1392–1397, 1993.
434. R. C. Tausworthe. Random numbers generated by linear recurrence modulo two. *Mathematics of Computation*, 19:201–209, 1965.
435. S. Tezuka. Walsh-spectral test for GFSR pseudorandom numbers. *Communications of the ACM*, 30(8):731–735, August 1987.
436. S. Tezuka. Random number generation based on the polynomial arithmetic modulo two. Technical Report RT-0017, IBM Research, Tokyo Research Laboratory, October 1989.
437. S. Tezuka. Lattice structure of pseudorandom sequences from shift-register generators. In O. Balci, editor, *Proceedings of the 1990 Winter Simulation Conference*, pages 266–269. IEEE Press, Piscataway, NJ, 1990.
438. S. Tezuka. A new family of low-discrepancy point sets. Technical Report RT-0031, IBM Research, Tokyo Research Laboratory, January 1990.
439. S. Tezuka. Polynomial arithmetic analogue of Halton sequences. *ACM Transactions on Modeling and Computer Simulation*, 3:99–107, 1993.
440. S. Tezuka. A generalization of Faure sequences and its efficient implementation. Technical Report RT0105, IBM Research, Tokyo Research Laboratory, 1994.
441. S. Tezuka. *Uniform Random Numbers: Theory and Practice.* Kluwer Academic Publishers, Norwell, MA, 1995.
442. S. Tezuka. Polynomial arithmetic analogue of Hickernell sequences. In H. Niederreiter, editor, *Monte Carlo and Quasi-Monte Carlo Methods 2002*, pages 451–459. Springer, New York, 2004.
443. S. Tezuka. On the necessity of low-effective dimension. *Journal of Complexity*, 21:710–721, 2005.
444. S. Tezuka. Discrepancy between QMC and RQMC. *Uniform Distribution Theory*, 2:93–105, 2007.
445. S. Tezuka and H. Faure. I-binomial scrambling of digital nets and sequences. *Journal of Complexity*, 19:744–757, 2003.
446. S. Tezuka and P. L'Ecuyer. Efficient and portable combined Tausworthe random number generators. *ACM Transactions on Modeling and Computer Simulation*, 1(2):99–112, 1991.
447. S. Tezuka and P. L'Ecuyer. An analysis of add-with-carry and subtract-with-borrow generators. In J. J. Swain, D. Goldsman, R.C. Crain, and J. R. Wilson, editors, *Proceedings of the 1992 Winter Simulation Conference*, pages 443–447. IEEE Press, Piscataway, NJ, 1992.

448. S. Tezuka, P. L'Ecuyer, and R. Couture. On the add-with-carry and subtract-with-borrow random number generators. *ACM Transactions on Modeling and Computer Simulation*, 3(4):315–331, 1994.
449. S. Tezuka and T. Tokuyama. A note on polynomial arithmetic analogue of Halton sequences. *ACM Transactions on Modeling and Computer Simulation*, 4:279–284, 1994.
450. J. P. R. Tootill, W. D. Robinson, and D. J. Eagle. An asymptotically random Tausworthe sequence. *Journal of the ACM*, 20:469–481, 1973.
451. H. F. Trotter and J. W. Tukey. Conditional Monte Carlo for normal samples. In H. Meyer, editor, *Symposium on Monte Carlo Methods*, pages 80–88. John Wiley and Sons, New York, 1956.
452. J. Tsitsiklis and B. Van Roy. Regression methods for pricing complex American-style options. *IEEE Transactions on Neural Networks*, 12:694–703, 2001.
453. B. Tuffin. On the use of low-discrepancy sequences in Monte Carlo methods. Technical Report No. 1060, I.R.I.S.A., Rennes, France, 1996.
454. B. Tuffin. A new permutation choice in Halton sequences. In P. Hellekalek, G. Larcher, H. Niederreiter, and P. Zinterhof, editors, *Monte Carlo and Quasi-Monte Carlo Methods 1996*, volume 127 of Lecture Notes in Statistics, pages 427–435. Springer-Verlag, New York, 1998.
455. B. Tuffin. Variance reduction order using good lattice points in Monte Carlo methods. *Computing*, 61:371–378, 1998.
456. J. G. van der Corput. Verteilungsfunktionen: I, II. *Proceedings of the Nederlandse Akademie van Wetenschappen*, 38:813–821, 1058–1066, 1935.
457. B. Vandewoestyne and R. Cools. Good permutations for deterministic scrambled Halton sequences in terms of L_2-discrepancy. *Journal of Computational and Applied Mathematics*, 189:341–361, 2006.
458. F. J. Vázquez-Abad. RPA pathwise derivative estimation of ruin probabilities. *Insurance: Mathematics and Economics*, 26:269–288, 2000.
459. F. J. Vázquez-Abad and D. Dufresne. Accelerated simulation for pricing Asian options. In D. J. Medeiros, E. F. Watson, J. S. Carson, and M. S. Manivannan, editors, *Proceedings of the 1998 Winter Simulation Conference*, pages 1493–1500. IEEE Press, Piscataway, NJ, 1998.
460. E. Veach. *Robust Monte Carlo methods for light transport simulation.* PhD thesis, Stanford University, 1997.
461. D. Wang and A. Compagner. On the use of reducible polynomials as random number generators. *Mathematics of Computation*, 60:363–374, 1993.
462. S. S. Wang. Discussion on the paper "Understanding Relationships Using Copulas" by E. Frees and E. Valdez. *North American Actuarial Journal*, 3:137–141, 1999.
463. X. Wang and K.-T. Fang. The effective dimension and quasi-Monte Carlo integration. *Journal of Complexity*, 19:101–124, 2003.
464. X. Wang and F. J. Hickernell. Randomized Halton sequences. *Mathematical and Computer Modelling*, 32:887–899, 2000.
465. X. Wang, C. Lemieux, and H. Faure. A note on Atanassov's discrepancy bound for the Halton sequence. Technical report, Department of Statistics and Actuarial Science, University of Waterloo, 2008.
466. X. Wang and I. H. Sloan. Why are high-dimensional finance problems of low effective dimension? *SIAM Journal on Scientific Computing*, 27:159–183, 2005.
467. Y. Wang and F. J. Hickernell. An historical overview of lattice point sets. In K.-T. Fang, F. J. Hickernell, and H. Niederreiter, editors, *Monte Carlo and Quasi-Monte Carlo Methods 2000*, pages 158–167. Springer, New York, 2001.
468. T. Warnock. Computational investigations of low discrepancy point sets. In S. K. Zaremba, editor, *Application de la théorie des nombres à l'analyse numérique*, pages 319–343. Academic Press, New York, 1972.

469. G. W. Wasilkowski. Integration and approximation of multivariate functions: Average-case complexity with isotropic Wiener measure. *Journal of Approximation Theory*, 77:212–227, 1994.
470. G. W. Wasilkowski. Average case complexity. *Journal of Complexity*, 12:257–272, 1996.
471. S. Wegenkittl. *Generalized ϕ-divergence and frequency analysis in Markov chains.* PhD thesis, University of Salzburg, 1998.
472. W. J. Welch, R. J. Buck, J. Sacks, H. P. Wynn, T. J. Mitchell, and M. D. Morris. Screening, predicting, and computer experiments. *Technometrics*, 34:15–25, 1992.
473. W. Whitt. Bivariate distributions with given marginals. *The Annals of Statistics*, 4(6):1280–1289, 1976.
474. B. A. Wichmann and I. D. Hill. An efficient and portable pseudo-random number generator. *Applied Statistics*, 31:188–190, 1982. See also corrections and remarks in the same journal by Wichmann and Hill, 33: 123 (1984); McLeod 34: 198–200 (1985); Zeisel 35: 89 (1986).
475. B. A. Wichmann and I. D. Hill. Building a random number generator. *Byte*, 12(3):127–128, 1987.
476. G. A. Willard. Calculating prices and sensitivities for path-dependent derivatives securities in multifactor models. *Journal of Derivatives*, 5:45–61, Fall 1997.
477. J. R. Wilson. Antithetic sampling with multivariate inputs. *American Journal of Mathematical and Management Sciences*, 3:121–144, 1983.
478. J. L. Wirch and M. R. Hardy. A synthesis of risk measures for capital adequacy. *Insurance: Mathematics and Economics*, 25:337–347, 1999.
479. H. Woźniakowski. Average case complexity of multivariate integration. *Bulletin (New Series) of the American Mathematical Society*, 24:185–194, 1991.
480. H. Woźniakowski. Average case complexity of linear multivariate problems. Part 1: Theory; part 2: Applications. *Journal of Complexity*, 8:337–372, 373–392, 1992.
481. R. Wu and M. C. Fu. Optimal exercise policies and simulation-based valuation for American-Asian options. *Operations Research*, 51:52–66, 2003.
482. C. Xing and H. Niederreiter. A construction of low-discrepancy sequences using global function fields. *Acta Arithmetica*, 73:87–102, 1995.
483. S. K. Zaremba. La méthode des bons treillis pour le calcul des intégrales multiples. In S.K. Zaremba, editor, *Application de la théorie des nombres à l'analyse numérique*, pages 39–116. Academic Press, New York, 1972.
484. P. Zinterhof. Über einige abschätzungen bei der approximation von funktionen mit gleichverteilungsmethoden. *Österreichischen Akademie der Wissenschaften Mathematisch-Naturwissenschaf Sitzungsberichte II*, 185:121–132, 1976.
485. cg.scs.carleton.ca/~luc/rnbookindex.html.
486. csrc.nist.gov/rng/.
487. en.wikipedia.org/wiki/mersenne_twister.
488. lib.stat.cmu.edu/designs/owen.html.
489. mint.sbg.ac.at.
490. mint.sbg.ac.at/hintlib/.
491. parallel.bas.bg/~emanouil/sequences.html.
492. random.mat.sbg.ac.at/news/seedingtt800.html.
493. support.microsoft.com/kb/828795.
494. www.cs.columbia.edu/~ap/html/information.html.
495. www.cs.uwaterloo.ca/~paforsyt/agon.pdf.
496. www.iro.umontreal.ca/~simardr/ssj/index.html.
497. www.iro.umontreal.ca/~simardr/testu01/tu01.html.
498. www.math.uwaterloo.ca/~clemieux/flfactors.html.
499. www.mathworks.com. See documentation on the function `rand`.
500. www.multires.caltech.edu/software/libseq/.
501. www.netlib.org/toms/659.
502. www.research.att.com/~njas/oadir/index.html.
503. www.stat.fsu.edu/pub/diehard/.

Index

acceptance-rejection, 20, 46–48, 55
admissible integers, 165
ANOVA
  components, 232, 240, 262
  decomposition, 210, 214–229, 325
antithetic variates, 89–101, 109, 136, 273–274, 299
arbitrage, 249
array-RQMC, 210, 240, 263, 312, 317
asymptotically random, 77
average dimension, 221

Babenko-Zaremba index, 193
baker transformation, 197, 240
bank example, 13–19, 88, 96, 100, 105, 107, 111, 116, 120, 129, 137, 205, 235
batch means (within MCMC), 304
Bayesian inference, 303
Bernoulli polynomial, 194
Black-Scholes-Merton formula, 250, 251, 280, 298
bootstrapping, 25, 38, 136, 225
borehole function, 321
Box-Muller algorithm, 50
Brownian bridge, 33, 222–225, 240, 299
  generalized, 223, 256, 262
Brownian motion, 42, 55, 222, 248, 257, 260, 262, 278
  geometric, 248, 258
burn-in period, 304

Cholesky decomposition, 53, 56, 223, 253, 295, 298
coefficient of determination, 109
common random numbers, 58, 107, 132–135, 138, 281, 299
complete market, 249, 256
completely uniformly distributed sequence, 307
component-by-component, 152, 240, 245
conditional Monte Carlo, 119–125, 136, 225, 231, 263, 279
conditional tail expectation, 294, 300
control variable, 101–110, 136, 230–231, 273, 299
  external, 107, 135
  internal, 107
  multiple, 108
copula models, 43, 53–54, 56
coupling from the past, 303, 310
Cranley-Patterson rotation, 204
crude Monte Carlo, 12
curse of dimensionality, 9

delta-gamma approximation, 295
detailed balance condition, 307, 309
diaphony, 194, 198
  dyadic, 195
    weighted, 196
digital net
  dual space, 191
digital sequence
  Sobol', 262
digital net, 155
  (t, k, s)-net, 156, 329, 333
  dual space, 211
  scrambled net, 12, 221
  shift net, 174
digital sequence, 155
  (t, s)-sequence, 156
  Faure, 154, 156, 161–163, 177
    numerical results with, 270
  generalized Faure, 169
    numerical results with, 271

generalized Niederreiter, 167
generalized Sobol', 167, 221
Niederreiter, 163–164
Niederreiter-Xing, 168
polynomial arithmetic analogue of Halton sequence, 168
Sobol', 78, 146, 154, 157–161, 198, 217, 229, 239, 299
numerical results with, 237, 270, 281, 287, 293, 297
dimension distribution, 221
dimension-stationary, 177, 199, 262
direction numbers, 157, 159, 166, 271
discrepancy
L_2 discrepancy, 183, 199
(definition of) low-discrepancy sequence/point set, 143
extreme, 182
generalized L_2 discrepancy, 186
isotropic, 182
star, 142, 153–157, 162, 165, 181, 308
weighted L_2, 242
mean-square, 245
discretization, 255, 280
Euler, 257
dual pricing method, 286–287

effective dimension, 216–222, 226, 228, 240, 326
effective sample size, 315
efficiency, 89, 98, 117, 122, 129, 133, 232, 331
elementary intervals, 76, 156, 188
empirical CDF, 25, 38, 140, 233, 298, 300
equidistribution, 75–80, 156, 188, 206
maximal, 77
equity-linked contract, 31–34, 39
exact sampling, 310–312
exclusive-or, 64, 157
experimental design
balanced incomplete block design, 332
experimental design, 301, 322
factorial design, 322
OA-based Latin hypercube design, 329, 333
space-filling design, 323
exponential twisting/tilting, 114, 265, 296

financial model
Heston, 257, 262, 281
jumps, 260, 264
lognormal, 31, 37, 42, 247, 278, 289, 290, 298
regime switching, 258, 264
variance gamma, 260, 298
finite difference, 282
formal Laurent series, 163, 173, 178, 338–339
Fourier series, 191, 192, 211
Freivald's algorithm, 35
fully projection-regular, 144, 148, 158, 177, 198, 328
fundamental theorem of option pricing, 249

generating matrices, 155–164, 167–170, 206–208, 213
Gillespie's algorithm, 27–31
τ-leap approach, 39
Girsanov's theorem, 278
global illumination problem, 112
global sensitivity indices, 215, 217, 240, 322, 329, 332
good lattice points, *see* lattice–Korobov point set
Gray code, 159, 169
greeks, 282, 288
delta, 282, 289, 295
gamma, 282, 289, 295

Halton sequence, 153–154, 162, 165, 167, 198, 212, 239
generalized, 165, 198, 239
numerical results with, 237, 271
numerical results with, 270
Hammersley point set, 154
hedging, 288
hit-and-miss method, 35

importance sampling, 110–119, 230, 265, 295, 299, 313
weighted, 115, 131, 313
infinitesimal perturbation analysis, 115, 276, 290, 299
inversion, 16, 20, 35–37, 41, 44–46, 55–56, 94, 132, 230, 256, 324

jackknifing, 105, 137

Kalman filter, 313, 318, 333
Karhunen-Loève expansion, 224
Koksma-Hlawka inequality, 184, 196, 214, 242

Laguerre polynomial, 285
Latin hypercube sampling, 136, 208, 233, 323
Latin supercube sampling, 209, 317
lattice, 70, 147

basis, 70, 72, 147
copy rule, 148, 198
determinant, 147
dual, 72, 191, 193, 211
extensible, 150
extensible Korobov, 151, 198
integration, 147
invariants, 147
Korobov point set, 145, 148, 175, 193, 228, 235, 239, 309, 333
numerical results with, 232, 268
polynomial, *see* polynomial lattice
rank, 147
rank-1, 148, 239, 240, 245
shortest vector, 72, 193
structure (for PRNG), 67, 70–73
least-squares Monte Carlo, 285
likelihood ratio, 112, 277, 313
method, 291, 299
Lindley's equation, 14
Longstaff and Schwartz algorithm, 285
LP_τ-sequence, 78, 154

Markov Chain Monte Carlo, 303–312
martingale, 249, 286
measure
absolutely continuous, 113
equivalent martingale, 249
risk-neutral probability, 249, 256, 282, 294
Metropolis-Hastings algorithm, 303, 305–310, 312, 333
acceptance probability, 306, 309
multiple-try, 309, 333
MinT, 157, 174
moment-matching method, 282, 299
Moro's algorithm, 44
mortgage-backed security, 217, 255, 268, 269, 299

naive Monte Carlo, 12
numerical examples, *see* bank example, greeks, option (American, Asian), ruin probability, stochastic activity network, value-at-risk

occupation time, 259
option
American, 283–288
Asian, 217, 224, 236, 250, 255, 262, 273, 278, 292, 298
Bermudan, 283
continuation value, 284
digital, 224
European, 247
path-dependent, 248
path-independent, 248
put-call parity, 299
rainbow, 252, 256, 298
order statistics, 25, 50, 233
orthogonal array, 324, 326–329
OA-based Latin hypercube design, 329, 333
ordered, 174, 329
strength, 326

P_α, 194
weighted, 193, 216
percentile estimation, *see* quantile estimation
perfect sampling, *see* exact sampling
periodization, 196
Sidi's transformation, 197, 240
polar method, 50
polynomial lattice, 170–174
extensible, 173
integration lattice, 172, 196
Korobov, 178, 229
numerical results with, 262, 268, 281
polynomial LCG, 178
polynomial version of Hickernell sequences, 174
rank-1, 171, 173
Salzburg Tables, 172
primitive
element, 61, 85, 176
polynomial, 62, 64, 157, 164, 167, 337
principal components, 223
probabilistic Monte Carlo algorithm, 35
product rule, 6–9
projection
definition of, 144
quality of, 143, 154–169, 177, 216–217, 262, 328
use in quality measures, 186–187, 194–196, 228–229
propagation rule, 199
proposal distribution, 305, 314
pseudorandom number generator
$\mathbb{F}_2$-linear, 65, 193, 229
add-with-carry, 66, 75
bad, 24, 57, 81
combined, 62, 63, 65
combined Tausworthe, 178, 199, 229, 262
cycle, 60, 69, 85, 179
explicit inversive congruential, 67, 86
generalized feedback shift register, 64
jumping ahead, 60, 85

lagged-Fibonacci, 62, 66
linear congruential, 23, 61, 175, 229, 308
linear feedback shift register, 64, 76, 79, 85
Mersenne-Twister, 24, 65, 76
mid-square method, 58
MRG32k3a, 24, 63, 86
multiple recursive, 62
multiplicative (linear) congruential, 61, 85
nonlinear, 67
period, 24, 59, 64, 175, 179, 308
randomness, 57, 68
RANDU, 24, 61, 71, 86
seed, 24, 38, 59, 69, 175
subtract-with-borrow, 66, 74–75
Tausworthe, 64, 178
tempering, 65

$(q_1, \ldots, q_s)$-partition, 156, 189, 206
quantile estimation, 25, 294, 298
quasi-regression, 226

radical-inverse function, 145, 150, 173
Radon-Nikodym, 277
randomization
digital shift, 205–206
scrambling, 206–208, 213, 239
shift, 176, 204–205
rare event, 111, 294
ratio estimate, *see* weighted importance sampling
recurrence-based point set, 62, 175–180, 193, 239, 308–309
regenerative
epochs, 265
process, 265
simulation, 135, 266–268
repeatability, 58, 60
residual resampling, 318
resolution, 76, 86, 159, 192, 229
Richtmyer sequence, 145, 152
ruin probability, 115, 264–268
storage process, 265
surplus process, 264
Russian roulette, 112

sampling plan, 323, 333
permuted-columns, 331
substituted-columns, 331
score function, *see* likelihood ratio method
scrambled net, *see* digital net
self-financing, 289
sensitivity analysis, 132, 321, 329
sequential Monte Carlo, 312–320
bootstrap filter, 112, 316
filtering distribution, 312, 314
particles, 313, 317
properly weighted sample, 313, 333
sequential importance sampling, 315
shift net, *see* digital net
Sidi's transformation, *see* periodization
Simpson's rule, 5
Sobolev space, 243
spectral test, 71, 86, 193, 229
weighted, 195
splitting, 104, 112, 137, 230
statistical test, 24, 68, 80–84
p-value, 80, 82
birthday spacings test, 84, 86
collisions test, 82–84
dense case, 83
negative entropy test, 82
overlapping test, 84
Pearson chi-square test, 81, 84
serial test, 81
sparse case, 83
type I error, 80
stochastic activity network, 96–99, 106, 117, 122–125, 133, 137, 232
stochastic approximation, 276–278
stochastic differential equation, 248, 257, 258, 264–265, 279
stochastic mesh method, 284, 286
stochastic volatility, 258, 279
stopping time, 19, 283
stratification, 125–131
optimal allocation, 126, 137
post stratification, 127
proportional allocation, 126, 137
surrogate function, 321, 332
synchronization, 96, 107, 135
systematic resampling, 318

t-value, 78, 86, 156–161, 167, 169, 174, 191–193, 199, 207, 229
TailVar, *see* conditional tail expectation
time-reversible, 333
tractability, 152, 229, 241
trapezoidal rule, 5–9, 34, 141, 143

unbounded dimension, 19, 21, 30, 176, 205, 239, 260, 263, 311
uniformly directed cutset, 122
uniformly distributed sequence, 184

value-at-risk, 233, 293–298, 300

van der Corput sequence, 145, 150, 153, 165
variance of RQMC estimator
  estimate, 202–203
  formulas, 211–214
variation
  in the sense of Hardy and Krause, 184
  in the sense of Vitali, 185
  infinite, 186, 197

Walsh series, 188–196, 211–213
  multi base, 342
weighted Monte Carlo, 110

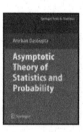